Environmental Microbiology for Engineers

Environmental Microbiology for Engineers

Volodymyr Ivanov
Nanyang Technological University
Singapore

CRC Press is an imprint of the
Taylor & Francis Group, an **informa** business

CRC Press
Taylor & Francis Group
6000 Broken Sound Parkway NW, Suite 300
Boca Raton, FL 33487-2742

Printed in the United States of America on acid-free paper
10 9 8 7 6 5 4 3 2 1

International Standard Book Number: 978-1-4200-9234-9 (Hardback)

Library of Congress Cataloging-in-Publication Data

Ivanov, Volodymyr.
Environmental microbiology for engineers / Volodymyr Ivanov.
p. cm.
Includes index.
ISBN 978-1-4200-9234-9 (alk. paper)
1. Microbial ecology. 2. Bioremediation. I. Title.

QR100.I93 2010
579--dc22 2010025789

Visit the Taylor & Francis Web site at
http://www.taylorandfrancis.com

and the CRC Press Web site at
http://www.crcpress.com

I would like to dedicate this book to my dear wife Elena Stabnikova, who is a research partner in all my projects; co-supervisor of my graduate students; and companion in all my travels, adventures, and businesses.

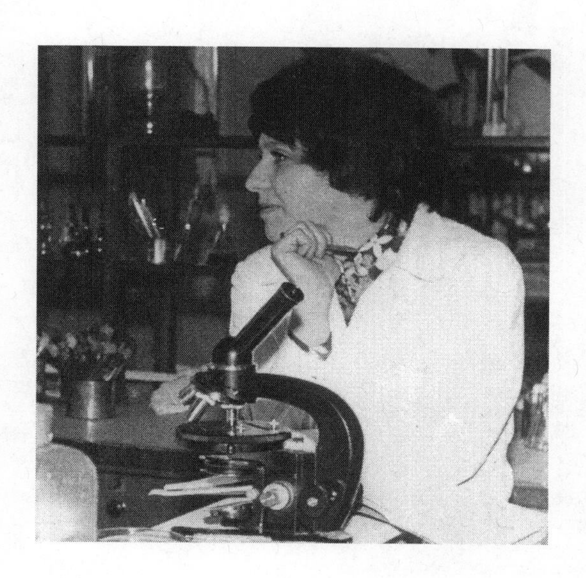

Contents

Preface

Microorganisms play an important role in the protection of humans, animals, plants, air, water, soil, and engineering systems from chemical or biological pollution, deterioration, and corrosion, and in the restoration of polluted and degraded environments. This is an area of civil and environmental engineering but the engineering solutions must be based on the relevant knowledge of microbiology.

This book is intended for undergraduate and graduate students at civil and environmental engineering schools and universities. It covers such essential topics as diversity and functions of microorganisms in environmental engineering systems and environmental bioengineering processes, applied microbial genetics and molecular biology, microbial biodegradation of organic substances, public health microbiology for environmental engineers, microbiology of water and wastewater treatment, and biotreatment of solid waste and soil bioremediation, as well as essential topics for civil engineering processes such as biodeterioration, biocorrosion, and biocementation. The objective is to provide a practical understanding of microorganisms and their functions in the environment and in environmental engineering systems.

To enhance this understanding of microbial functions in engineering systems, I have provided a tutorial bank, a quiz bank, and an exam question bank, as well as the solutions to these tutorial and exam questions in the solutions manual. The principal intention of this solutions manual is to help readers formulate answers to quantitative bioengineering and microbiological questions and to perform design calculations.

Author

Volodymyr Ivanov received his MEng degree in industrial biotechnology from the Ukrainian University of Food Technologies, Kiev, Ukraine, and his PhD degree in microbiology and biotechnology from the Institute of Microbiology and Virology of the Ukrainian Academy of Sciences, Kiev, Ukraine.

He was head of the biotechnological department in this institute; professor of microbiology and biotechnology at Kiev National University (Kiev, Ukraine), at Chulalongkorn University (Bangkok, Thailand), and at Gwangju Institute of Science and Technology (Gwangju, South Korea); and has been teaching environmental microbiology and environmental biotechnology to undergraduate and graduate students at Nanyang Technological University, Singapore, for the last 10 years. He has published over 130 papers, 8 patents, and 15 book chapters and monographs in different areas of industrial and environmental microbiology and biotechnology.

1

Microorganisms

System

A system is an entity formed by interactions between its components. It is surrounded by an environment and is separated from it by a boundary. The boundary depends on the type of the system and can be of physical, chemical, biological, or social origins. The set of interactions between the components constitutes the structure of the system (Figure 1.1).

Self-Organized System

An engineering system is controlled by outside elements: engineering design, construction, management, and automatic control. A self-organized system is one which is organized by internal interactions and permanent feedbacks between the components. A complex chemical structure can be self-formed by atomic interactions and molecular self-assembly (Figure 1.2a). A complex biological structure can be self-formed by diffusion-limited growth and aggregation of cells (Figure 1.2b). Molecular self-assembly is the formation of 3D structure of proteins, which are catalysts of all the biochemical reactions in a cell, and deoxyribonucleic acid (DNA, a molecule of genetic information) by hydrogen bonds and hydrophobic forces between the monomers of these polymers. Self-assembly of DNA and protein molecules is a major tool in nanoengineering assembling.

Life as Self-Organized Growth of Biomass

Life can be defined as a process of self-organized growth, maintenance, and decay of biomass on Earth. A biomass is a mass of organisms that are self-replicating life units. Life also includes many other essential features:

- Exchange of materials and energy with the environment
- Permanent decrease of entropy (increase of complexity) in living systems
- Formation of specific biochemical substances

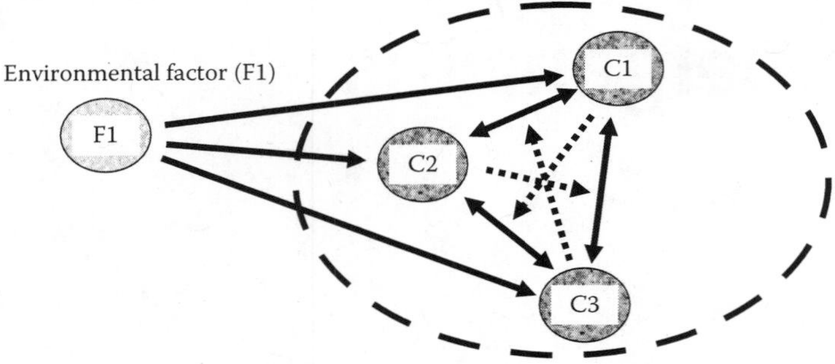

Figure 1.1 Interactions in a system. The arrows show interactions between components. Dotted arrows show the influence of the component on the interactions between other components. Influence of one component on the interactions between the environmental factor and other components are not shown.

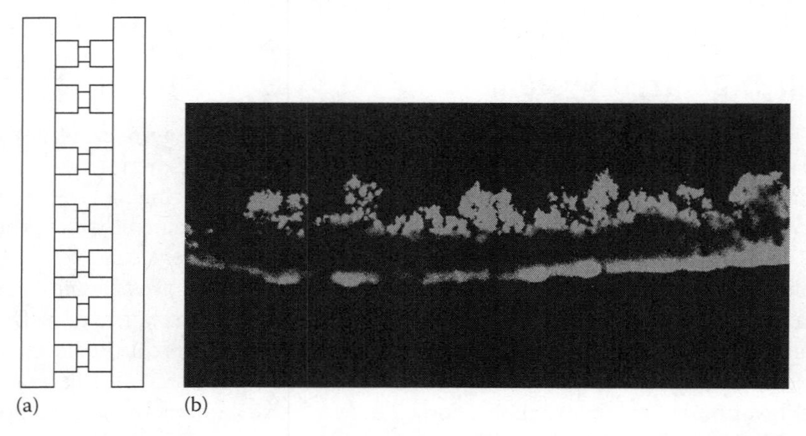

Figure 1.2 (a) DNA double strands are formed as a helix due to hydrogen bonds between N, H, and O atoms of the bases of the two strands of DNA. (b) The folded structure of microbial biofilm due to diffusion-limited cell growth and aggregation. Gradients of oxygen and nutrients, and products of life activity determine the structure of the microbial biofilm used for wastewater treatment.

- Specific biotransformations of these substances
- Specific interactions between organisms
- Adaptation, evolution, reproduction, and use of the life program

Every living system includes living components separated by a boundary from the environment, which is a source of substances and energy for biomass growth and maintenance. The biomass components organize the self-feeding and

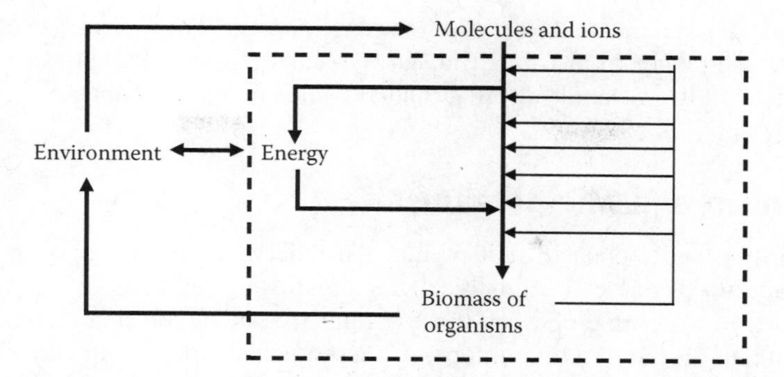

Figure 1.3 Life as a process of growth and maintenance of biomass using energy and substances from the environment. Arrows show the flow of materials and energy; dotted line shows the boundary of a living system.

self-growth of the biomass (Figure 1.3). Therefore, life, in terms of activity, can be described as the self-organized growth of biomass:

$$\frac{dX}{dt} = f(X)$$

where
X is biomass
dX/dt is growth of biomass

Cell

There are two types of cells: a prokaryotic (bacterial) cell, which is relatively small and simple in structure, and a eukaryotic cell, which is bigger, more complex, and contains intracellular organelles (nucleus, mitochondria, chloroplast, and others) performing specific functions (Figure 1.4). All cells and organelles are separated from the environment by a lipid membrane, which is not permeable to polarized substances and ions. Many cells are covered with a wall that offers

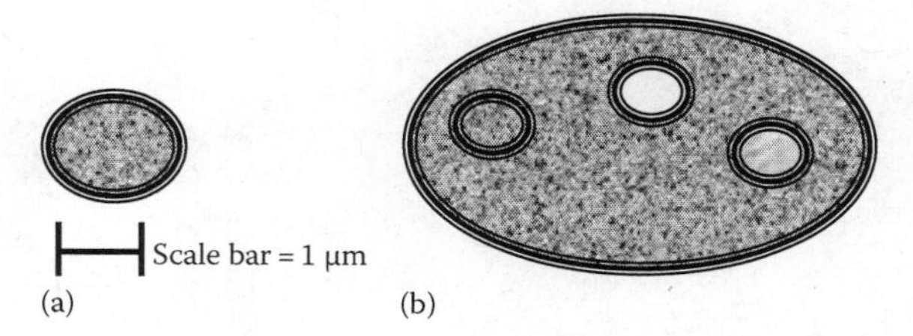

Figure 1.4 Schematic representation of (a) prokaryotic and (b) eukaryotic cells covered with a wall and a membrane. The eukaryotic cell contains different organelles covered with a membrane.

mechanical protection to the cell from the environment. The typical size of a prokaryotic cell ranges from 1 to 10 μm. The typical size of a eukaryotic cell ranges from 10 to 100 μm. Details of cell structure can be found in Chapter 5.

Organism and Microorganism

Organism is self-replicated unit of life. An unicellular organism is an organism made up of one cell. A multicellular organism is made up of many cells. Microorganisms or microbes are those organisms that are too small to be visualized without the aid of a microscope. A microorganism may be an unicellular or a multicellular organism.

Size of Microorganisms

Microorganisms range in size from 1 μm to approximately 100 μm (Table 1.1). Biogenic reproductive particles such as prions and viruses are below this range. Most multicellular plants and animals are above 100 μm, and thus are visible without the aid of a microscope. Note that only approximate sizes are listed in Table 1.1. Individual objects can be smaller or bigger than the sizes listed in Table 1.1.

Biology

Biology (from Greek *bio* = life + *logos* = speech, knowledge) is the science of life. It studies the dynamics and statics of living systems as well as the interactions

TABLE 1.1 Size of Biomolecules, Biogenic Particles, and Organisms

Size	Typical Group of the Objects	Method of Experimental Visualization
0.1 nm	Atom	Atomic force microscope (AFM)
1 nm	Monomer	AFM
10 nm	Biopolymer, prion	Transmitting electron microscope (TEM)
100 nm	Virus	TEM
Microorganisms		
1 μm	Prokaryotes (bacteria and archaea)	TEM and scanning electron microscopes, light microscope
10 μm	Fungi	Light microscope
100 μm	Algae and protozoa	Light microscope
1–10,000 mm	Animals and plants	Visible without a microscope

Note: Units of size are as follows: 1 angstrom = 1×10^{-10} m; 1 nanometer = 1 nm = 1×10^{-9} m; 1 micrometer = 1 μm = 1×10^{-6} m; 1 millimeter = 1 mm = 1×10^{-3} m.

between them and the environment. Living systems are called biological systems. The elementary unit of life and of every biological system is the cell.

Microbiology

Microbiology is a branch of biology devoted to the study of microorganisms (microbes) that include both unicellular and multicellular organisms. These microorganisms are not visible without the aid of a microscope because they are smaller than 70 μm (the resolution limit of human vision). There are many microbiological disciplines such as industrial microbiology, medical microbiology, veterinary microbiology, agricultural microbiology, and environmental microbiology, which are specified by their objects of study.

Environmental Microbiology

Environmental microbiology studies microbes in the lithosphere, hydrosphere, and atmosphere, as well as in environmental engineering systems used for the biotreatment of water, wastewater, solid wastes, soil, and gas. Approximately 70% of existing environmental engineering technologies are based on applications of microorganisms. Environmental microbiology also comprises such important civil engineering topics as microbial-induced biocorrosion, biodeterioration of construction materials, microbiological quality of indoor air, and microbial improvement of geotechnical properties of soil.

Why Microbes Are Used in Environmental Engineering

1. Microbes have the highest cell surface to cell volume ratio (Table 1.2). Rates of biochemical reactions and biodegradation as well as the growth rate are proportional to S/V ratio. So, microbes with the highest S/V ratio have the highest rate of biodegradation, which is often a major goal of environmental engineering processes.
2. Microorganisms can perform diverse biochemical reactions, so they are useful in biotransformation and the degradation of environmental pollutants (Table 1.3).

TABLE 1.2 Cell Surface to Volume Ratio (S/V) of Spherical Cells

Organism	Diameter (μm)	S/V
Bacteria	1	6
Fungi	10	0.6
Protozoa	100	0.06

Note: S/V for spherical cell $= (\pi D^2)/(\pi D^3/6) = 6/D$.

TABLE 1.3 Processes Performed by Microorganisms

Process	Conventional Reaction (S_i Is a Substance)	Example of Application
Oxidation	$S_1 - e^- \rightarrow S_2$	Oxidation of ammonium (nitrification) in bioremoval of ammonium from wastewater
Reduction	$S_1 + e^- \rightarrow S_2$	Reduction of nitrate (denitrification) in bioremoval of nitrate from wastewater
Degradation (mineralization)	$S \rightarrow CO_2$	Removal of toxic pollutants from water, soil, and gas
Transformation	$S_1 \rightarrow S_2$	Removal of toxic pollutants from water, soil, and gas
Hydrolysis	$(S_1)_n + nH_2O \rightarrow nS_2$	Biodegradation of biopolymers in agricultural and food-processing wastes
Biosynthesis	$nS_1 \rightarrow S_2$	Growth of biomass
Biorecognition	$S + E \rightarrow [SE]$	Microbial biosensors

Organisms of Importance in Environmental Microbiology

All organisms are composed of cells. On the basis of the cell type, organisms can be classified as prokaryotes or eukaryotes. Prokaryotic cells are relatively simple in structure; they lack a true nucleus covered by a membrane. The structure of a eukaryotic cell is more complex because it contains organelles that serve as compartments for performing special metabolic functions. The objects of environmental microbiology are prokaryotes such as bacteria and archaea, and eukaryotes such as microscopic fungi, microscopic algae, and protozoa (unicellular animals). Viruses are not organisms but they are studied under environmental microbiology because they reproduce in organisms.

Viruses

A virus is a biological particle that reproduces in numerous copies inside a live prokaryotic or eukaryotic host cell, and then releases from that cell. An extracellular virus particle does not perform any biochemical reactions by itself. Therefore, a virus has no such essential features of organism as self-reproduction and self-feeding. However, it is capable of self-organized reproduction and biomass growth inside a host cell. The type of host cell is usually strictly specific for each type of virus. Typical virus sizes range from 20 to 200 nm. Viruses contain a single type of nucleic acid, either DNA or RNA, coated usually with a protein envelope. Virology is a specialized science studying viruses.

Subviral Particles

There are also other simpler than virus particles, such as viroids, virusoids (satellites), and prions, which also reproduce in a host cell. A viroid is a circular, single-stranded RNA without a protein coat. A virusoid (a satellite) is similar to a viroid in structure but requires a helper virus for successful reproduction. Prions are protein molecules that reproduce in numerous copies inside a living cell, and then release from the cell.

Importance of Viruses in Environmental and Civil Engineering

Viruses are important in environmental and civil engineering because of the following reasons:

1. Viruses must be removed from indoor air to prevent dispersion inside a building by the ventilation system.
2. Viruses must be removed or destroyed during water and wastewater treatment.
3. Viruses of bacteria (bacteriophages) can infect and degrade the bacterial communities used in environmental engineering.
4. Detection of bacteriophages is used to monitor the microbial pollution of the environment.
5. Viruses may be used as a vector (carrier) of genes in artificial or natural genetic recombination (gene exchange between organisms).

Prokaryotes

Prokaryotes are microorganisms with prokaryotic-type cells. The most common cell shapes are spherical and rod shaped. They consist of two phylogenetic groups: bacteria and archaea. The typical size of cells ranges from 1 to 3 μm but there are cells smaller than 0.5 μm or bigger than 50 μm. Prokaryotes being the smallest among organisms have the highest cell surface to cell volume ratio and the highest growth rate (on medium with carbohydrates), and are most active in the biosynthesis or degradation of organic matter. The range of conditions suitable for life of prokaryotes is significantly wider than that of eukaryotes (Table 1.4).

Bacteria and Archaea

There are two groups of prokaryotes: bacteria and archaea. Bacteria are able to utilize either light (phototrophs), inorganic substances (lithotrophs), or organic substances (heterotrophs) as an energy source for growth and reproduction. They are adapted to a wide range of environmental conditions and are major degraders of dead organic matter on earth. Archaea comprises prokaryotes that are similar to bacteria with regard to cell structure and size but differ from them in numerous molecular biological properties such as nucleotide sequence of rRNA

TABLE 1.4 Approximate Range of Conditions Suitable for Life of Different Microbial Groups

Environmental Factors	Approximate Range of Conditions Suitable for Life				
	Bacteria	**Archaea**	**Fungi**	**Algae**	**Protozoa**
Source of energy	Light Organic Inorganic	Light Organic Inorganic	Organic	Light	Organic
Source of carbon	Organic CO_2	Organic CO_2	Organic	CO_2	Organic
Redox potential (eH) (mV)	−150 (Anaerobic) +150 (Aerobic)	−200 (Strictly anaerobic) +100 (Aerobic)	>0 (Aerobic conditions for majority)	>0 (Aerobic)	>0 (Aerobic)
Temperature (°C)	−10 to +110	−10 to +110	0–50	0–40	15–40
pH	2–11	2–9	4–9	6–9	6–8

and chemical structure of the cell membrane and cell wall. This group includes extremophiles, i.e., prokaryotes living at one or two of the following extreme environmental conditions:

- Absence of oxygen and low redox potential (methanogens)
- Low (<3) pH of the environment (*Thermoplasma*)
- High (>90°C) temperature (hyperthermophiles)
- High (>10% NaCl) salinity (halophiles)

Importance of Prokaryotes for Environmental and Civil Engineering

1. Bacteria are used for the biosynthesis of useful substances in biotechnological industry, for the production of probiotics in medicine and agriculture, for the treatment of wastewater and soil bioremediation in environmental engineering, and for bioclogging and biocementation in civil engineering.
2. Many bacteria are harmful to human, animal, and plant health. Therefore, the removal or killing of harmful bacteria in water, wastewater, air, or solid waste is an important task in environmental engineering.
3. Some species of bacteria cause corrosion of metals and deterioration of materials. Prevention of bacteria-caused corrosion of metals and biodeterioration of materials are essential tasks in civil engineering.
4. Methanogenic archaea are used in anaerobic biodegradation of organic wastes. Other groups of archaea do not yet have industrial scale applications in civil or environmental engineering.

Fungi

Fungi are eukaryotic micro- or macroorganisms. They assimilate only chemical energy from organic substances selectively transported into the cell through the cell membrane. The typical diameter of the cell is about 10 μm. Cells often combine in branched filaments, called hyphae, which then combine in a network called mycelium (Figure 1.5).

Unicellular fungi are called yeasts (Figure 1.6).

The mycelium creates the visible fruit bodies of fungi called mushrooms (Figure 1.7).

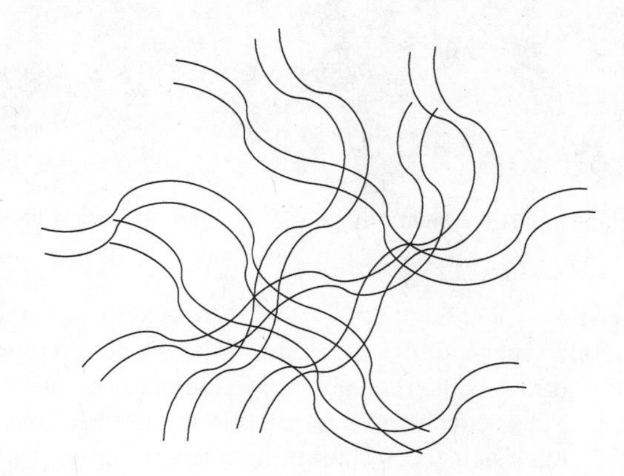

Figure 1.5 Mycelium of fungi.

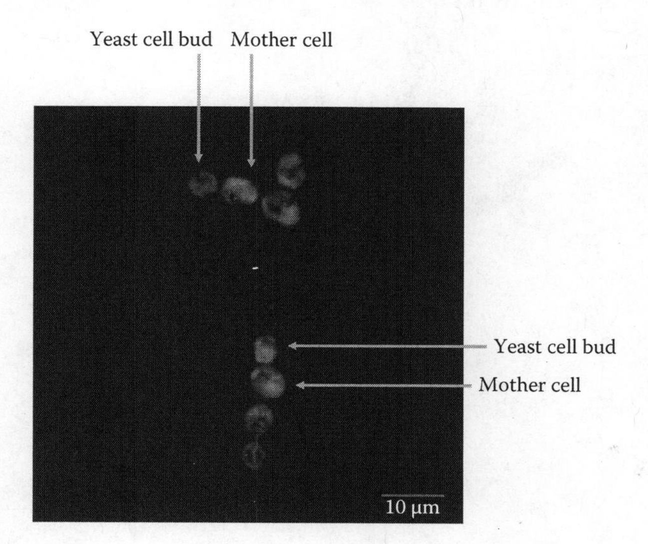

Figure 1.6 (See color insert following page 292.) Yeast cells.

Figure 1.7 (See color insert following page 292.) Cultivation of edible mushrooms on saw dust.

Mycelial fungi are typical habitants of soil and are active degraders of organic monomers and biopolymers. Fungi digest insoluble organic matter by secreting digestive enzymes outside cells, then absorbing the soluble nutrients. Most fungi are saprophytes, i.e., they consume dead organic matter. Molds are aerobic filamentous fungi growing on surfaces containing nutrients (Figures 1.8 and 1.9).

There are groups called slime molds (myxomycetes) and water molds (oomycetes), which are actually protozoans and not fungi.

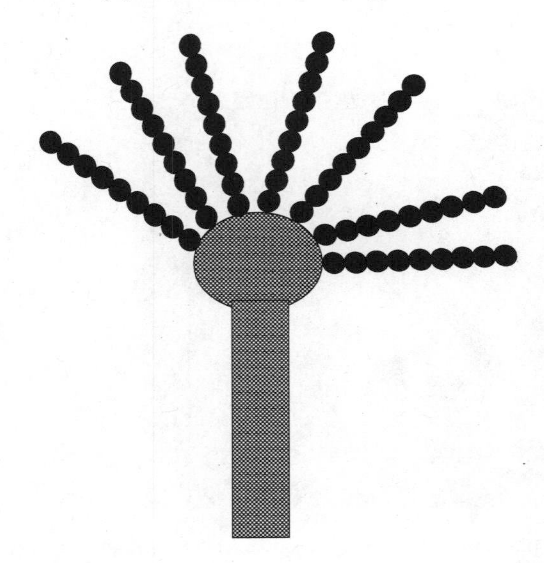

Figure 1.8 Spores of mold.

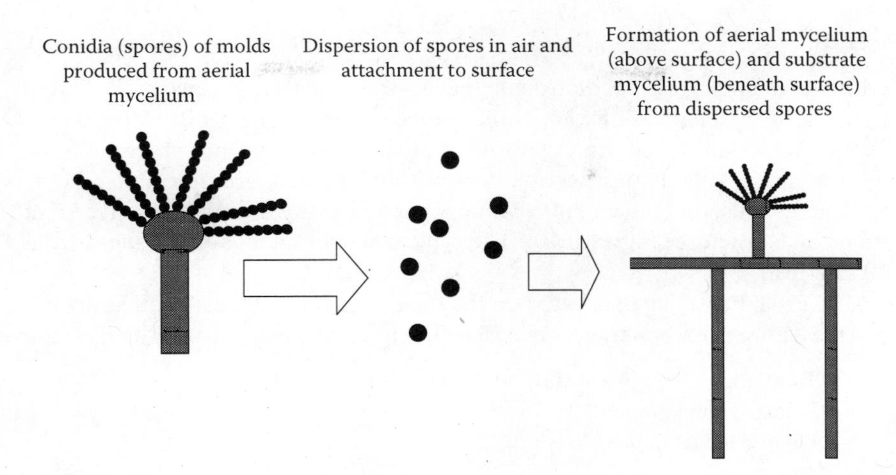

Figure 1.9 Mold.

Importance of Fungi for Environmental and Civil Engineering

1. Fungi are used in environmental engineering mainly for the composting (biotransformation) of organic wastes and biodegradation of toxic organic substances in soil. Industrial applications of fungi include production of antibiotics, enzymes (biocatalysts of chemical reactions), and cultivation of edible mushrooms (Figure 1.7). These value-added substances can be produced by fungi from solid organic waste.
2. One of the largest applications of unicellular fungi called yeasts is the production of ethyl alcohol and alcohol-containing wine, liquors, and beer. Production of fuel alcohol from carbohydrates is a rapidly developing industry. This industry is closely related to environmental engineering when cellulose-containing food, and agricultural or wood-processing wastes are used for the production of ethanol fuel.
3. Fungi are active destructors of wood. Therefore, protection of wood construction materials from fungal deterioration is essential for civil engineering. Another aspect important for civil engineering is the formation of spores by filamentous microscopic fungi known as molds. Molds have a surface mycelium and aerial hyphae that contain spores (conidia), which are often toxic for human and can cause allergy and asthma attacks. If the indoor ventilation is not sufficient and the wall, floor, and ceiling materials, and different coatings are suitable for growth of the molds, they release spores into the indoor air, causing different diseases. This release of the spores by the molds could be an essential factor for the so-called sick building syndrome, a set of the diseases that appear in improperly designed, constructed, and maintained buildings.

Algae

Algae are eukaryotic, phototrophic micro- or macroorganisms assimilating light energy and carbon dioxide as the source of carbon. They often float on the surface of water. Algae are organisms that carry out oxygenic photosynthesis with water serving as the electron donor. Many of them are unicellular, some are filamentous, and others are colonial. Unicellular algae are highly diverse in shape and structure (Figure 1.10). The typical size of an algal cell ranges from 10 to 20 μm.

The grouping of algae is based on the type of chlorophyll, cell wall structure, and the nature of carbon reserve material. The following major groups of algae are

- Green algae
- Golden-brown algae
- Algae without cell walls
- Dinoflagellates
- Red algae
- Brown algae

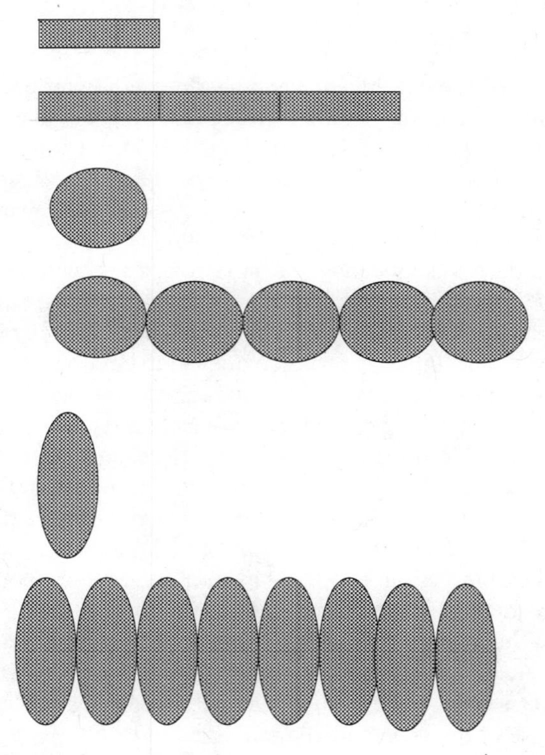

Figure 1.10 Some common shapes of algal cells.

Figure 1.11 (See color insert following page 292.) Pollution of fresh water due to excessive growth of algae and cyanobacteria caused by pollution of water with nitrogen and phosphorous. Excessive growth of phototrophs in reservoirs can clog water filters and give reservoir water a bad taste and smell.

Algae should not be confused with cyanobacteria, which are bacteria but are sometimes called as "blue-green algae," an obsolete and incorrect term.

Importance of Algae for Environmental Engineering

1. Algae remove nutrients from water.
2. Excessive growth of algae deteriorates water quality in coastal area, rivers, lakes, and reservoirs (Figure 1.11).
3. Some algae are fast growing in polluted waters and produce toxic compounds, this being the reason for the so-called red tides.
4. The spectrum of the species of microscopic algae present in natural water is used as an indication of water quality.
5. Pigments and unsaturated fatty acids can be manufactured as by-products from algae.
6. Algae can be used for the production of fuel using waste nutrients and light energy by the accumulation of lipids in the cells.

Protozoa

Protozoa are eukaryotes; these unicellular animals are capable of absorbing soluble organic food or ingest and digest other microbes or food particles inside the cell. The major type of animal nutrition is to engulf food particles and then to digest them inside the organism. The typical size of a protozoan cell ranges from

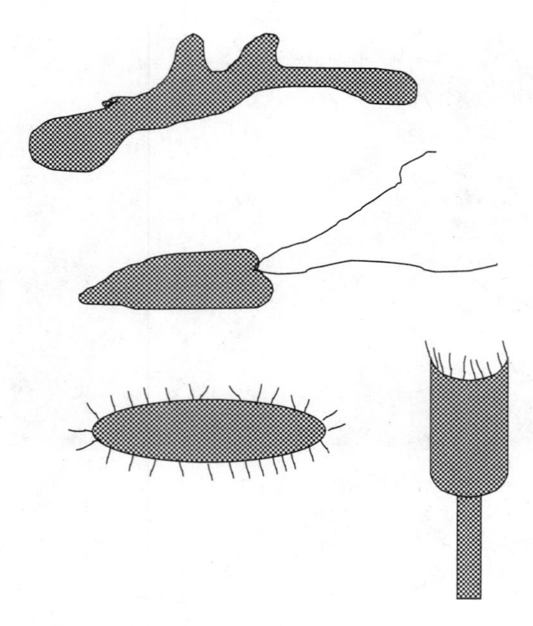

Figure 1.12 Diverse shape of protozoan cells.

10 to 50 μm. The cells produce cysts under adverse environmental conditions. These cysts are resistant to desiccation, starvation, and disinfection.

The four major groups of protozoa are distinguished by their motility mechanism (Figure 1.12):

1. Amoeba: move by means of false feet
2. Flagellates: move by means of flagella
3. Ciliates: use cilia for locomotion
4. Protozoa: no means of locomotion

Importance of Protozoa for Environmental Engineering

1. Protozoa consume suspended bacteria and bacteria from activated sludge during wastewater treatment, thus removing suspended solids and improving the quality of the treated wastewater.
2. Monitoring changes in the protozoan community is used to control the operating conditions of an aerobic wastewater treatment plant.
3. Disinfection of water polluted with the cysts of pathogenic protozoa is a technically difficult problem in environmental engineering.

2

Static Biochemistry

Static and Dynamic Biochemistry

Biochemistry can be defined as the chemistry of life. This science is conventionally classified into static biochemistry, studying the structure of the molecules participating in biochemical reactions, and dynamic biochemistry, studying the changes of these molecules during life processes.

Chemical Features of Life

The major chemical features of life are as follows:

1. Such elements as H, C, N, O, P, and S are major elements of live matter or biomass. Carbon is the most important element. It chemically bonds to other atoms in a number of ways to produce biomolecules.
2. Many molecules in the biomass are specific stereoisomers, that is, molecules having the same structural formulas, but one is the mirror image of the other, just as the left hand is a mirror image of the right hand.
3. The same types of biomolecules can be found in all forms of life, ranging from bacteria to humans.
4. The biomolecules include such groups of monomers as monosaccharides, amino acids, nucleotides, which are used for the synthesis of polysaccharides, proteins, and nucleic acids (RNA and DNA), respectively.
5. All biochemical reactions are accelerated by protein catalysts called enzymes.
6. Nucleic acids carry genetic information and are used as tools for the biosynthesis of enzymes.

An explanation of the chemical features of life is given in this chapter.

Balance of Elements

The balance of only four atoms—C, H, O, and N—is accounted for in the biotechnological calculations. For example, the most common empirical formula of biomass is $CH_{1.7}O_{0.5}N_{0.12}$. By the mass conservation law, elements do not disappear in biochemical reactions. So, all the equations must be balanced by elements. For example, aerobic biosynthesis of biomass from glucose (empirical formula $C_6H_{12}O_6$) can be described by the balanced equation:

$$C_6H_{12}O_6 + 8.1O_2 + 0.36NH_3 \rightarrow 3CH_{1.7}O_{0.5}N_{0.12} + 3CO_2 + 4.35H_2O$$

Covalent Bond

Sharing of electrons between atoms constitutes a chemical bond. If electrons are shared equally, the bond is covalent, usually shown as C–C, C–N, C–O (solid lines represent covalent bond). The covalent bond can also be a double or triple bond, shown as C=C, C=O, or, C=C, respectively.

Organic Compounds and Functional Groups

Carbon atoms combined by covalent bonds form a great variety of organic compounds, which are chains, rings, spheres, or tubes. The skeleton of a carbon chain or ring has specific attachments called functional groups. Major functional groups participating in biochemical reactions are

1. The hydroxyl group, –OH, is present in alcohols, R–OH, and carbohydrates, $(CHOH)_n$, where R is the organic radical and n is the number of carbon atoms in molecule. The hydroxyl group can be oxidized to carbonyl group, H–C=O.
2. The carbonyl group present in aldehydes, R–C(O)H, can be oxidized to the carboxyl group, HO–C=O.
3. The carboxyl group is present in carboxylic organic acids, RCOOH. Carboxylic acids are weak acids. The carboxylic group can be oxidized to CO_2 by decarboxylation, for example

$$RCOOH \rightarrow CO_2 + RH$$

4. The ester group, –COO–R, is formed from hydroxyl and carboxyl groups during dehydration, that is, removal of water molecule. The addition of a water molecule, hydrolysis, can yield an organic acid and alcohol:

$$R_1\text{–}COO\text{–}R_2 + H_2O \leftrightarrow R_1COOH + R_2OH$$

5. The amino group, $-NH_2$, is present in amino acids, $HOOC\text{–}R\text{–}NH_2$, and organic amines, RNH_2.
6. The sulfhydryl group, –SH, is present in some amino acids.
7. The phosphate group, $-PO_4^{3-}$, is present in different compounds, for example, phospholipids, nucleotides, and nucleic acids.

Polarity of Chemical Bond

During the sharing of electrons in covalent bonding, the atom with greater electronegativity draws the electron pair closer to it, forming a polar covalent bond. The dipole moment of the chemical bond is multiplication of the partial charges of bonding atoms with a distance between them. Unshared electron pairs make large contributions to dipole moments. For example, unshared electrons on the oxygen atom in the water molecule contributes to the dipole moment of water molecule, H_2O.

Water

Water is the major component of all biochemical reactions. The content of water in cells ranges from 70% to 90%. The water molecules, H–O–H, have a weak negative charge at the oxygen atom and a weak positive charge at the hydrogen atoms. Due to this, the water molecule is slightly polar. Therefore, all ionic and polar molecules are soluble in water due to interactions between the dipole moments of water and polar molecules. Nonpolar molecules, such as lipids or hydrocarbons, are not soluble in water.

pH of Solution

Water is ionized into hydrogen ions (H^+) and hydroxide ions (OH^-):

$$H_2O \leftrightarrow H^+ + OH^-$$

$[H^+]\,[OH^-] = 10^{-14}$ in pure water at 25°C, where $[H^+]$ and $[OH^-]$ are molar concentrations of protons and hydroxide ions. More exactly, there must be considered not the molar concentrations but the molar activities of hydronium ions, H_3O^+, and hydroxide ions, which could be replaced by the molar concentrations at low concentrations. The molar concentration of protons in pure water $= 10^{-7}$. The concentration of protons is usually expressed in terms of $pH = -\log[H^+]$. The pH of pure water is 7, which is neutral pH. When acids are dissolved in water, the pH of the solution is decreased. Addition of base increases the pH of solution. Examples of pH in nature and engineering systems are given in Table 2.1.

Intermolecular Forces

Intermolecular forces are significantly weaker than covalent bonds but are extremely important in biological systems, creating 3D structures of the molecular aggregates. These forces are

1. The hydrogen bond is the electrostatic interaction between any electronegative atom and a hydrogen atom, which is bonded to another electronegative atom. The most common hydrogen bonds in biological

TABLE 2.1 Values of pH in Nature and Engineering Systems

pH	Examples from Nature and Engineering
12	Solution of lime, $Ca(OH)_2$, used to increase pH in water treatment
11	Solution of ammonia
10	Surface of concrete (depends on age)
9	Water of soda lake, soap
8	Sea water
7	Pure water, intracellular pH, freshwater
6	Rain water
5	Water of acid rain, skin
4	Tomato juice, beer
3	Vinegar, orange juice
2	Lemon juice, stomach acid
1	Mine drainage water, alum (coagulant for water treatment)

systems are in the groups −OH···O− and −OH···N− (dotted lines represent hydrogen bonds). The force of the hydrogen bond is approximately 10% of the force of the C−H covalent bond. Numerous hydrogen bonds between parts of biopolymer molecules ensure their self-folding and 3D structure.

2. The force of dipole–dipole interaction between molecules is approximately 1% of the covalent bond force. This interaction could be important in microbial cell aggregation and adherence of microbial cell to other surfaces.
3. London forces (other terms are van der Waals forces or dispersion forces) are due to the weak (approximately 0.1% of the force of covalent bond) attractions between temporary multipoles in molecules at close range without permanent multipole moments. These forces ensure interaction between nonpolar molecules. London forces participate in biopolymer folding, in biorecognition of the molecules, that is, in the mutual orientation of the neighboring molecules, and in the attachment of microbial cells to surfaces.

Hydrophobic Substances and Hydrophobic Forces

Hydrophobic substances are nonpolar and not soluble in water. Hydrophobic molecules are repelled by molecules of water and aggregated together (Figure 2.2a). This aggregation is called as hydrophobic interaction. The aggregation of lipids around cell cytoplasm forms a cell barrier, the cell membrane, which isolates cell contents from the environment and prevents the free exchange of polar molecules between cell and environment (Figure 2.1a).

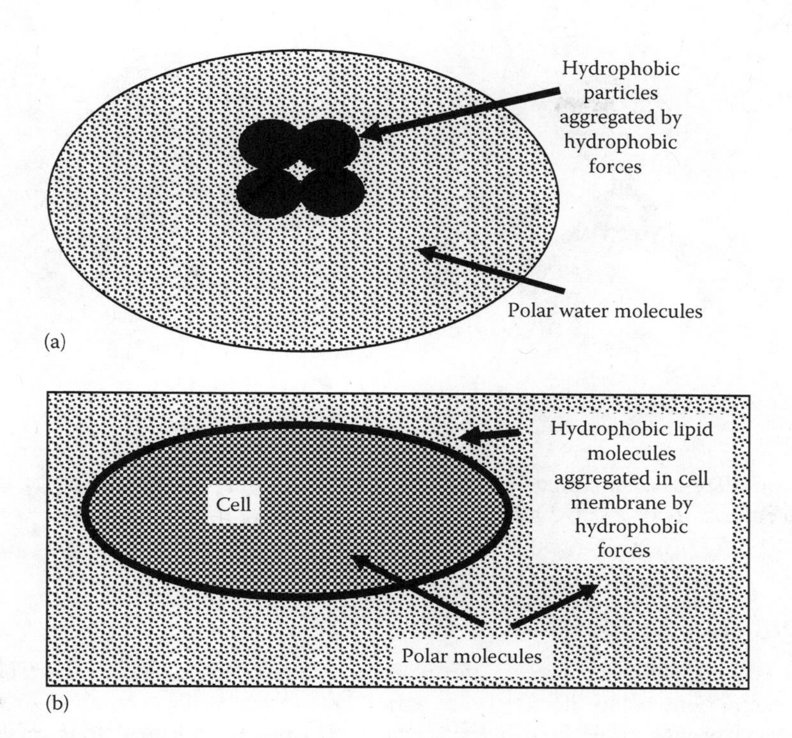

Figure 2.1 Hydrophobic interactions between the droplets (a) or molecules (b) of a hydrophobic substance.

Stereoisomers

A very essential feature of many biological molecules is that they exist in the cell, only as one stereoisomer. When four different moieties are bound to a carbon atom, such molecules may exist as stereoisomers. Such molecules have the same structural formulas, but one is the mirror image of the other, just as the left hand is a mirror image of the right hand (Figure 2.2). Optical isomers of carbohydrates and amino acids are given the designations D and L. D-Sugars predominate in biological systems, whereas amino acids in cell are in L-form.

Biological Monomers and Polymers

Polymers are major components of cell, typically constituting 90% of dry cell matter. The other 10% is made up of Na^+, K^+, HPO_4^{2-}, Cl^- ions, and low-molecular-weight organic compounds. These compounds are mainly monomers, repeating units of the biopolymers. The major cell monomers are monosaccharides, amino acids, and nucleotides, which form, due to the formation of covalent and other bonds, polysaccharides, proteins, and nucleic acids, respectively. Lipids also can

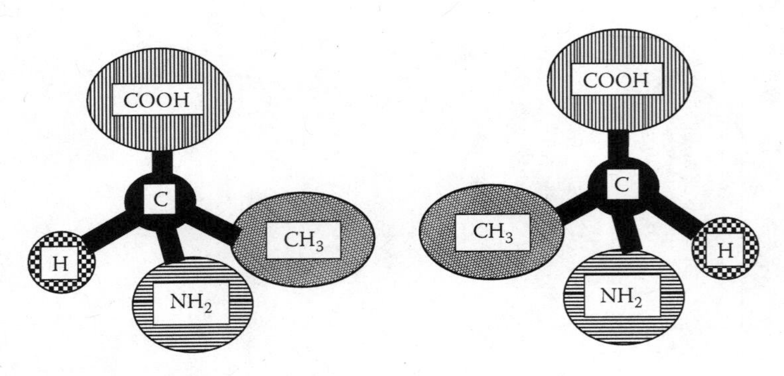

Figure 2.2 Stereoisomers of the amino acid alanine.

be considered as monomers producing membranes due to aggregation of lipid molecules by hydrophobic forces.

Monosaccharides

Monosaccharides are carbohydrates with the common empirical formula CH_2O, an exemption being deoxyribose. Every monosaccharide contains several hydroxyl groups, −OH, and one carbonyl aldehyde, −CHO, or ketone, −C=O, group. Monosaccharides contain from three to seven atoms of carbon and are called, respectively, triose (glyceraldehyde, for example), tetrose, pentose (ribose and deoxyribose, for example), hexose (glucose and fructose, for example), and septose. They exist in linear form and ring form (Figure 2.3).

The major functions of monosaccharides in cells include

1. Providing a source of energy
2. Assimilation of CO_2 using energy of light, that is, they are the primary product of photosynthesis: $6CO_2 + 6H_2O + \text{energy of light} \rightarrow C_6H_{12}O_6 + O_2$
3. Functioning as a key component in the synthesis of DNA (deoxyribose) and RNA (ribose)
4. Functioning as a key component in the synthesis of cell walls of many microorganisms

Oligosaccharides and Polysaccharides

Molecules of monosaccharides combine through ether bonds, −O−, with the removal of a water molecule, as shown in the reaction between two monosaccharides:

$$R_1\text{–OH} + R_2\text{–OH} \leftrightarrow R_1\text{–O–}R_2 + H_2O$$

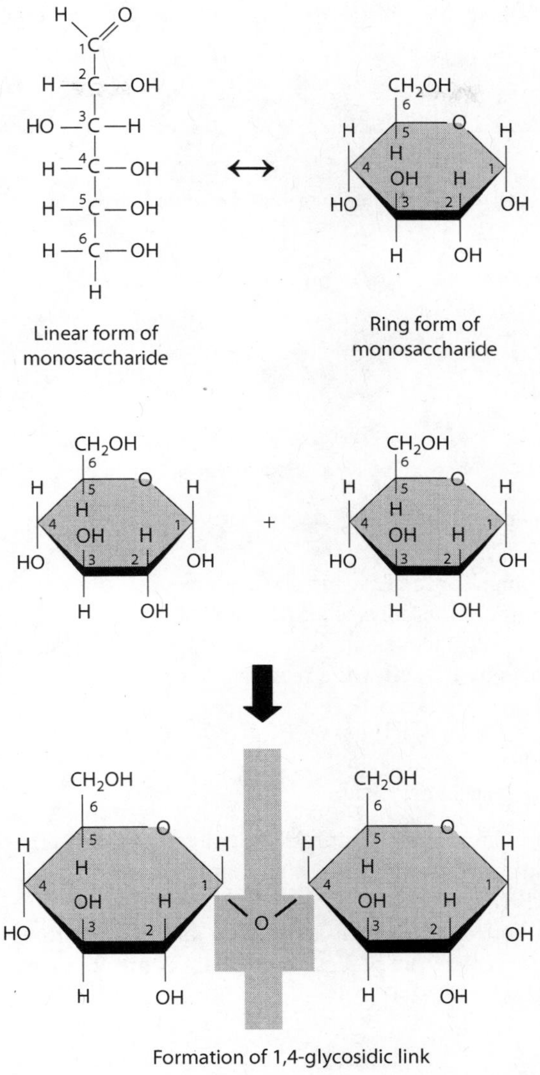

Linear form of monosaccharide

Ring form of monosaccharide

Formation of 1,4-glycosidic link

Figure 2.3 Monosaccharides.

The bond between two monosaccharides is called a glycosidic bond (Figure 2.3). The products of this reaction are either oligosaccharides containing a few monomer units or polysaccharides containing thousands of monomer units combined in a linear chain or branching chains. The glycosidic bond can be either α′- or β-bond, depending on the directions of the substitute groups on the carbons flanking the ring oxygen (Figure 2.4). Glycosidic bonds usually connect the first and fourth atoms of C in the monomers (α′-1,4- and β-1,4 bonds) that are in the

ά-1,4 Glycosidic bonds combining monomers of glucose in the chain of glycogen and starch

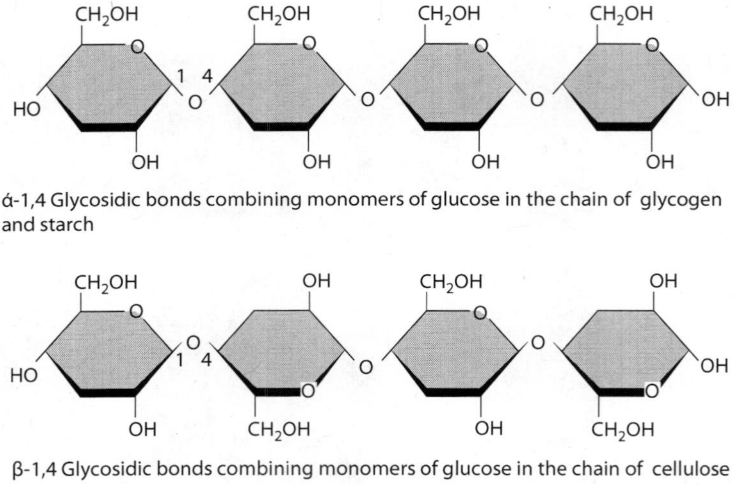

β-1,4 Glycosidic bonds combining monomers of glucose in the chain of cellulose

Figure 2.4 Polysaccharides.

polysaccharide chain, or the first and sixth atoms of C (α′-1,6- and β-1,6 bonds) at the branching point of the polysaccharide chains.

Polymerization of glucose and depolymerization of polysaccharide by hydrolysis can be shown by the following reaction:

$$nC_6H_{12}O_6 \leftrightarrow (C_6H_{10}O_5) + nH_2O$$

The important polysaccharides are

1. Cellulose, which is a structural component of cell walls of plants and fungi, synthesized from glucose monomers connected with β-1,4 glycosidic bonds (Figure 2.4)
2. Glycogen and starch, which are carbon and energy stores of microbial cells, synthesized from glucose monomers connected with α′-1,4 glycosidic bonds (Figure 2.4)

Lipids and Membranes

Lipid molecules in cells are joined together, but not by covalent bonds. Thousands of lipid molecules are joined together in lipid membranes or granules due to hydrophobic forces. The types of lipids are as follows:

1. Triglycerides (fats), which are typically three long-chain fatty acids (with 14–20 atoms of C), linked to a glycerol molecule by an ester bond.
2. Phospholipids, which consist of glycerol linked to two long-chain fatty acids and one phosphate group connected to an organic compound, usually choline.

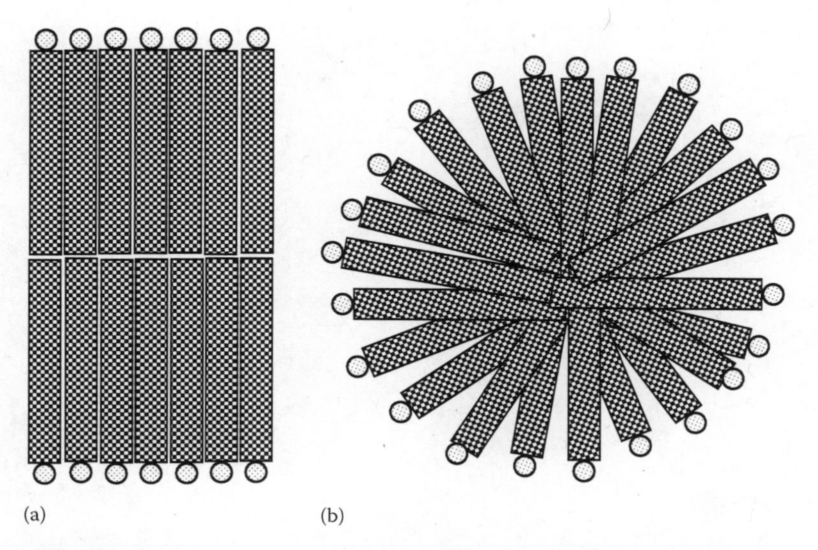

Figure 2.5 Structure of (a) lipid membrane and (b) lipid micelle. ◎ Hydrophilic part of the molecule. ▬ Hydrophobic part of the molecule.

3. Sterols, which are hydrophobic substances with a common structure of four fused carbon rings.
4. Waxes, which are long-chain fatty acids linked to long-chain alcohols or carbon rings.

Lipid Membranes

The carboxylic acid head of fatty acid is polar and hydrophilic, while the long carbon chain of fatty acid is nonpolar and hydrophobic. Therefore, in aqueous solution, fatty acids form membranes or micelles that shield the hydrophobic carbon chain from the polar water molecules (Figure 2.5). Aggregated molecules of lipids, containing hydrophobic and hydrophilic parts, form a bilayer lipid membrane surrounding the cell and cellular organelles (Figure 2.5). Usually, phospholipids are used to form a bilayer cell membrane, separating cell cytoplasm or organelle matrix from surrounding medium (Figure 2.6).

Membrane Melting and Freezing

The long carbon chain of fatty acids in triglycerides or phospholipids may either be saturated (like an alkane) or unsaturated with one or more double bonds (like an alkene) (Figure 2.6). The melting temperature of the lipid structure, membrane, or micelle depends on the content of unsaturated fatty acids. In natural fatty acids, a double bond forms *cis*-isomer, where both parts of the isomer are

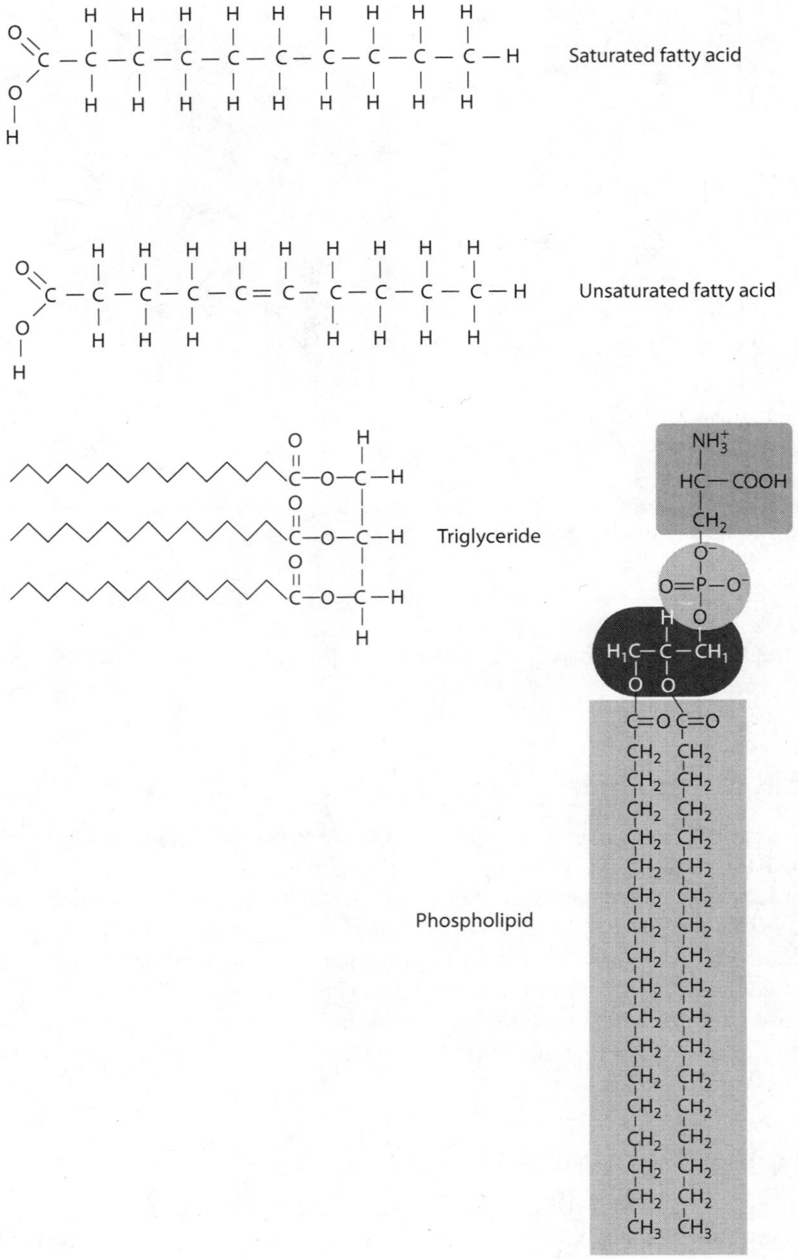

Saturated fatty acid

Unsaturated fatty acid

Triglyceride

Phospholipid

Figure 2.6 Lipids and long-chain fatty acids.

on the same side of double bond. This decreases intermolecular hydrophobic interaction between neighboring triglyceride or phospholipid molecules, causing decrease of the melting/freezing temperature. When the phospholipid membrane of a cell freezes, it cannot function effectively as a cell barrier. Therefore, microorganisms living at low temperatures have a high content of unsaturated fatty acids in their cell membranes.

Amino Acids and Proteins

There are 20 amino acids used for protein synthesis (Table 2.2).

Each amino acid contains at least one amino group, $-NH_2$, and at least one carboxyl group, $-COOH$ (Figure 2.7). Natural amino acids are optically active molecules (an exception being glycine), because four different groups are attached to the central carbon atom: an amino group, a carboxylic group, a hydrogen atom, and a group that is specific for each amino acid.

Molecules of amino acids combine by a peptide bond, $-CO-NH-$, with the removal of a water molecule, as shown in the reaction between two amino acids (Figure 2.7):

$$R_1\text{–}CH(NH_2)COOH + R_2\text{–}CH(NH_2)COOH$$

$$\rightarrow R_1\text{–}CH(NH_2)CO\text{–}NH\text{–}CH(COOH)\text{–}R_2 + H_2O$$

The products of this reaction are peptides containing a few amino acids or proteins containing hundreds of amino acids.

Structure of Protein Molecule

There are four levels of protein structure (Figure 2.8):

1. The primary structure is the sequence of amino acids connected by peptide bonds.
2. The secondary structure is the bonding pattern of the amino acids (e.g., helix, sheet, etc.) created by the hydrogen bonds.
3. The tertiary 3D structure consists of the domain, where the sheets or helixes fold on each other and become stable due to hydrogen, hydrophobic, disulfide bonds, and salt bridges.
4. The quaternary structure consists of several protein molecules connected together as one unit.

3D self-folded tertiary and quaternary structures are created by the following forces:

1. Hydrophobic interactions between hydrophobic amino acids
2. Strong $-S-S-$ bonds between SH-containing amino acids in the molecule

TABLE 2.2 Amino Acids

Amino Acid and Its Code in the Written Sequence of Protein	Structure
Alanine (Ala)	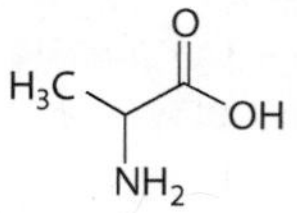
Arginine (Arg)	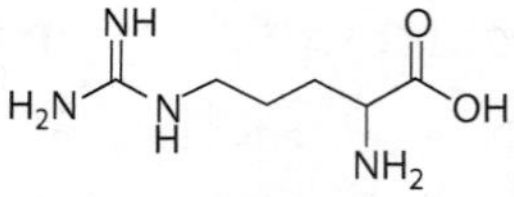
Asparagine (Asn)	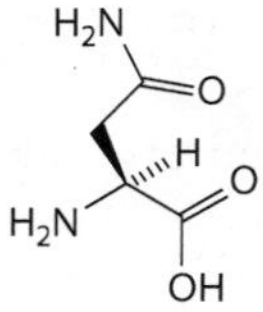
Aspartic acid (Asp)	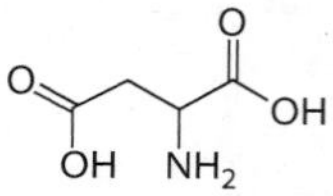
Cysteine (Cys)	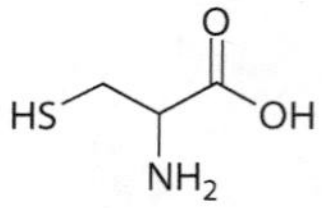
Glutamic acid (Glu)	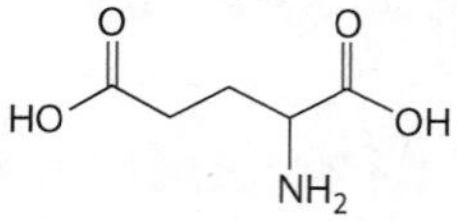
Glutamine (Gln)	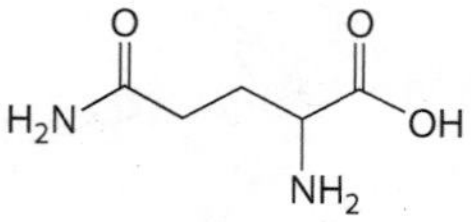
Glycine (Gly)	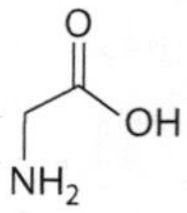
Histidin (His)	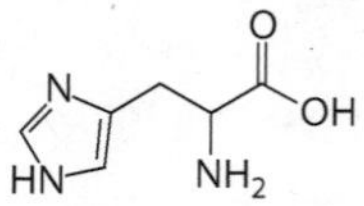
Isoleucine (Ile)	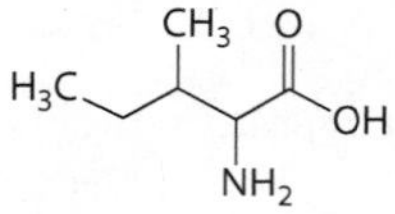

TABLE 2.2 (continued) Amino Acids

Amino Acid and Its Code in the Written Sequence of Protein	Structure
Leucine (Leu)	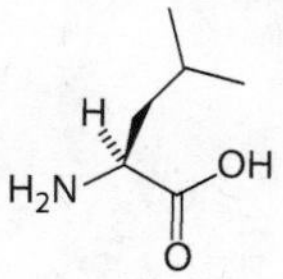
Lysine (Lys)	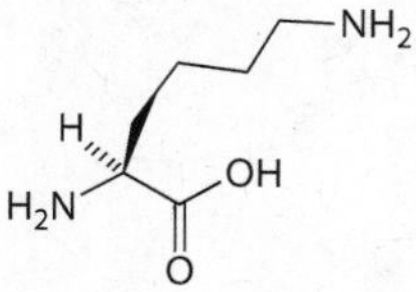
Methionine (Met)	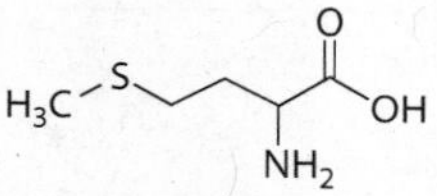
Phenylalanine (Phe)	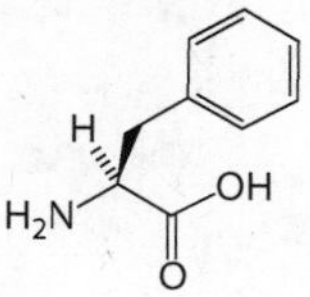
Proline (pro)	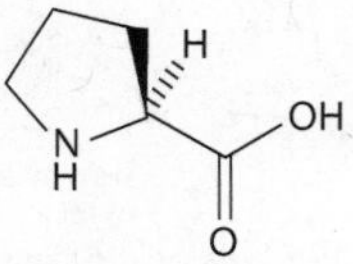
Serine (Ser)	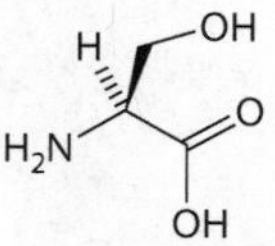
Threonine (Thr)	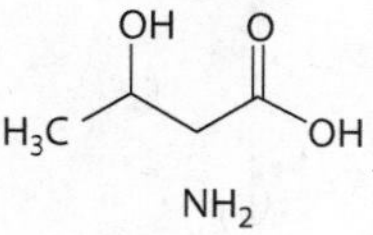
Tryptophan (Trp)	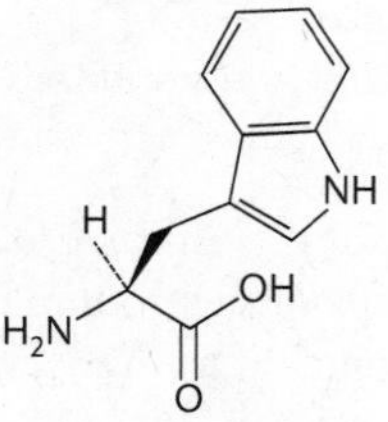

(continued)

TABLE 2.2 (continued) Amino Acids

Amino Acid and Its Code in the Written Sequence of Protein	Structure
Tyrosine (Tyr)	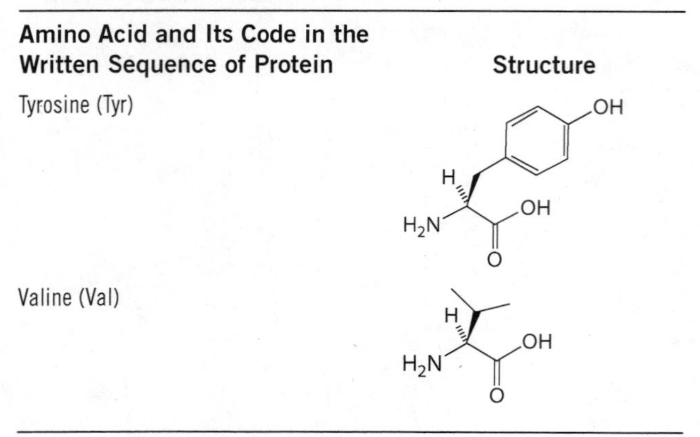
Valine (Val)	

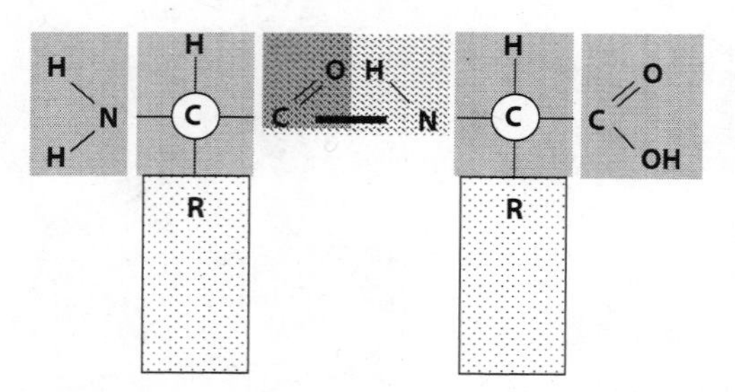

Figure 2.7 Amino acid and peptide.

3. –COO–Ca–OOC– and –COO–Mg–OOC– bridges between free carboxylic groups of dicarbonic amino acids
4. Hydrogen bonds –OH···O– and –OH···N–

This 3D-folded sequence of amino acids determines the biological activity of protein in cell.

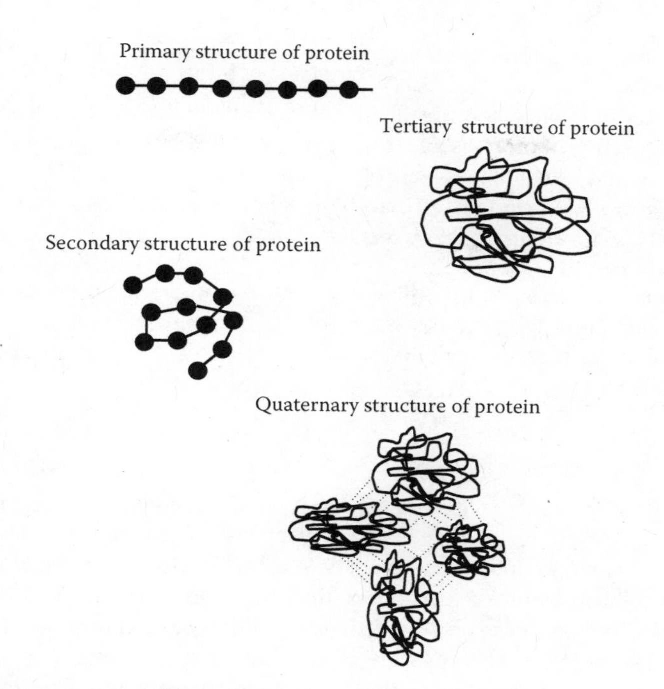

Figure 2.8 Levels of protein structure.

Globular and Fibrous Proteins

There are two types of tertiary structures: globular proteins (globe = sphere), which are usually water soluble, and fibrous proteins, which are insoluble in water. Self-folding in globular proteins is performed so that the hydrophobic parts of amino acids are directed toward the interior of the protein molecule, whereas the hydrophilic parts are bound outward, allowing dipole–dipole interactions with water molecules, ensuring solubility of the protein molecule. The majority of globular proteins are enzymes, protein catalysts of biochemical reactions. Fibrous proteins form long protein filaments due to self-folding with hydrophobic groups of amino acids directed outwards, which ensures insolubility of molecule in water. Fibrous proteins perform structural functions in the cell or organism, connecting different parts, covering the organism, or contracting fibers in muscles.

Denaturation of Proteins

Denaturation is a change of the secondary, tertiary, or quaternary structures of proteins following the inactivation of the proteins. The environmental engineering aspects of denaturation are as follows:

1. Thermal denaturation of proteins in pasteurization and sterilization processes
2. Denaturation of proteins by chemical oxidants in disinfection process
3. Denaturation of proteins due to destruction of hydrogen bonds by organic solvent and salts: toxic effects of organic solvents and salts
4. Denaturation of proteins caused by reaction of amino acids with heavy metals: toxic effect of heavy metals

The denaturation can be reversible if the effect of the denaturing agent is gentle and the action is short. In this case, the 3D structure of protein can be self-restored. Globular proteins tend to denature more easily than fibrous proteins.

Enzymes

The majority of proteins are enzymes, catalysts of all biochemical reactions in a cell. Being a catalyst, an enzyme does not disappear after a reaction, but regenerates in every reaction and starts up a new reaction cycle. An enzyme decreases the energy of activation of a reaction, thus increases the rate of the reaction (Figure 2.9). To reproduce, a prokaryotic cell requires several thousand biochemical reactions. So, several thousands of specific enzymes must be synthesized in a prokaryotic cell to catalyze these biochemical reactions. The number of

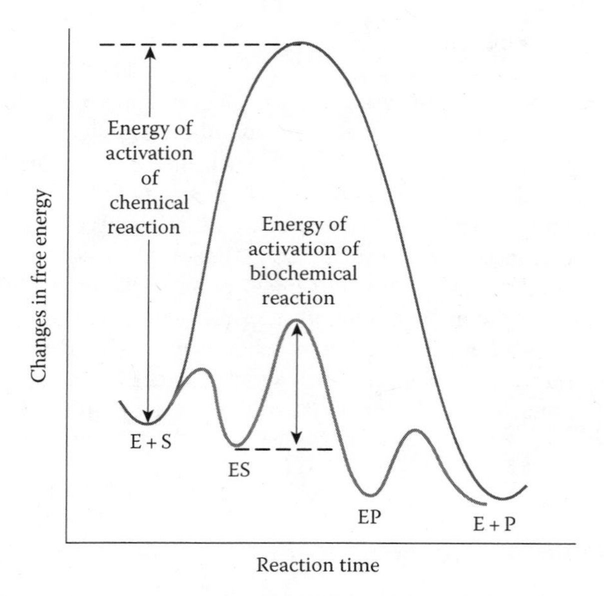

Figure 2.9 Catalytic action of enzyme due to decrease in activation energy. E, enzyme; S, substrate; ES, enzyme–substrate complex; EP, enzyme–product complex, P, product.

biochemical reactions and, respectively, the number of enzymes in an eukaryotic cell could be 10 times bigger than that in a prokaryotic cell.

Coenzymes

It is the part of an enzyme that binds to a substrate molecule and is termed the active center of enzyme. Any changes in the structure of the active center caused by physical or chemical factors terminates enzymatic activity. Very often, the active center of an enzyme is a cofactor (coenzyme), which is a simple organic molecule able to attach/detach to an enzyme molecule. Examples of these cofactors (coenzymes) are NAD^+ and $NADP^+$ (oxidized nicotinamide adenine dinucleotide and oxidized nicotinamide adenine dinucleotide phosphate), FMN (flavine mononucleotide), and FAD (flavin adenine dinucleotide), which play an important role in oxidation–reduction reactions.

Nucleotides

Nucleotides are monomers of deoxyribonucleic acids (DNA) and ribonucleic acids (RNA), as well as cofactors in cell metabolism and control. The cofactors playing an important role in energetic metabolism are adenosine triphosphate (ATP) and nicotinamide adenine dinucleotide (NAD^+). Each nucleotide contains a nitrogen-containing base, phosphate, and sugar (Figure 2.10).

The combination of a nitrogen-containing base with sugar is called a nucleoside. Nucleotides also exist in activated forms containing two or three phosphates, called nucleotide diphosphates or triphosphates, respectively. If the sugar in a nucleotide is deoxyribose, the nucleotide is called a deoxynucleotide; if the sugar is ribose, the term ribonucleotide is used.

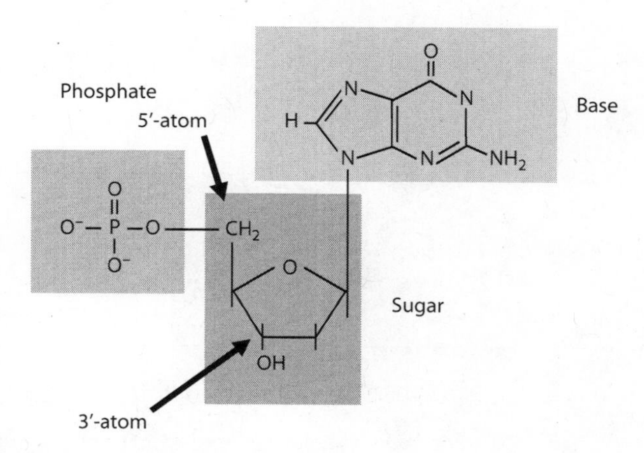

Figure 2.10 Structure of nucleotide.

The sugar in nucleotides is ribose in RNA or deoxyribose in DNA. The four bases of DNA are adenine (A), guanine (G), cytosine (C), and thymine (T). The same bases, with the exception of thymine, are used in RNA, where Uracil (U) replaces thymine as the fourth base. Adenine and guanine are derivatives of purine, but cytosine, thymine, and uracil are derivatives of pyrimidine.

Oligonucleotides

Nucleotide molecules are linked by phosphodiester bonds, which are double ester bonds between the phosphorus atom in a phosphate group PO_4^{3-} and two other molecules of ribose or deoxyribose (R–OH) connected with base (B_i) over two ester bonds:

$$B_1\text{–O–R–OH} + \text{OH–R–O–}B_2 + HPO_4^{2-}$$
$$\rightarrow B_1\text{–O–R–O–}PO_2^{-}\text{–O–R–O–}B_2 + H_2O$$

The products of this reaction are either oligonucleotides containing a few nucleic acids or polynucleotides containing hundreds of nucleic acids. A polynucleotide chain has no branches. Each strand of nucleic acid is made up of a sugar–phosphate backbone with different purine or pyrimidine bases joined by N-glycosidic bonds to another repeating sugar–phosphate backbone (Figure 2.11). Oligonucleotides containing from 8 to 20 nucleotides are used in environmental engineering as specific probes for the detection and identification of microorganisms.

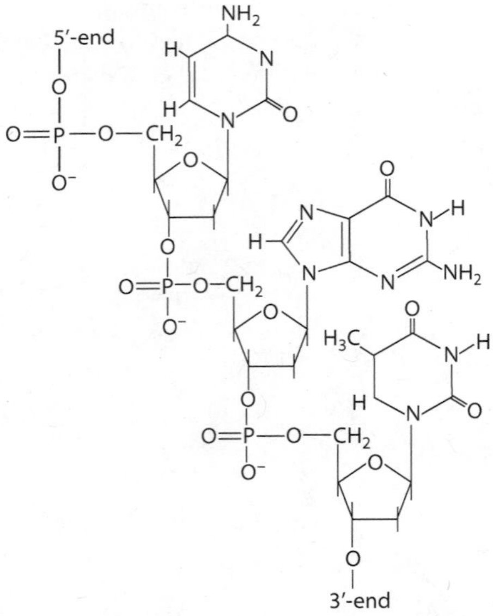

Figure 2.11 Structure of nucleic acid strand.

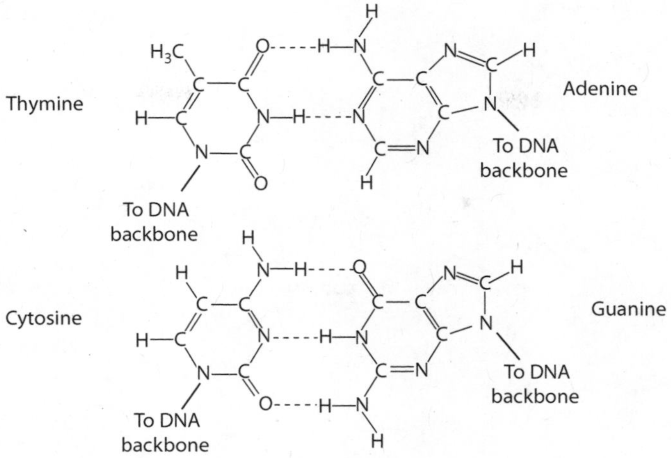

Figure 2.12 Base pairing in DNA. Dotted lines show hydrogen bonds between bases.

DNA Structure

DNA consists of two sugar–phosphate backbones, which are held together by hydrogen bonds between the bases on the two strands. The absence of hydroxyl group at 2′ atom of C in deoxyribose makes the DNA backbone more stable in comparison with RNA. Hydrogen bonds in DNA ensure the specific binding between the pairs of the bases: adenine pairs with thymine only and cytosine pairs with guanine only (Figure 2.12).

As a consequence of AT and GC base pairing in DNA, there are always the same number of A and T residues and G and C residues. The content of G+C residues (=100% − content of A+T residues) is an identification character for specific DNA.

Due to the strict stereochemical pairing of bases connected to the deoxyribose–phosphate backbone of DNA, two strands of DNA having complementary sequences of nucleotides wrap around each other to form a double helix, where the sugar–phosphate backbone is on the outside and the bases are in the middle (Figures 2.13 and 2.14). The outer phosphate groups give a high negative charge to DNA molecule, so DNA is always bound by cations or positively charged proteins.

The strand polarities are opposite each other. The sequence of one strand of DNA is customarily read from 5′ to 3′ atom of deoxyribose in the strand backbone, for example, 5′-ATTCTGCACCCGT-3′, where A is adenine, T is thymine, C is cytosine, and G is guanine.

DNA Melting

The strands in the double helix of DNA are held together by hydrogen bonds between the nearest bases and by hydrophobic interactions between the stacked

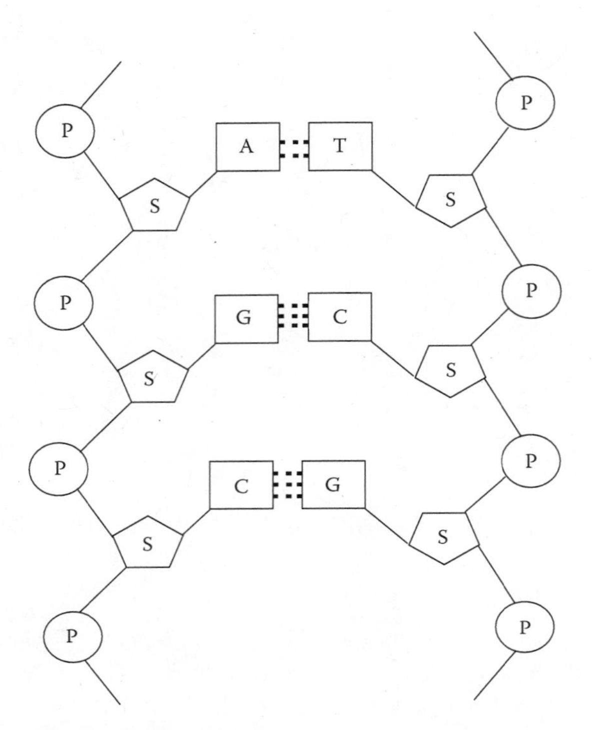

Figure 2.13 Structure of DNA.

ones on top of the other bases. These weak interactions can be disrupted by heat, as well as by acidic or basic conditions, or substances that destroy hydrogen bonds. If hydrogen bonds are destroyed, the DNA strands are separated. This separation is referred to as DNA melting.

There are two intermolecular hydrogen bonds in an AT base pair and three intermolecular hydrogen bonds in a GC base pair. This difference in number of hydrogen bonds in AT base pairs in comparison with GC base pairs determines the greater stability of DNA structures with a higher GC base pair content. This explains why DNA of thermophilic microorganisms living at temperatures higher than 50°C has higher content of G+C than DNA of mesophilic microorganisms living at 20°C–30°C.

DNA as a Carrier of Genetic Information

The sequence of nucleotides in DNA carries genetic information, which is an inherited program of cell life. During DNA replication, two complementary strands in the molecule of DNA are separated and each strand is used to produce a complementary strand of DNA so that new two copies of DNA molecule are synthesized. Thus DNA replication is making copies of the genetic information of a cell (Figure 2.15).

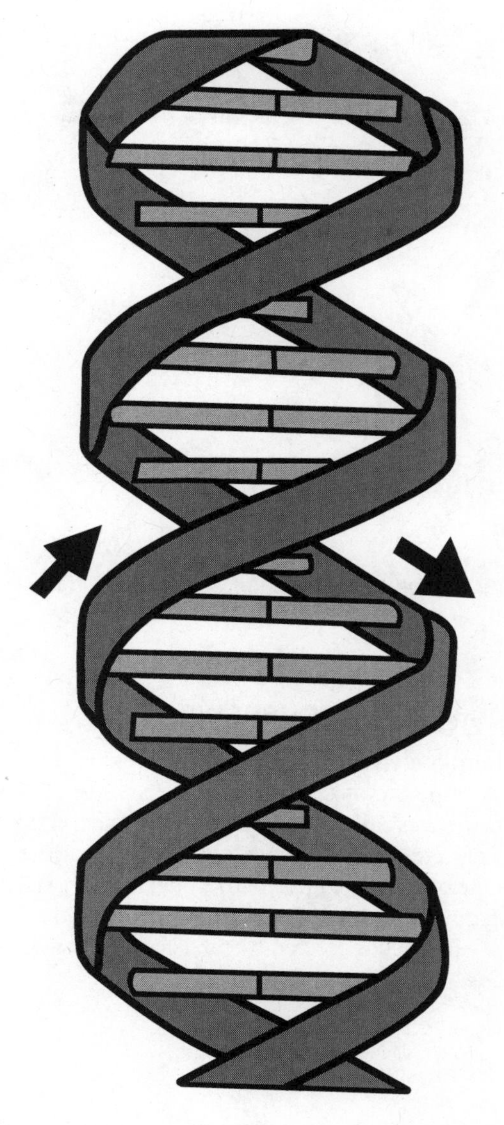

Figure 2.14 DNA Double helix. Arrows show the opposite direction of DNA synthesis in complementary strands.

The information stored in the DNA double helix is redundant. If one base in the strand is missed or changed, the complementary base on the opposite strand still contains the information, so the missed base on one strand of DNA can be repaired. If it is not repaired, the change in the DNA nucleotide sequence, called mutation, will be inherited.

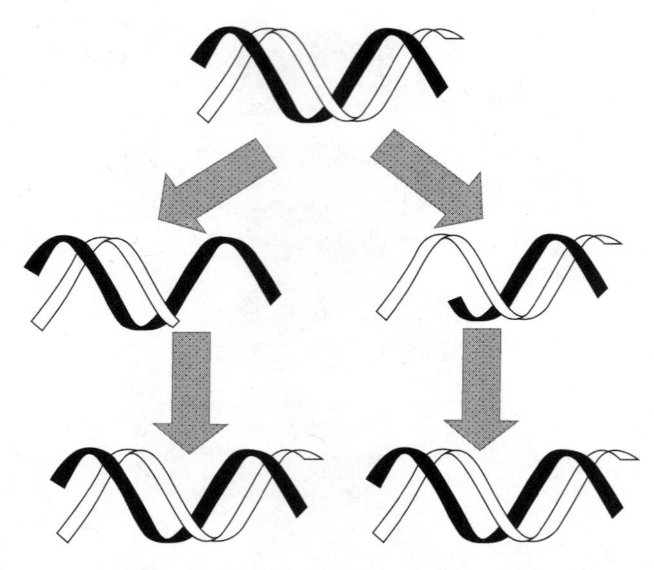

Figure 2.15 DNA replication as a copying of two new strands of DNA on two existing strands of DNA.

DNA Sequences and Sequencing

There are several million base pairs in DNA of prokaryotes and several billion base pairs in DNA of eukaryotes. For many of them, the DNA sequences have been determined (a process called DNA sequencing) and are stored in the national, international, and private databases, many of which are accessible by the scientific community. One of the most informative public database is GenBank (http://www.ncbi.nlm.nih.gov/Genbank/index.html), an annotated collection of all publicly available DNA sequences created by the National Center for Biotechnology Information of USA.

RNA

RNA is usually single-stranded but there are double-stranded sites inside the molecule. There are four types of RNA, which are important in converting the DNA nucleotide sequence in a gene to a sequence of amino acids in a protein:

1. Messenger RNA (mRNA)
2. Transfer RNA (tRNA)
3. Ribosomal RNA (rRNA)
4. Micro RNA (miRNA) and other small RNAs

Messenger RNA

Messenger RNA (mRNA) is a copy of the information carried by a gene (a sequence of nucleotides coding information for the synthesis of one enzyme, for example) on the DNA. mRNA is a sequence of nucleotides complementary to the gene sequence. mRNA is produced by a process called as transcription. The role of transcribed mRNA is to transfer the portion of information from DNA to the translation tools of cell.

Transfer RNA

Transfer RNA (tRNA) is a molecule that binds a specific amino acid and transfers it to the sequence of mRNA so that the sequence of nucleotides in mRNA is translated into a sequence of amino acids. For this translation, nucleotide triplets in a specific part of tRNA molecule, called the anticodon, complementarily pairs with a triplet of nucleotides (called codon) in the mRNA molecule. There are 64 ($=4^3$) combinations of 4 nucleotides in triplets.

Genetic Code

Due to complementary pairing of nucleotide triplets on tRNA and mRNA, the sequence of 64 trinucleotides of mRNA is translated as the sequence of 20 amino acids. tRNA is a molecule with the same structure in all organisms (4 arms and 3 loops with different functions) containing 75–95 nucleotides.

Ribosomal RNAs and Translation

Ribosomal RNA (rRNA) is a component of ribosomes, the particles synthesizing proteins in the cell. The ribosomes of eukaryotes and prokaryotes contain different proteins and different rRNA molecules, as shown in Table 2.3.

16S rRNA

The nucleotide sequence of the 16S rRNA and the sequence of related gene of 16S rRNA are widely used in environmental microbiology for the detection and identification of prokaryotes in environmental engineering systems. It is because the gene of rRNA is the least variable gene in prokaryotic cell. Therefore, the genes of 16S rRNA of several thousands of prokaryotic species were sequenced and the sequences are stored in GenBank and other databases, for example, in Ribosomal Database Project (http://rdp.cme.msu.edu/).

TABLE 2.3 Content of Prokaryotic and Eukaryotic Ribosomes

Organism	Size of Ribosome	Large Subunit (and RNA Molecules)	Small Subunit (and RNA Molecule)
Prokaryote	70S	50S (5S rRNA, 23S rRNA)	30S (16S rRNA)
Eukaryote	80S	60S (5S rRNA, 5.8S rRNA, 28S rRNA)	40S (18S rRNA)

Note: S denotes Svedberg unit that is used in analytical ultracentrifugation of polymers. Bigger particles sediment faster and have higher Svedberg values. The S units of the ribosome cannot be a sum of S units of the subunits because they represent measures of sedimentation rate that depends not only on mass but also on the shape of the molecule.

Functions of rRNA in Ribosome

Ribosomal RNA molecules make up at least 80% of the RNA molecules found in a typical cell. rRNAs molecules play several roles in translation performed by the ribosomes:

1. A catalytic role in protein synthesis.
2. A recognition role. For example, the 16S rRNA of a prokaryotic ribosome binds to a start codon in mRNA and is involved in the correct positioning of both the mRNA and the tRNA; the rRNA molecules have a structural role forming the scaffold on which the ribosomal proteins assemble.

MicroRNA

MicroRNAs, 19–25 nucleotides in length, are involved in the regulation of transcription and translation in eukaryotes. These molecules are partially complementary to one or more mRNA molecules, and their main function is to inhibit protein translation. mRNA can also initiate degradation of the mRNA molecule, thus regulating the translation of the protein coded by specific mRNA. Small inhibitory RNA (siRNA) are similar to microRNA but inhibition by siRNA requires an exact match to the target mRNA. These and other small RNA molecules are considered as tools that control production of specific proteins in the cell.

Summary on the Functions of Monomers and Polymers in Cell

The summary of the functions of monomers and polymers in cell structure is shown in Table 2.4.

The groups of these functions are as follows:

1. Formation of cell structure
2. Biological energy production and storage

TABLE 2.4 Functions of Monomers and Polymers in Microbial Cell Structure

Cell Part	Type of Substance	Function of Substance
Capsule around cell	Polysaccharides	Cell protection against starvation, desiccation, or intoxication
Cell flagella, cilia, pili, and fimbriae	Proteins	Cell motion, attachment, interactions
Cell wall	Polysaccharides Proteins	Mechanical rigidness and shape of cell
Cell membrane	Lipids Proteins	Selective permeability of substances into and out of cell
Cell cytoplasm	Monosaccharides	Sources of energy and carbon for biosynthesis
	Polysaccharides	Storage of energy and carbon for biosynthesis
	Amino acids	Monomers for protein synthesis Osmotic pressure regulation
	Proteins	Enzymes
	Nucleotides	Energy storage and transfer
	DNA	Storage of genetic information
	RNAs	Tools for cell control and protein synthesis
	Ions	Osmotic pressure regulation Energy production
	Inorganic cell inclusions	Polyphosphate—storage of phosphorus Sulfur granules—storage of metabolic product Gas in gas vesicles (bubbles) inside cell

TABLE 2.5 Monomers and Polymers in Microbial Cell Functions (Involvement in Cell Structure Is Shown in Table 2.4)

Cell Function	Type of Substance	Function of Substance
Energy production and storage	Monosaccharides	Biological energy production
	Polysaccharides	Sources of energy and carbon storage
	Membrane lipids	Energy generation
	Storage lipids	Sources of energy and carbon storage
Cell biosynthesis and reproduction	Proteins	Enzymes
	DNA	Store of genetic information
	RNA	Tools of transcription and translation
	Polyphosphate	Storage of phosphorus
Cell behavior	Ions	Sources of energy for cell movement
	Monosacccharides	Osmoprotectors
	Polysaccharides	Storage of energy for survival
	Amino acids	Osmoprotectors

3. Cell biosynthesis and reproduction
4. Cell behavior (locomotion, survival, and taxis)

The summary of the role of monomers and polymers in cell functions is shown in Table 2.5.

3

Dynamic Biochemistry

Biochemistry is the chemistry of life. Dynamic biochemistry studies the changes in molecules during life processes. These coordinated changes in molecules in life systems are called metabolism (from Greek *metabole*, change). Substances produced by metabolism are termed metabolites.

Levels of Metabolism

Metabolism involves coordinated biochemical reactions at different levels. These levels, in order of complexity, are

- Molecule + enzyme
- Linear, branched, or cycling metabolic pathways and polyenzyme complexes
- Cell compartment or organelle of microbial cell
- Metabolic block of microbial cell
- Metabolism of whole microbial cell
- Metabolism of microbial population
- Metabolism of microbial ecosystem
- Microbial metabolism of whole biosphere

Biochemical Reaction

The fundamental unit of metabolism is the biochemical reaction. All biochemical reactions in cells are catalyzed by protein molecules called enzymes:

$$S + E(+C) \leftrightarrow [ES] \leftrightarrow [EP] \leftrightarrow P + E(+C) \tag{3.1}$$

where

E is the enzyme
S is the substrate
ES is the unstable enzyme–substrate complex
EP is the unstable enzyme–product complex
P is the product

The enzyme is regenerated in this reaction and starts up a new reaction cycle. The coenzyme, C, is an active part of enzyme, but it is often weakly bound with the enzyme.

Major Features of Enzymes

The major features of enzymes are as follows:

1. Enzymes are protein catalysts of biochemical reactions.
2. Enzymes possess high specificity—they increase the rate of specific biochemical reactions.

Mechanism of Enzymatic Catalysis

The constant k of chemical reaction rate depends on energy of activation, which is defined by the Arrhenius equation:

$$\ln k = \left(-\frac{E_a}{R}\right) \times \left(\frac{1}{T}\right) + \ln A, \qquad (3.2)$$

where
E_a is the activation energy of chemical reaction
R is the gas constant (8.31 J/K mol)
T is the absolute temperature (K)
A is the frequency factor for reaction

The activation energy is the minimum amount of energy required to initiate a chemical reaction. An enzyme decreases the energy of activation of a reaction significantly by stereochemical arrangement of substrate inside the 3D structure of protein, which is favorable for initiation of the reaction. Decrease of activation energy increases the rate of the reaction (Figure 2.9). Typically, enzyme-catalyzed biochemical reactions take place thousands or even million times faster than the same chemical reactions occurring without enzyme.

Specificity of Enzymes

Bacterial cells require several thousands of biochemical reactions to reproduce. So, several thousands of specific enzymes must be synthesized in the cell to catalyze these biochemical reactions.

The specificity of an enzyme is created by stereochemical specificity of the binding between active center of enzyme and substrate.

Inactivation of Enzymes

Any changes in the 3D structure of enzyme due to thermal or chemical denaturation leads to a decrease of catalytic activity of enzyme and eventual termination. Any changes in the structure of the active center or in the coenzyme, caused by physical or chemical factors, also terminate enzymatic activity. An important environmental engineering example is the inactivation of SH-containing active centers of enzymes by ions of heavy metals reacting with SH group and the resulting inactivation of enzyme activity. Oxidants and organic solvents are also strong inhibitors of enzymes due to effecting chemical changes of active center or changes in 3D structure of protein.

Classification of Enzymes

The trivial name of an enzyme includes the type of substrate, the type of catalyzed reaction, and ends with "ase." For example, an enzyme catalyzing the oxidation of alcohol by removal of hydrogen has the trivial name alcohol dehydrogenase. To identify thousands of enzymes, the enzyme classification (EC) numbers and names are used. Six classes of enzymes are

1. *Oxidoreductases* catalyze oxidation-reduction reactions. Oxidoreductases are important enzymes for environmental engineering because they catalyze oxidation, which is commonly used for the biodegradation of organic substances. Some subclasses of oxidoreductases important for environmental engineering are shown in Table 3.1.
2. *Transferases* catalyze transfers of different chemical groups. They are most important in biosynthesis of cellular substances and regulation of cellular processes.
3. *Hydrolases* catalyze hydrolysis reactions where a molecule is split into two or more smaller molecules by the addition of water. These enzymes

TABLE 3.1 Some Subclasses of Oxidoreductases

Subclass	Catalyzed Reaction	Importance for Environmental Engineering
Dehydrogenases	Hydrogen transfer from the substrate to NAD^+	Biodegradation of organic substances
Oxidases	Hydrogen transfer from the substrate to molecular oxygen producing hydrogen peroxide as a by-product	Biodegradation of xenobiotics
Peroxidases	Oxidation of a substrate by hydrogen peroxide	Primary step in biodegradation of lignin
Oxygenases	Oxidation of a substrate by molecular oxygen with water as by-product	Primary step in biodegradation of aliphatic and aromatic hydrocarbons

TABLE 3.2 Some Subclasses of Hydrolases

Subclass	Catalyzed Reaction	Importance for Environmental Engineering
Cellulases	Hydrolysis of cellulose to monomers	Biodegradation of cellulose-containing wastes; production of biofuel
Amylases	Hydrolysis of starch	Primary step in biodegradation or biotransformation of starch-containing wastes
Proteases	Hydrolysis of proteins	Primary step in biodegradation or biotransformation of protein-containing wastes
Ribonucleases (RNAases)	Hydrolysis of ribonucleic acids (RNAs)	Primary step in biodegradation or biotransformation of RNA-containing wastes
Deoxyribonucleases (DNAases)	Hydrolysis of deoxyribonucleic acid (DNA)	Primary step in biodegradation or biotransformation of DNA-containing wastes
Phosphatases	Removal of phosphate groups from organic compounds	Step in removal of phosphate from organic compounds
Lipases	Hydrolysis of lipids to glycerol and long-chain fatty acids	Primary step in biodegradation of lipids

are important for environmental engineering because they catalyze breakdown of biopolymers to monomers. Some subclasses of hydrolases are shown in the Table 3.2.

4. *Lyases* catalyze the cleavage of C–C, C–O, C–S, and C–N bonds by means other than hydrolysis or oxidation. These enzymes catalyze some reactions of biodegradation of organic compounds.
5. *Isomerases* catalyze rearrangements of isomers, for example optical isomers.
6. *Ligases* catalyze the reactions of biosynthesis where two molecules are joined.

The EC number of an enzyme includes four digits. For example, the EC number of catalase is EC1.11.1.6. The first digit indicates class, and subsequent digits indicate subclasses and sub-subclasses. In scientific and technical literature, it is required to provide not only the trivial names of enzymes but also their EC numbers and names.

Environmental Applications of Microbial Enzymes and Coenzymes

Some enzymes are extracellular, ejected from the cell into the medium to hydrolyze biopolymers prior to the consumption of their monomers. However, the majority of enzymes are intracellular ones. Extracellular enzymes, separated from cells, and intracellular enzymes, extracted from cells and separated from cell debris and cellular components, can be used in environmental engineering as catalysts for the following purposes:

- Pretreatment of organic wastes by hydrolysis of cellulose using different cellulases, or lipids using different lipases
- Production of biosensors of organic substances utilizing the strong specificity of enzyme–substrate interactions
- Performance of all molecular–biological analysis and processes, including PCR, DNA sequencing, and construction of genetically modified microbial strains

Coenzymes often are not synthesized by microorganisms living in symbiotic or parasitic relationships with humans, animals, and plants. So, these substances, known as vitamins, must be added in the media for cultivation of symbiotic or parasitic (pathogenic) microorganisms. Such media are used in monitoring of the environment for microorganisms.

Control of Individual Biochemical Reaction

All biochemical reactions are catalyzed by enzymes. The rate of a biochemical reaction is controlled by

1. The content of enzyme in cell or concentration of extracellular enzyme in the medium, (X_e), mol of enzyme g^{-1} of biomass or g of enzyme L^{-1}. The content of intracellular enzyme in cell or extracellular enzyme in medium is changed by the turning on or off of the transcription of the relative gene.
2. The substrate concentration affects the rate of enzymatic reaction according to the Michaelis–Menten equation:

$$V = \frac{V_{max} S}{(S + K_s)}, \tag{3.3}$$

where
V and V_{max} are velocities
S is the concentration of substrate
K_s is the constant of half saturation for substrate

3. Different inhibitors (I) decrease enzymatic velocity. There are noncompetitive and competitive inhibitors of enzyme activity. Nonspecific enzyme inhibitors such as acids, bases, alcohols, and heavy metals change the structure of the active center of enzyme and the secondary,

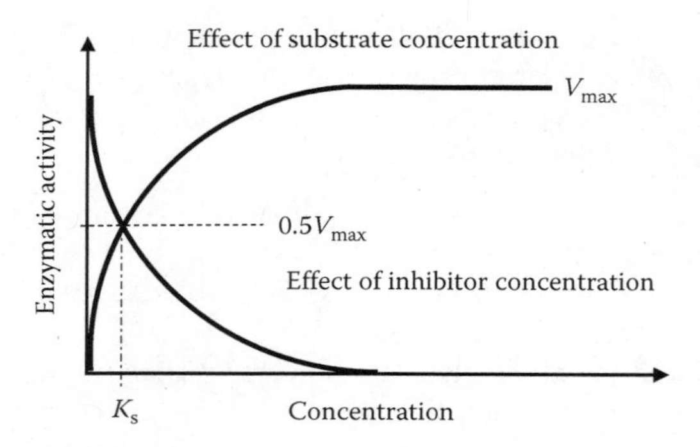

Figure 3.1 Control of biochemical reaction.

tertiary, and quaternary structure of enzymes. The noncompetitive inhibition of enzymatic activity can be described by the following equation:

$$V = \frac{V_{max}S}{\left[(S+K_s)/(1+I/K_i)\right]}, \tag{3.4}$$

where

- V and V_{max} are current and maximum velocities, respectively
- S is concentration of substrate
- K_i is the inhibition constant describing the affinity of the inhibitor for the enzyme
- K_s is the substrate concentration at half of the maximum velocity (V_{max}) of the reaction
- I is concentration of inhibitor

4. Specific inhibitors are the competitors of substrates for active center of enzyme. Competitive inhibition of enzymatic activity can be described by the equation (Figure 3.1)

$$V = \frac{V_{max}S}{\left[(S+K_s)/(1+I/K_i)\right]}, \tag{3.5}$$

where the symbols are same as in the paragraph above.

The velocity of reactions is used for the quantification of enzyme through enzymatic activity, in terms of the enzyme unit (U), which is the amount of enzyme that catalyzes the conversion of 1 μmol of substrate per minute at optimal conditions. Empirical enzyme units are used in engineering practice, for example, velocities of the changes in optical density or viscosity of the medium, or increase

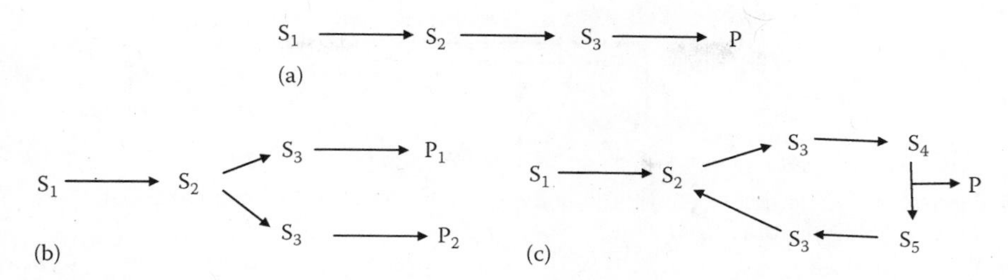

Figure 3.2 (a) Linear, (b) branched, and (c) cyclic biochemical pathways. S_1, ..., S_5 are substrates as well as intermediate products and P, P_1, P_2 are final products of the biochemical pathways.

of product concentration in the medium. The rate of biochemical reaction can be given as mol of substrate g^{-1} of biomass s^{-1} or as mol of substrate L^{-1} s^{-1}.

Control of a Group of Biochemical Reactions

Biochemical reactions are integrated in linear, branched, or cycling metabolic pathways and polyenzyme complexes (Figure 3.2).

In eukaryotic cells, a specific and bigger group of biochemical reactions is performed in the cell organelle. There are groups of biochemical reactions, usually connected with multistep oxidation, transfer of electrons, and energy generation, that occur in the membrane (membrane-dependent enzymes) in both prokaryotic and eukaryotic cells.

The rate of the slowest reaction, usually the first reaction, determines the rate of the sequence of cycle of the reactions ("bottle neck" rule). The rate control in the sequence or cycle is performed by cells due to (1) precursor activation and feedback inhibition and (2) control of enzyme synthesis and degradation.

Regulation of Enzymatic Activity in the Sequence of Enzymes

The rates of the sequence of enzymes are regulated by two mechanisms:

1. When the concentration of the first substrate in the sequence of enzymes is increased, the rate of the biochemical pathway is also increased. It can be performed by activation of the last enzyme or all enzymes of the pathway by the first substrate (Figure 3.3). This mechanism is called as precursor activation.
2. When the concentration of the final product of the pathway is increased, the rate of the biochemical pathway is decreased by inhibition of the first enzyme or all enzymes of the pathway (Figure 3.3). This mechanism is called as feedback inhibition by final product.

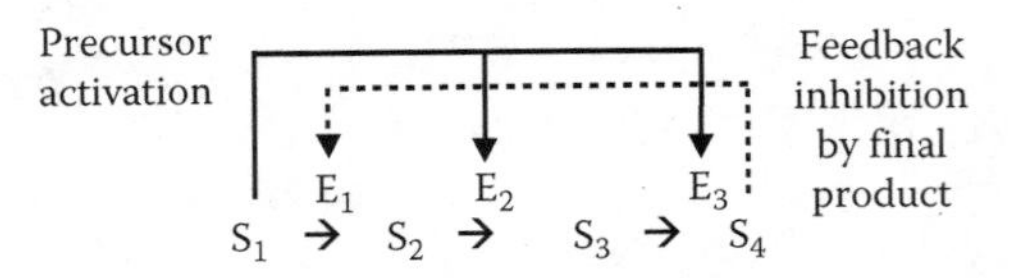

Figure 3.3 Regulation of enzymatic activity. In the sequence of enzymes. S_1–S_4 are substrates; E_1–E_3 are enzymes.

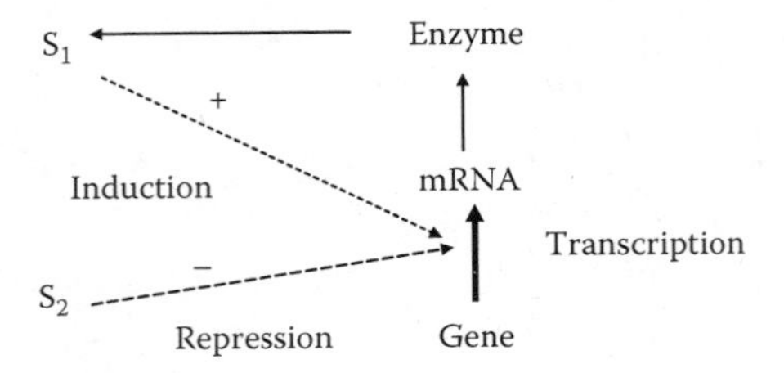

Figure 3.4 Regulation of enzyme synthesis.

Regulation of Enzyme Synthesis and Degradation

The effective control mechanism that regulates the content of enzyme in the cell is the turning on or off of the transcription of the relative gene. The appearance of new substrate in the medium may induce the synthesis of the enzyme needed to metabolize this substrate. The substrate, which is metabolized faster or with higher efficiency, can repress the synthesis of the enzymes catalyzing metabolism of another substrate (Figure 3.4).

This effect, called catabolite repression, is often revealed at the level of cell metabolism when glucose and other substrates are present in medium. In this state, enzymes oxidizing glucose are synthesized, but the synthesis of enzymes for oxidation of another substrate is repressed. However, the synthesis of these enzymes starts when glucose gets consumed from the medium. This mechanism is important for environmental engineering because biodegradation of slowly degraded organic compounds in wastewater or polluted soil can be repressed by the addition of glucose or some other rapidly metabolized carbohydrates.

Metabolic Blocks

Biochemical pathways are grouped in metabolic blocks that perform different functions in cell metabolism (Figure 3.5).

The function of pretreatment reactions is the hydrolysis of biopolymers, oxidation of hydrocarbons, and phosphorylation of monosaccharides before transport of organic compounds through cell membrane.

The specific transport of substances through cell membrane, against their concentration gradients, is performed by protein-based nanoscale gates using biological forms of energy.

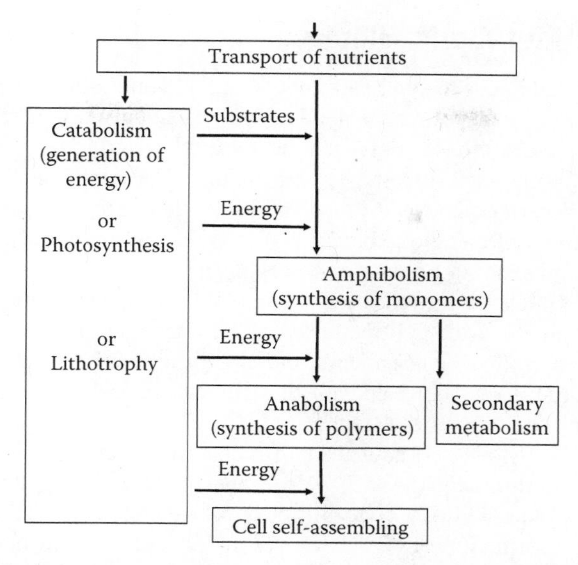

Figure 3.5 Pretreatment reactions within the cell (hydrolysis, oxidation, and phosphorylation).

Biological forms of energy for all energy-consuming reactions in cells are produced by the following reactions:

1. Catabolism, which is the generation of biological forms of energy using oxidation of organic matter accompanied by the formation of substances used for the production of monomers.
2. Photosynthesis, which is the generation of biological forms of energy using the energy of sunlight accompanied by the assimilation of CO_2 and the formation of substances for the production of cell biomass.
3. Lithotrophy, which is the generation of biological forms of energy using the oxidation of inorganic substances accompanied by the assimilation of CO_2 and the formation of substances used for the production of monomers.

Amphibolism is the block of biochemical reactions involved in the synthesis monomers.

Anabolism is the block of biochemical reactions involved in the synthesis of polymers from monomers. Synthesized polymers are self-assembled and formed within cells.

Microorganisms have a specific block of biochemical reactions termed as secondary metabolism. These biochemical reactions produce substances that are not essential for cell formation and reproduction, but can help cells to survive or to colonize their environment, for example, by the production of antibiotics, toxins, or autoregulators.

Control of Metabolic Blocks

A prokaryotic cell needs to synthesize between 500 and 5,000 different enzymes to replicate itself. Eukaryotic cells need to synthesize between 5,000 and 50,000 different enzymes. All these enzymes must be synthesized in temporal order and in optimal quantities within cells. The mechanisms of control of enzyme activity have been described above. At the same time, there are some mechanisms to control the activity of metabolic blocks acting at the level of cell, cell aggregates, multicellular organism, or population of organisms.

Transport of nutrients into the cell and metabolites out of the cell is performed by specific channels in the cell membrane. The rates of transport through these channels are controlled by the membrane electric potential, i.e., the difference in electric charge across both sides of the membrane, and by the concentration gradient of ions and organic substances.

There are also whole-cell activators of metabolic blocks, for example, cyclic adenosine monophosphate (cAMP) activating supply of energy, cyclic guanosine monophosphate (cGMP) controlling ion channels in cell, and nitric oxide (NO) activating formation of cGMP. cAMP and cGMP are constantly produced from AMP and GMP, respectively, by cyclases and are degraded by phosphodiesterases; thus, the activators of cyclases or inhibitors of phosphodiesterases, for example, caffeine, affect levels of cAMP and cGMP in cells. These levels affect the rates of metabolic blocks not directly but by the activation of protein kinase enzymes, which modify the activity of other enzymes by their reversible phosphorylation, i.e., by the addition of phosphate groups to the enzyme molecule, changing hydrophobicity of its parts, 3D structure, and enzyme activity or protein–protein interactions.

Temporal Control of Cell Metabolism

Metabolic activity within cells varies with time. Nonspecific factors causing these changes include pH, temperature, concentration of oxygen, concentration of nutrients, and specific, repeating patterns of metabolic changes of cells that depend on

- Period of cell cycle (time from one cell division to other)
- Period of life cycle (time from origin to death of cell)

This control is performed by the regulation of DNA transcription. Several mechanisms controlling temporal changes of cell metabolism are known:

1. Adenine or cytosine methylation of DNA, which changes the production of specific enzymes.
2. Acetylation and methylation of histones, which are positively charged proteins, covering negatively charged DNA. These transformations

of histones change the transcription of specific genes and synthesis of related enzymes.

3. Synthesis of small RNAs that interfere with the transcription of specific genes and synthesis of related enzymes.

Control of Metabolism in Microbial Populations, Aggregates, and Ecosystems

Whole-cell activators of metabolic blocks like cAMP, cGMP, and NO are used to change metabolism in cell population or aggregates. Therefore, NO, an important signaling molecule, can be used in engineering practice to prevent formation of or disperse microbial biofilms on the surface of biomedical devices or on membrane surfaces in reverse osmosis facilities or membrane bioreactors. NO is an unstable free radical. It can be produced by many aerobic and microaerophilic bacteria. In animals, NO production is a stress response and can lead to killing of infectious microorganisms because of direct inhibition with NO and NO-related formation of radicals, which are toxic for bacteria.

Quorum Sensing

Both prokaryotic and eukaryotic microorganisms can produce different low-molecular-weight regulators of metabolism using as the signals to synchronize the physiological reactions of all cells in a population or ecosystem. The sense of these intercellular signals is to prepare cells of population for starvation, to fast growth, to repair of damages, or to mating.

Communication between bacterial cells, called quorum sensing, is used for coordination of gene expression in cells of population. When the concentration of cells and produced signaling molecules are increased to some level due to cell aggregation, signaling molecule (inducer) binds with the receptor in cell and activates transcription of certain genes, including those for inducer synthesis. Different molecules are used as signals of cell aggregation: peptides in Gram-positive bacteria; autoinducer AI-2 in both Gram-positive and Gram-negative bacteria; and acyl-homoserine lactones (AHL, Figure 3.6), the most common signaling molecules for Gram-negative bacteria.

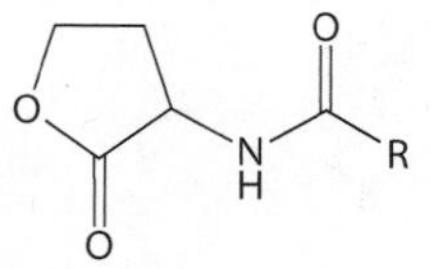

Figure 3.6 Quorum-sensing signaling molecule: *N*-acyl homoserine lactone. R varies by chain lengths (4–18 carbon atoms) and some other minor chemical changes.

Microorganisms use quorum sensing to coordinate cell aggregation and formation of biofilms. Therefore, it is important for environmental and civil engineering bioprocesses with the involvement of microbial biofilms and microbial aggregates, including different biofilm reactors, bioclogging, membrane biofouling, interaction of microorganisms with plants and animals, and sedimentation of bacterial cells in wastewater treatment.

Control of Metabolism in Biosphere

All living creatures on the earth are interdependent and can be considered as one global ecosystem usually called biosphere or Gaia ("land" or "earth," from the ancient Greek). The common subsystems in this ecosystem are components of soil, freshwater, seawater, and atmosphere. Soil subsystems are relatively isolated, the subsystems of the oceans are changing slowly because of their big mass, but the masses of atmospheric components are relatively low in comparison with masses of ocean components; therefore biochemical dynamics of such atmospheric components as H_2O, CO_2, and CH_4 determine the metabolism in biosphere through average temperatures of soil and ocean surfaces and intensity of photosynthesis (CO_2 assimilation) following the supply of organic compounds produced by photosynthetic organisms into the biosphere.

4

Biooxidation and Bioreduction

Oxidation–reduction is a way of producing biological forms of energy that can be used for all cell activities. Oxidation–reduction is widely used for the biodegradation of wastes and the bioremediation of environment in environmental engineering.

Oxidation–Reduction

Oxidation is the loss of electron(s) or hydrogen. Reduction is the gain of electron(s) or hydrogen. Electrons are donated by one compound (electron donor is oxidized) and are accepted by another compound (electron acceptor is reduced). The examples of coupled oxidation–reduction reactions are

Oxidation of carbohydrate and reduction of oxygen

$CH_2O + O_2 \rightarrow CO_2 + H_2O$ Complete oxidation–reduction

$CH_2O \rightarrow 4e^- + C^{4+} + H_2O$ Electron donor is oxidized Oxidation

$O_2 + 4e^- \rightarrow 2O^{2-}$ Electron accepton is reduced Reduction

Oxidation of ferrous ion and reduction of oxygen

$4Fe^{2+} + O_2 + 2H_2O \rightarrow 4Fe^{3+} + 4OH^-$ Complete oxidation–reduction

$Fe^{2+} \rightarrow Fe^{3+} + e^-$ Oxidation of iron (II) Oxidation

$O_2 + 4e^- \rightarrow 2O^{2-}$ Reduction of oxygen Reduction

$H_2O \leftrightarrow H^+ + OH^-$ Dissociation of water

Oxidation of ferrous ion and reduction of nitrate

$5Fe^{2+} + NO_3^- + 6H^+ \rightarrow 5Fe^{3+} + 0.5N_2 + 3H_2O$ Complete odixation–reduction

$Fe^{2+} \rightarrow Fe^{3+} + e^-$ Oxidation of iron (II) Oxidation

$NO_3^- + 5e^- \rightarrow 0.5N_2$ Reduction of oxygen Reduction

$H_2O \leftrightarrow H^+ + OH^-$ Dissociation of water

Oxidation Number

The oxidation number is a charge on the atom in a compound after electrons have been transferred or shared in bonding. Three rules to calculate the oxidation number are as follows:

1. The oxidation number of hydrogen in compounds is +1 and 0 for H_2.
2. The oxidation number of oxygen in compounds is –2; it is 0 for O_2 and –1 for H_2O_2.
3. The oxidation number of other elements is determined as the charge on the compound – (+1) × (number of hydrogen atoms in compound – (–2) × number of oxygen atoms in the compound) divided by number of atoms of element in compound. The examples of calculations of oxidation numbers of elements in different compounds participating in biochemical oxidation–reduction are shown in Table 4.1.

Oxidation of Carbon

The highest reduced state of carbon, –4, is found in methane, CH_4, and the highest oxidation state, +4, is found in carbon dioxide, CO_2. A total of eight electrons are transferred during the oxidation of methane to carbon dioxide:

TABLE 4.1 Oxidation Numbers of Elements in Different Compounds Participating in Biochemical Oxidation–Reduction

Minimum Oxidation State (Reduced Compound—Donor of Electrons)	Intermediate Oxidation State (Reduced or Oxidized Compound—Can Be Donor or Acceptor of Electrons)	Maximum Oxidation State (Oxidized Compound—Acceptor of Electrons)
CH_4 (methane) oxidation number of C = –4 = 0 (no charge) – 4(+1)	$C_6H_{12}O_6$ (glucose) oxidation number of C = 0 = [0 – 12(+1) – 6(–2)]/6	CO_2 oxidation number of C = +4 = 0 (no charge) – 2(–2)
H_2O oxidation number of O = –2 = 0 – (+2)		O_2 Oxidation number of O = 0
NH_4^+ Oxidation number of N = –3 = +1 (charge of NH_4^+ ion) – 4 × (+1)	N_2 Oxidation number of N is 0	NO_3^- Oxidation number of N = +5 = –1 (charge of NO_3^- ion) – 3(–2)
H_2S oxidation number of S = +2 = 0 (no charge) – (–2)	S oxidation number of S = 0	SO_4^{2-} oxidation number of S = +6 = –2 (charge of SO_4^{2-} ion) – 4(–2)
Fe^{2+}	Fe^0	Fe^{3+}

$$CH_4 \rightarrow 8e^- + CO_2$$

Carbon in all other organic compounds has an intermediate oxidation number from −4 to +4. The number of electrons donated during the oxidation of carbon = oxidation number of carbon in product of oxidation – oxidation number of carbon in electron donor, for example:

Oxidation number of C in ethanol, C_2H_5OH, is $[0 - (+6) - (-2)]/2 = -2$

Oxidation number of C in carbon dioxide produced from ethanol is +4

$$C_2H_5OH \rightarrow 2CO_2 + 6e^-$$

Oxidation–Reduction during Fermentation

Reactions such as intramolecular oxidation–reduction of organic compounds, called fermentation, also occur. Some atoms of carbon in organic molecules, during fermentation, are oxidized to CO_2 and other atoms are reduced to a more reduced state; the molecule then splits to more reduced and more oxidized molecules than the initial molecule. An example of intramolecular oxidation of glucose to carbon dioxide and ethanol is

$$CH_2OH\text{–}CHOH\text{–}CHOH\text{–}CHOH\text{–}CHOH\text{–}CH_2OH \rightarrow 2CO_2 + 2CH_3CH_2OH$$

Oxidation number of C in glucose = 0

Oxidation number of C in carbon dioxide = +4

Oxidation number of C in ethanol = −2

Energy

Energy is the ability to do work. If energy is released in a reaction, the change of free energy, ΔG (calculated as the free energy in the products minus the free energy in the reactants), is less than zero (Figure 4.1). There are physical, chemical, and biological forms of energy.

Biological work includes the biosynthesis of necessary monomers and polymers, transport of nutrients from medium into cell, intracellular transport of metabolites, and cell or organism movement. The cell has mechanisms to capture energy in the specific biological forms suitable to carry out biological work.

Two Sources of Biological Forms of Energy

There are two sources of biological forms of energy: the energy of visible light (physical form of energy) and the energy of reduced organic or inorganic compounds (chemical form of energy). The related terms are phototrophy (use of light energy for cell growth) and chemotrophy (use of chemical energy for cell growth).

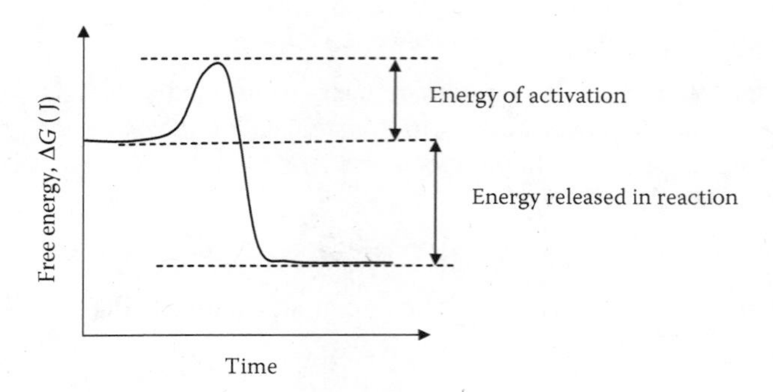

Figure 4.1 Release of energy in a biochemical reaction.

Chemotrophy

Chemotrophy can be chemoorganotrophy (use of energy of reduced organic compounds) or chemolithotrophy (use of energy of reduced inorganic compounds). Chemoorganotrophy and chemolithotrophy are often replaced by the simplified terms heterotrophy and lithotrophy, respectively (Figure 4.2).

Donors of electrons in chemotrophy are

- Organic carbon (heterotrophy)
- Inorganic compounds such as H_2, S^{2-} and S^0, CO, Fe^{2+}, NH_4^+, and NO_2^- (lithotrophy)

Terminal acceptors of electrons can be

- Organic carbon (fermentation)
- Inorganic compounds such as SO_4^{2-}, Fe^{3+}, Mn^{4+}, NO_3^-, NO_2^-, and CO_2 (anaerobic respiration)
- Oxygen and oxygen radicals (aerobic respiration)

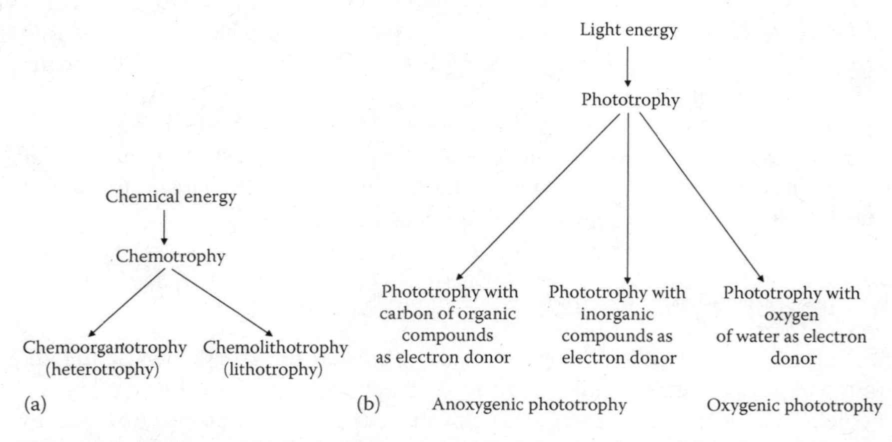

Figure 4.2 Sources of biological energy for (a) chemotrophy and (b) phototrophy.

Chemotrophic Oxidation–Reduction Reactions

Typical oxidation–reduction reactions with transfer of *n* electrons from donor to acceptor, used for the generation of biological forms of energy during chemotrophy, are as follows:

C^d (organics) + C^d (organics) $\rightarrow C^{(d-n)} + C^{(d+n)}$ Fermentation

$H_2 + 2Fe^{3+} \rightarrow 2H^+ + 2Fe^{2+}$ Anaerobic respiration

$2CH_2O$ (organics) + $SO_4^{2-} \rightarrow 2CO_2 + S^{2-} + 2H_2O$ Anaerobic respiration

$5CH_2O$ (organics) + $4NO_3^- + 4H^+ \rightarrow 5CO_2 + 2N_2 + 7H_2O$ Anaerobic respiration

CH_2O (organics) + $O_2 \rightarrow CO_2 + 2H_2O$ Aerobic respiration

$NH_4^+ + 2O_2 \rightarrow NO_3^- + 2H^+ + H_2O$ Aerobic respiration

Chemotrophy: Fermentation

Fermentation is intramolecular oxidation–reduction. There are many types of fermentations, as shown below:

Fermentation of glucose to ethanol (ethanol fermentation) is used for the production of fuel from starch or cellulose-containing wastes:

$$C_6H_{12}O_6 \rightarrow 2CH_3CH_2OH + 2CO_2$$

Fermentation of glucose to lactic acid (homolactic fermentation) is used in food industry and for the production of polylactic acid, which is biodegradable plastic:

$$C_6H_{12}O_6 \rightarrow 2CH_3CH(OH)COOH$$

Fermentation of glucose to butyric acid (butyric acid fermentation) is an important reaction of anaerobic biodegradation of organic wastes:

$$C_6H_{12}O_6 \rightarrow CH_3CH_2CH_2COOH + 2CO_2 + 2H_2$$

Fermentation of lactic acid to propionic and acetic acids (propionic fermentation) is used in food industry (cheese production) and is an important step in anaerobic biodegradation of organic wastes:

$$3C_3H_6O_3 \rightarrow 2C_3H_6O_2 + CH_3COOH + CO_2$$

Chemotrophy: Anaerobic (Anoxic) Respiration

Anaerobic (anoxic) respiration is the oxidation of substances other than the oxygen terminal acceptors of electrons, such as NO_3^-, Fe^{3+}, SO_4^{2-}, and CO_2. Biologically available energy gained during respiration is larger than the energy produced during fermentation. Anoxic respiration is typical only for prokaryotes. Different electron acceptors are used for energy generation by specific physiological groups of prokaryotes, including

1. Nitrate (NO_3^-) and nitrite (NO_2^-), used by denitrifying bacteria (denitrifiers)
2. Succinate, used in anoxic respiration with fumarate as electron donor
3. Sulfate (SO_4^{2-}), used by sulfate-reducing bacteria
4. Sulfur (S), used by sulfur-reducing prokaryotes
5. Ferric ions (Fe^{3+}), used by iron-reducing bacteria
6. Ions of different oxidized metals, used as acceptors of electrons
7. Carbon dioxide (CO_2), used by methanogens

Chemotrophy: Aerobic Respiration

Aerobic respiration is the oxidation of substances with oxygen as the terminal acceptor of electrons. Almost all organic compounds and many inorganic compounds can be biologically oxidized by oxygen:

$$C_6H_{12}O_6 \text{ (glucose)} + 6O_2 \rightarrow 6CO_2 + 6H_2O \text{ chemoorganotrophy}$$

$$4Fe^{2+} + O_2 + 4H^+ \rightarrow 4Fe^{3+} + 2H_2O \text{ chemolithotrophy}$$

Phototrophy

Donors of electrons in phototrophy can be

1. Organic carbon
2. Reduced inorganic compounds (S^{2-}, Fe^{2+}; it is possible to predict the discovery that donors of electrons such as NH_4^+ or even N_2 can also be used in photosynthesis)
3. Oxygen of water (O^{2-})

The terminal acceptor of electrons in phototrophy is CO_2.

Oxygenic and Anoxygenic Phototrophic Oxidation–Reduction Reactions

Typical oxidation–reduction reactions with transfer of electrons from donor to acceptor, used for the generation of biological forms of energy during anoxygenic photosynthesis, are as follows:

$$CH_3CH_2OH + CO_2 + \text{light} \rightarrow 3CH_2O \text{ (biomass)}$$

$$H_2S + CO_2 + \text{light} \rightarrow CH_2O \text{ (biomass)} + S + H_2O$$

Anoxygenic photosynthesis carried out by purple and green photosynthetic bacteria. Oxygen is not generated during this type of photosynthesis.

A typical oxidation–reduction reaction with transfer of electrons from donor to acceptor, used for the generation of biological forms of energy during oxygenic photosynthesis, is as follows:

$$H_2O + CO_2 + \text{light} \rightarrow CH_2O\ (\text{biomass}) + 0.5O_2 + H_2O$$

Oxygen is generated during this type of photosynthesis. Oxygen generated by oxygenic bacteria and plants and accumulated in the atmosphere is a major condition for life on earth because of two reasons: (1) the protection of terrestrial life from ultraviolet (UV) radiation and (2) the majority of eukaryotes are obligate aerobes generating biological energy by aerobic respiration and can live only in the presence of oxygen.

Electron-Transfer Chain

Transfer of electrons from donor to acceptor in cells is performed in several steps in cytoplasm with the formation of the redox mediator, reduced nicotinamide adenine dinucleotide, NADH ($+H^+$), which is a form of biological energy functioning like a universal fuel of cells. It is further stepwise oxidized in the electron-transport chain, which is a sequence of numerous redox mediators arranged inside membranes (cytoplasmic membrane, mitochondrial inner membrane, and thylakoid membrane).

Proton-Motive Force

Energy is released in portions during stepwise oxidations–reductions in the electron transfer chain and is accompanied by the transfer of protons from the inside to the exterior space of the membrane. This transfer of protons creates gradients of electric field (positive outside membrane) and concentration of protons on different sides of membrane (Figure 4.3).

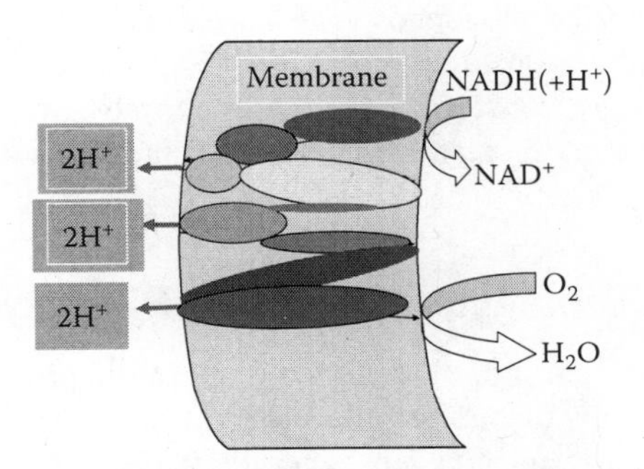

Figure 4.3 (See color insert following page 292.) Formation of proton-motive force on membrane due to electron transfer from donor to acceptor and vectorial transfer of protons to membrane outside.

Accumulation of protons on one side of the membrane forms a proton-motive force, which is a form of biological energy.

2H⁺ ADP + P_i ATP Membrane

Figure 4.4 (See color insert following page 292.) Oxidative phosphorylation. ADP, adenosine diphosphate nucleotide; ATP, adenosine triphosphate nucleotide; P_i, inorganic phosphate.

Oxidative Phosphorylation

The controlled passage of protons back across the membrane through specific membrane proteins is used to drive ion transport into the cell, cell motility, or adenosine triphosphate nucleotide (ATP) synthesis, which is like a monetary unit for cell, a universal form of biological energy. ATP is synthesized by a membrane-bound enzyme, ATP-synthase, from adenosine diphosphate nucleotide (ADP) and inorganic phosphate (P_i) using portion of energy accumulated as proton-motive force on membrane (Figure 4.4):

$$ADP + P_i + \text{portion of energy} \rightarrow ATP$$

Synthesis of ATP via generation of a proton-motive force by electron-transport chain is called oxidative phosphorylation.

Biological Forms of Energy

The most common form of biological energy in cells is ATP. A large amount of free energy is released and used for different biological functions when either one or two high-energy phosphate bonds in the molecule of ATP are hydrolyzed:

$$ATP \rightarrow ADP + P_i + \text{portion of energy}$$

$$ADP \rightarrow AMP + P_i + \text{portion of energy}$$

where AMP is adenosine monophosphate nucleotide. Other types of biological energy are proton-motive force ($\mu_{\Delta H^+}$), NADH ($+H^+$), and NADPH ($+H^+$). All forms of biological energy are convertible between themselves (Figure 4.5). Different forms of energy are used for performing different functions in cell (Table 4.2.).

Production of Biological Energy in Photosynthesis

Light energy is converted into biological energy in the process of photosynthesis. There are two steps of photosynthesis: light and dark reactions. The light reactions of anoxygenic (no oxygen generation) photosynthesis take place as follows:

$$\text{Energy of light} + H_2S \rightarrow \text{Energy of ATP} + \text{Energy of NADPH } (+H^+) + S$$

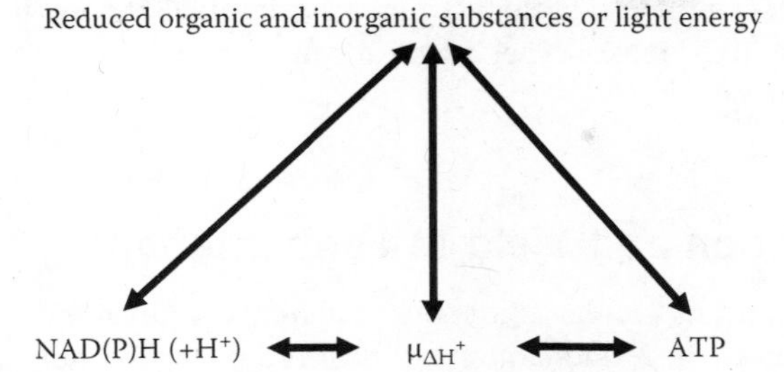

Figure 4.5 Conversion between different biological forms of energy.

TABLE 4.2 Functions of Major Forms of Biological Energy in Cell

Form of Biological Energy	Major Functions in Cell
ATP	Biosynthesis, transport of nutrients in cell, cell movement
Proton-motive force ($\mu_{\Delta H^+}$)	Formation of ATP, cell movement, ion transport
Reduced NAD, NADH (+H+)	Generation of ATP via oxidative phosphorylation (fuel of cell)
Reduced NADP, NADPH (+H+)	Reductions in cell

Light reactions of oxygenic (with oxygen generation) photosynthesis take place as follows:

$$\text{Energy of light} + H_2S \rightarrow \text{Energy of ATP} + \text{Energy of NADPH } (+H^+) + O_2$$

The light reactions are initiated by the absorption of a quantum of light by pigments. The major pigments are the chlorophylls. Other pigments such as carotenoids and phycobilins serve as accessory pigments absorbing and transferring the light energy to chlorophylls by fluorescence. Energy accumulated in ATP and NADPH (+H+) is used in dark reactions to reduce CO_2 into cell material:

$$6CO_2 + 18ATP + 12NADPH\ (+H^+) \rightarrow 6CH_2O\ (\text{biomass}) + NADP + ADP + P_i$$

ATP Yield of Oxidative Phosphorylation

Energy is released in portions during the transfer of electrons from reduced NAD to O_2 to synthesize up to 3 ATP. Theoretically, 38 moles of ATP can be synthesized from the oxidation of 1 mol of glucose ($C_6H_{12}O_6$):

$$C_6H_{12}O_6 + 38ADP + 38P_i \rightarrow 6CO_2 + 6H_2O + 38ATP$$

The real yield of ATP in oxidative phosphorylation varies from zero to three per two transferred electrons depending on species and conditions of oxidation. Calculation of the production of ATP in oxidative phosphorylation can be used in environmental engineering calculations of microbial production of biomass using coefficient 10.5 mol ATP per 1 g of dry microbial biomass of normal content (see Chapter 7).

ATP Production and Yield in Fermentation

Fermentation is intramolecular oxidation–reduction (Figure 4.6).

ATP is produced in fermentation by substrate phosphorylation, without the electron-transport chain. The fermentation products differ for various bacterial species and many of them have commercial importance. There are many types of fermentations, as follows:

- Fermentation of glucose to ethanol (ethanol fermentation) is used for the production of fuel from starch or cellulose-containing wastes:

$$C_6H_{12}O_6 \rightarrow 2CH_3CH_2OH + 2CO_2 + 2ATP$$

- Fermentation of glucose to lactic acid (homolactic fermentation) is used in food industry and for the production of polylactic acid, which is biodegradable plastic:

$$C_6H_{12}O_6 \rightarrow 2CH_3CH(OH)COOH + 2ATP$$

- Fermentation of glucose to butyric acid (butyric acid fermentation) is an important reaction of anaerobic biodegradation of organic wastes:

$$C_6H_{12}O_6 \rightarrow CH_3CH_2CH_2COOH + 2CO_2 + H_2 + 3ATP$$

Fermentation of lactic acid to propionic and acetic acids (propionic fermentation) is used in food industry (cheese production) and is an important step in anaerobic biodegradation of organic wastes:

$$3C_3H_6O_3 \rightarrow 2C_3H_6O_2 + CH_3COOH + CO_2 + ATP$$

Yield of ATP from 1 mol of glucose in ethanol fermentation is 19 times lower than maximum yield of ATP in oxidative phosphorylation. Respectively, the yield of biomass from glucose, which is based on oxidative phosphorylation, could be almost five times higher than the yield of biomass, which is based on fermentation process (see calculations in Chapter 7).

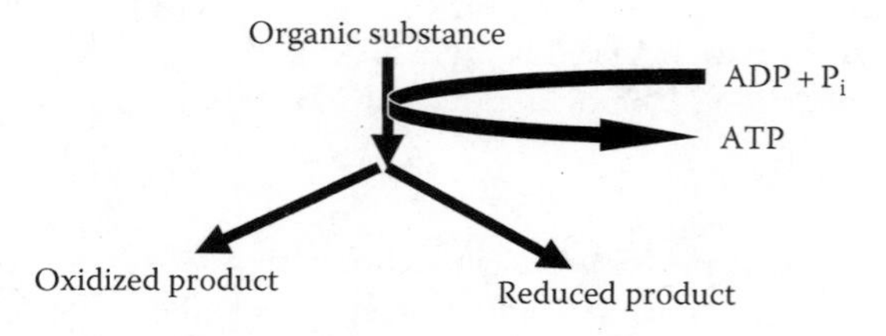

Figure 4.6 Chemical reaction(s) involved in fermentation.

ATP Yield of Lithotrophy

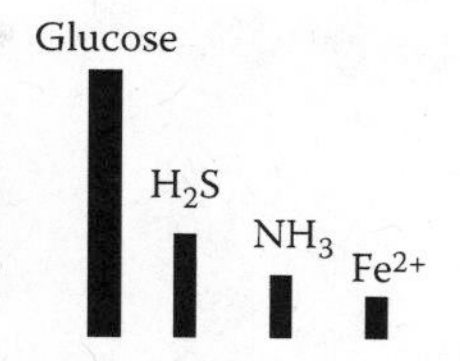

Figure 4.7 Comparison of ATP generation during oxidation of different electron donors by oxygen.

Chemolithotrophs are prokaryotes that can gain energy by oxidation of inorganic compounds. Most of the chemolithotrophs are autotrophs and aerobes that synthesize ATP by oxidative phosphorylation. However, the energy gained from oxidation of inorganic substances is usually significantly lower than energy produced in the oxidation of organic matter (Figure 4.7).

Long-Term Energy Storage in Cell

ATP, proton-motive force ($\mu_{\Delta H^+}$), NADH (+H^+), and NADPH (+H^+) are short-term energy stores in cell because of the high rate of their turnover in cells. For long-term energy storage, most cells accumulate osmotically inert storage polymers such as starch, glycogen, and poly-β-hydroxybutyrate. Fungi, plant, and animal cells can also accumulate lipids as stores of carbon and energy. Cells can accumulate up to 80% of biomass as storage compounds under conditions of energy source excess and shortage of nutrients for growth.

Relation to Oxygen and Generation of Energy

Relation to oxygen is one of the main features of microorganisms. Generation of biologically available energy in a cell is due to oxidation–reduction reactions. Oxygen is the most effective acceptor of electrons in the generation of energy from oxidation of substances but not all microorganisms can use it. The following groups of microorganisms differ in their relation to oxygen and generation of biological energy:

1. Obligate and tolerant anaerobic microorganisms producing biological energy by fermentation or anaerobic respiration
2. Facultative anaerobic bacteria, which are capable of producing biological energy by fermentation or anaerobic respiration if oxygen is absent or by aerobic respiration if oxygen is present
3. Microaerophilic bacteria and obligate aerobes producing biological energy by aerobic respiration only

5

Biodegradation

Biodegradation and Biotransformation

Biodegradation is a biochemical process in which complex compounds are broken down into simpler ones. Biodegradation is the most important process in wastewater treatment, solid waste treatment, and soil bioremediation.

Mineralization is the complete biodegradation of organic compounds to CO_2. The final aim of the majority of environmental engineering processes is the mineralization of organic pollutants.

Biotransformation is a biochemical process in which the chemical complexity of organic compounds is not changed significantly. Typical biotransformations are either the addition of a functional group, or the replacement of one functional group in the compound by another, or the oxidation/reduction of several carbon atoms in the compound. Biotransformations are preliminary and intermediate steps of biodegradation. An example of biotransformation is the incorporation of an oxygen atom in a hydrocarbon molecule prior to its transport and oxidation in a cell.

Biotransformation is not the complete biodegradation of toxic organic substances, but it could be important for detoxication of polluted water, soil, and solid wastes, because even small chemical changes in a substance can decrease its toxicity. However, in many cases, biotransformation of a toxic compound can increase its toxicity, so complete biodegradation is required for the treatment of polluted water, air, or soil.

Biochemical Reactions of Biodegradation

The biodegradation of organic compounds mainly involves the oxidation of substrate S to product P_1 with donation of n electrons to acceptor A forming product P_2:

$$S - ne^- \rightarrow P_1 \quad \text{and} \quad A + ne^- \rightarrow P_2, \quad \text{for example}$$

$$CH_2O - 4e^- + H_2O \rightarrow CO_2 + 4H^+ \quad \text{and} \quad O_2 + 4e^- + 4H^+ \rightarrow 2H_2O$$

An intermediate step of biodegradation could be the reduction of a substance or replacement of its functional group.

An important step in the degradation of a biopolymer, $(S_1)_n$, is its hydrolysis to monomers, S_2:

$$(S_1)_n + (n-1)H_2O \rightarrow nS_2.$$

Aerobic and Anaerobic Biodegradation

Biodegradative oxidation of organic substances can be performed by fermentation, anoxic, or aerobic respiration. Fermentation and anoxic respiration are anaerobic processes. Aerobic biodegradation is performed with involvement of oxygen as electron acceptor. The biodegradation of hydrocarbons and lignin requires incorporation of oxygen into molecules prior to further oxidation, so these substances are not degraded anaerobically. It is one reason for the accumulation of these substances under anaerobic conditions and formation of huge oil and coal deposits on the earth.

Biodegraded Substances

Natural biodegraded substances are (1) the components of dead biomass of microorganisms, plants, and animals; (2) organic wastes such as plant litter, animal and human feces, and urine; and (3) industrial, agricultural, aquacultural, and organic wastes from food processing.

Another group of organic substances to be biodegraded include (1) wastes and pollutants from oil-recovery and petrochemical industry; (2) wastes from chemical industry; and (3) organic pollutants of soil, water, and air.

Biodegradation Rates of Natural Substances

The components of dead biomass and organic wastes are degraded at different rates. The sequence of the biodegradation rates can be shown as follows: storage oligosaccharides > storage polysaccharides > RNA > proteins > DNA > structural polysaccharides > aliphatic and aromatic compounds > lipids > lignin. Storage polysaccharides have the fastest biodegradation rate and structural polysaccharides have a lower biodegradation rate. In this sequence, lignin has the slowest biodegradation rate.

Biodegradation Rates Correlate with Turnover Rates

The majority of cell components undergo turnover, i.e., they are degraded and synthesized again. The sequence of biodegradation rates correlates with the sequence of turnover rates of compounds in live cells. Storage polysaccharides, RNA, and proteins, which are actively involved in cell metabolism, have the highest turnover rates in cells. Structural compounds, like lipids, cellulose, and other cell wall polymers, which are not involved in the performance of the biochemical reactions, have the lowest turnover rate in cells and, consequently, the lowest rates of biodegradation.

Degradation of Storage Polysaccharides

The most quickly degraded substances are monosaccharides and oligosaccharides, for example, glucose and saccharose. Polysaccharides, which are used by cells as stores of energy and carbon, are usually polymers of glucose molecules connected by α-1,4 bonds.

Starch is the storage form of glucose in plant cells. Starch is a mixture of two polymers: amylose (linear polymer of glucose in starch) and amylopectin (branched polymer of glucose in starch with α-1,6 bonds in the points of branching).

Glycogen is the storage form of glucose in animal, fungal, and bacterial cells. Glycogen is a branched polymer similar to amylopectin. Starch and glycogen are split to mono- and oligosaccharides, or to linear chains of polysaccharides by enzymatic hydrolysis (Figure 5.1). There are many microbial enzyme amylases, which specifically hydrolyze starch. Bacteria and fungi usually secrete extracellular amylases to hydrolyze the polymer to monomers and oligomers that is shown by the equation

$$(C_6H_{10}O_5)_n + (n-1)H_2O \rightarrow nC_6H_{12}O_6$$

Biodegradation of Structural Polysaccharides

The function of structural polysaccharides is the support of cell integrity of bacteria, fungi, plants, and animals by the formation of a cell wall, or plant organism structure by formation of stem, branches, bark, and other structural elements.

Structural polysaccharides such as cellulose, hemicellulose, pectin, and chitin can be degraded by the hydrolytic action of microbial enzymes, while lignin

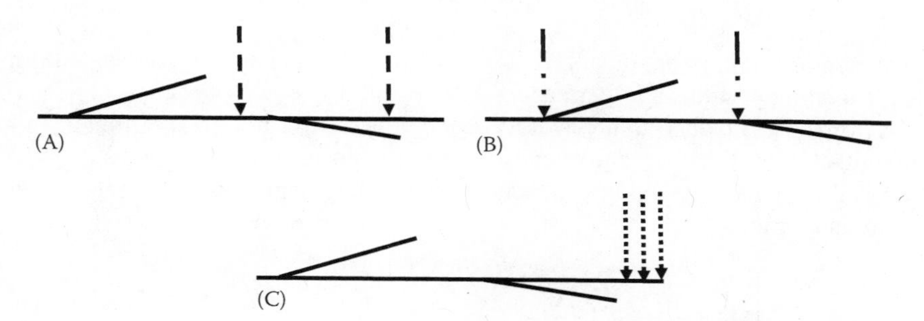

Figure 5.1 Action of different enzymes on biodegradation of starch or glycogen. (A) Random splitting of the chain, (B) splitting of the polymer in the points of branching, and (C) regular separation of monomers or oligomers from the end of the chain.

can be degraded only by bacteria and fungi using oxygen radicals produced by microbial cells.

The rate of microbial biodegradation is relatively high for pectin and hemicellulose ("soft glue" of cellulose fibers and plant cells), slow for cellulose and chitin (polymers conferring mechanical strength to organisms), and very slow for lignin ("strong glue" of cellulose fibers and plant cells).

Biodegradation of Hemicellulose and Pectin

Hemicellulose and pectin are heteropolymers, i.e., polymers of different monomers. They perform similar functions in plants, binding cellulose fibers and cells.

Hemicellulose is a branched polymer of pentoses (sugars with five atoms of carbon—xylose and arabinose) and hexoses (sugars with six atoms of carbon—glucose, mannose, and galactose). Hemicellulose has an amorphous structure and low resistance to acid hydrolysis. 3D molecules of hemicellulose bind together the microfibers of cellulose in the cell walls of plants. Hemicellulose can be degraded by many microorganisms that release the hydrolyzing enzymes, hemicellulases.

Pectin is a 3D polymer in the cell walls of nonwooded plants containing galactose, arabinose, xylose, and galacturonic acid esterified with methanol. The related hydrolyzing enzymes are pectinases.

Biodegradation of Cellulose

Cellulose is the major structural polysaccharide of plants. Cellulose is a linear polymer of glucose monomers connected by β-1,4 bonds. Cellulose is resistant to hydrolysis by chemical means. Only some species of bacteria and fungi can hydrolyze cellulose by excretion of extracellular enzymes, cellulases, to monomers and oligomers of glucose shown by the equation

$$(C_6H_{10}O_5)_n + (n-1)H_2O \rightarrow nC_6H_{12}O_6$$

Isolation of cellulose degraders from soil can be easily performed on the medium with solution of salts and filter paper.

Animals and plants cannot hydrolyze cellulose, but termites and ruminant animals such as cows, deer, and camels live in symbiosis with prokaryotes degrading cellulose and supplying products of incomplete degradation to insects and ruminant animals.

Biodegradation of Lignin

Lignin is a 3D polymer in cell walls of trees, comprising 25%–30% of dry mass of wood. It binds cellulose and other polymers in plants, and confers mechanical strength to the wood. Lignin has no defined primary structure because it is synthesized by radical-based biochemical reactions formed random C–C and C–O–C bonds between aromatic units. Therefore, it is hydrophobic and resistant to biodegradation. It is not degraded by hydrolysis as all other polymers are, but by using combined biological and chemical reactions causing random splitting of C–C and C–O bonds. The enzymes participating in biodegradation are called ligninases. Naturally biodegraded lignin from dead plants constitutes a major fraction of the organic material of soil called humus. Soil humus increases soil fertility by providing increased cation exchange capacity in the soil and expanding the capacity of moisture retention between flood and drought conditions.

White-Rot Fungi

Lignin biodegradation is slow and performed by only a limited number of microorganisms. White-rot fungi and some bacteria degrade lignin by a complex process of radical depolymerization. White-rot fungi possess a system, which allows them to disassemble lignin due to the formation of free radicals. Manganese peroxidase and lignin peroxidase are the most important enzymes in the biodegradation of lignin. Due to a nonspecific biodegradation mechanism, white-rot fungi can be also used for biodegradation of many other organic compounds. Many white-rot fungi are cultivated edible mushrooms, so after separation of fruit body and mycelium, the mycelium can be used for bioremediation of polluted soil (Figure 5.2). The majority of similar environmental engineering applications are with the white-rot fungus *Phanerochaete chrysosporium* (pronunciation: "fah-neh-roh-haete krih-zoh-sporium").

Biodegradation of Chitin

The structural polysaccharide of fungi, crustaceans, and insects is chitin, which is a polymer of *N*-acetylglucosamine monomers connected with β-1,4 bonds.

Figure 5.2 (See color insert following page 292.) Edible mushrooms containing lignin-degrading enzymes. They can be used for biodegradation.

It is hydrolyzed by chitinases, extracellular enzymes of microorganisms adapted to degrade dead biomass of fungi, crustaceans, and insects in nature.

Chitin is the main component of the cell wall of fungi and the exoskeletons of insects and crustaceans, for example, shrimps. It is a polymer of *N*-acetylglucosamine, which is a molecule of glucose connected with a –NH–$COCH_3$ group. The chemical structure of this polymer is similar to cellulose with the replacement of one hydroxyl group in the cellulose monomer by an acetylamino group. Chitin is degraded by extracellular enzymes, chitinases, produced by some species of bacteria and fungi and catalyzing hydrolysis of polymer in dead biomass of fungi, insects, and crustaceans.

Degradation of Proteins

The proteins of dead biomass are degraded through the digestion by protozoa or extracellular hydrolysis by bacteria and fungi, which secrete the enzymes proteases, quickly hydrolyzing protein to amino acids:

$$(\text{–HNRCO–})_n + H_2O \rightarrow H_2N\text{–}R\text{–}COOH$$

Some structural proteins, like keratin of hairs, are hydrolyzed slowly.

Degradation of Amino Acids

The deamination reaction produces ammonia and organic acids from amino acids:

$$H_2N\text{–}R\text{–}COOH + H_2O \rightarrow NH_3 + RCOOH$$

Subsequently, organic acids can be oxidized to CO_2.

The decarboxylation reaction produces biogenic amines from amino acids:

$$R\text{–}CH(NH_2)COOH \rightarrow RCH_2NH_2 + CO_2$$

The biogenic amines are toxic and are substances with foul odor, especially cadaverine and putrescine, which are produced from the amino acids lysine and ornithine:

$$H_2N\text{–}R\text{–}CH(NH_2)COOH \rightarrow H_2N\text{–}R\text{–}CH_2NH_2 + CO_2$$

Degradation of Nucleic Acids

The nucleic acids of dead biomass are degraded through digestion by protozoa or extracellular hydrolysis by bacteria and fungi, which produce the enzymes RNAses and DNAses that degrade macromolecules to smaller units capable of permeating the cell. Monomers are released from the cells and degraded finally to CO_2 and NH_3.

Degradation of Lipids

Lipids of fats and vegetable oils are hydrolyzed by lipases to glycerol and long-chain fatty acids. Lipids are insoluble in water, so they must be solubilized to the smallest droplets using natural or synthetic surfactants.

Lipid Glycerol Fatty acids

$$CH_2O\text{–}OOC\text{–}R_1 \quad CH_2OH \quad HOOC\text{–}R_1$$

$$CHO\text{–}OOC\text{–}R_2 + 3H_2O \rightarrow CHOH \quad HOOC\text{–}R_2$$

$$CH_2O\text{–}OOC\text{–}R_3 \quad CH_2OH \quad HOOC\text{–}R_3$$

Microbes can easily utilize the produced fatty acids and glycerol.

Biodegradation of Chemicals

Chemicals are degraded at different rates and often only under aerobic conditions. Aliphatic hydrocarbons are degraded quickly under aerobic conditions. Aromatic monocyclic and polycyclic hydrocarbons are degraded more slowly under aerobic conditions. Phenols and aromatic acids can be degraded anaerobically. Chlorinated aliphatic and aromatic hydrocarbons are degraded very slowly. Usually, the atoms of chlorine must be removed from the molecule prior to biodegradation. Some chemicals are completely nonbiodegradable.

Degradation of Aliphatic Hydrocarbons

The risk of oil pollution at sites of oil recovery during oil transport by tankers and pipelines, at oil refineries, and in oil storage is always high. One of the most effective ways to clean up polluted sites is the biodegradation of hydrocarbons.

Hydrocarbons are insoluble in water, so to be biodegraded they must be solubilized to the smallest droplets using mechanical stirring, synthetic surfactants, or surfactants synthesized by microorganisms.

The initial step of biodegradation involves the introduction of oxygen atoms from O_2 by monooxygenases, which incorporate one atom of oxygen to form hydroxyl group; the other atom of oxygen reacts with protons and electrons from reduced NAD to form water (Figure 5.3).

After this initial oxidation, hydrocarbons are converted into organic acids that can be metabolized to CO_2 and water as happens with the long-chain fatty acids of lipids.

$C_{15}H_{31}\text{—}CH_3 + NADH_2 + O\text{:}O$

↓

$C_{15}H_{31}\text{—}CH_2OH + NAD + H_2O$

↓ + NAD

$C_{15}H_{31}\text{—}CHO + NADH_2$

↓ + NAD + H_2O

$C_{15}H_{31}\text{—}COOH + NADH_2$

↓

Cyclic β-oxidation of fatty acids

Figure 5.3 Oxidation of aliphatic hydrocarbons.

Degradation of Aromatic Hydrocarbons

The initial step in the degradation of monocyclic and polycyclic aromatic hydrocarbons (PAH) often involves the introduction of oxygen atoms from O_2 by oxygenase enzymes. There are two types of these enzymes. Monooxygenases incorporate one atom of O_2 into the substrate as a hydroxyl group; the other generally combines with protons and electrons from reduced NAD to form water (Figure 5.4). The initial metabolites are oxygenated derivatives of the aromatic hydrocarbons. Dioxygenase incorporates both oxygen atoms of O_2 into the substrate (Figure 5.5). Initial metabolites are oxygenated derivatives of the aromatic hydrocarbons or the products of cleavage of aromatic ring.

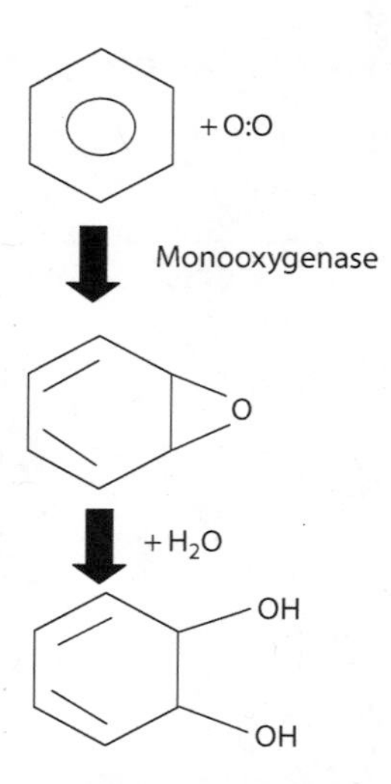

Figure 5.4 Oxidation of aromatic hydrocarbons by monooxygenation.

Biodegradation of BTEX Chemicals

Benzene, toluene, ethylbenzene, and xylenes (BTEX) are volatile monoaromatic hydrocarbons of crude petroleum and gasoline. Megatons of BTEX are also produced annually as the solvents and starting materials for the manufacturing of pesticides, plastics, and synthetic fibers. They are considered one of the major causes of environmental pollution because of widespread occurrences of leakage from underground petroleum storage tanks and spills at petroleum production wells, refineries, pipelines, and distribution terminals.

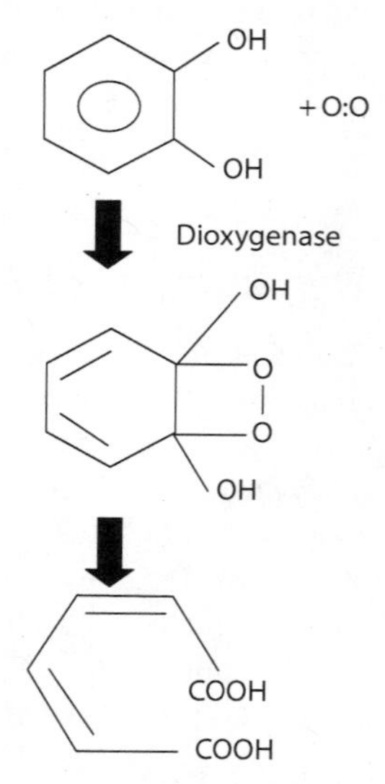

Figure 5.5 Oxidation of aromatic hydrocarbons by dioxygenation.

Bioremediation of Soil and Water Polluted by BTEX Chemicals

The BTEX chemicals (Figure 5.6) are components of gasoline and diesel and therefore are common pollutants of surface water, groundwater, and soil. BTEX can be degraded both aerobically and under anoxic conditions using ferric iron, sulfate, or carbon dioxide as terminal electron acceptors.

Xenobiotics

Thousands of organic compounds that do not exist in nature have been developed. Weakly degraded or nondegradable chlorinated and organophosphate insecticides, herbicides, fungicides, and different industrial, pharmaceutical, and military substances are called xenobiotics. Biodegradation of xenobiotics is an important part of industrial wastewater treatment and soil bioremediation.

A lot of xenobiotics are present in drinking water at extremely low concentrations, such as μg/L or even ng/L, but must be removed because of their potential negative health effects. Biodegradation of these compounds in nature and engineering systems for water treatment are important for maintenance of high water quality.

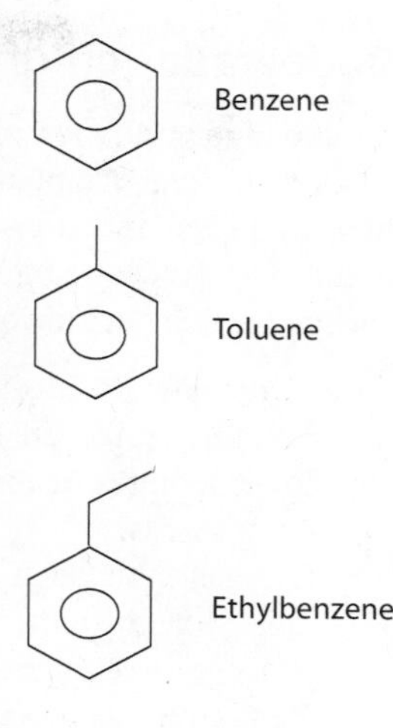

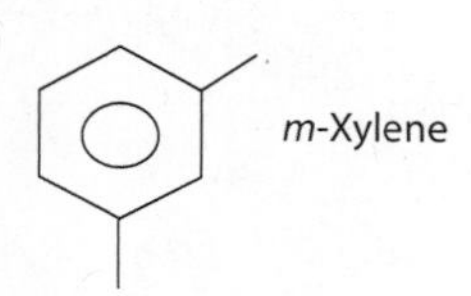

Figure 5.6 BTEX chemicals.

Plasmids of Degradation

Specific enzymes are used by cells in the degradation of xenobiotics. The genes of these specific enzymes of biodegradation are often located not in the bacterial chromosome but on mobile genetic elements such as plasmids (Figure 5.7). The multiplication (amplification) of the biodegradation genes and new effective combinations of genes for the enhanced biodegradation of several different toxic chemicals can be constructed in bacterial cells using genetic engineering techniques.

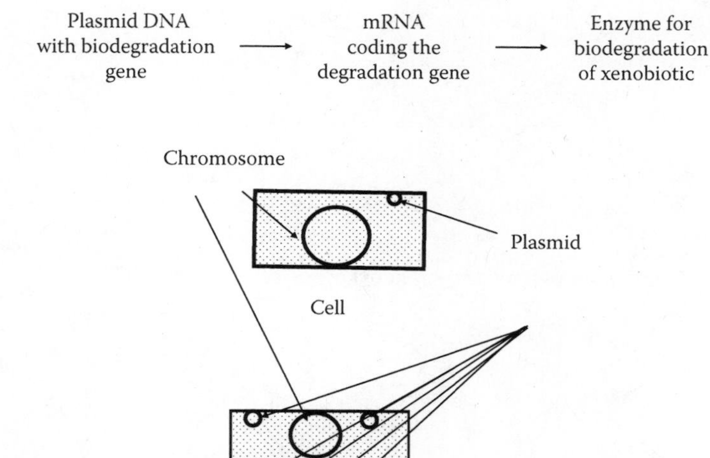

Figure 5.7 Plasmids of degradation.

Biodegradation of Chlorinated Hydrocarbons

Aerobic biodegradation of chlorinated hydrocarbons, for example, chlorophenols or polychlorinated biphenyls (PCBs) (Figure 5.8), includes the steps described above and some reactions carried out to return the organic chlorine to its mineral state. It is a process called dechlorination. There are different biochemical reactions of aerobic dechlorination:

- Hydrolytic dechlorination: when an atom of chlorine in an organic compound is replaced by the hydroxylic group of water (Figure 5.9). The intermediates of this process are oxygenated derivatives of the organic compounds.
- Oxygenolytic dechlorination: when oxygenase incorporates both oxygen atoms of O_2 into the substrate (Figure 5.10). The intermediates of this process are oxygenated derivatives of the organic compounds.
- Reductive dechlorination: when hydrogen produced during fermentation reacts with organic chlorine, detaching it from the organic molecule (Figure 5.11).

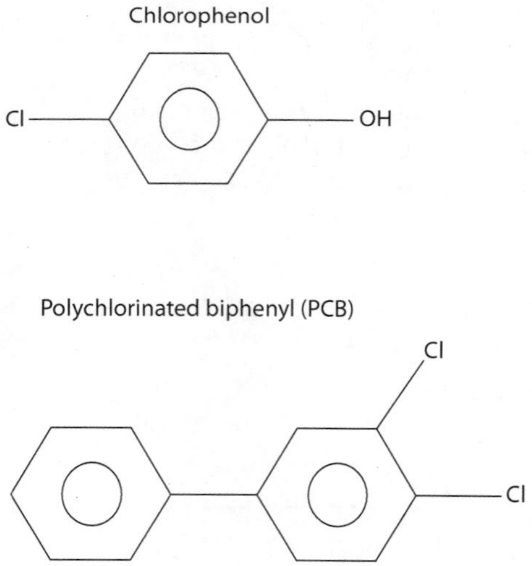

Figure 5.8 Chlorinated phenol and biphenyl.

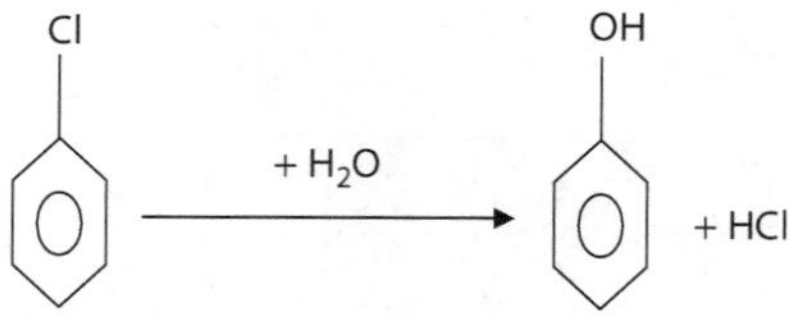

Figure 5.9 Hydrolytic dechlorination.

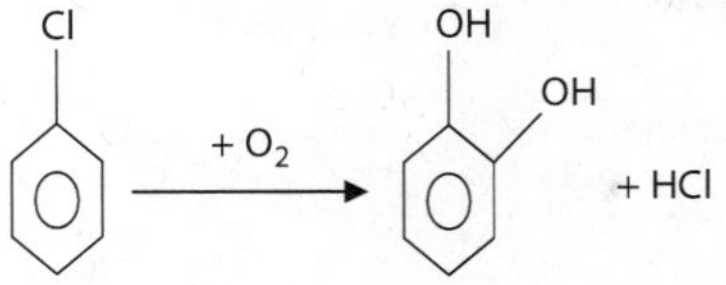

Figure 5.10 Oxygenolytic dechlorination.

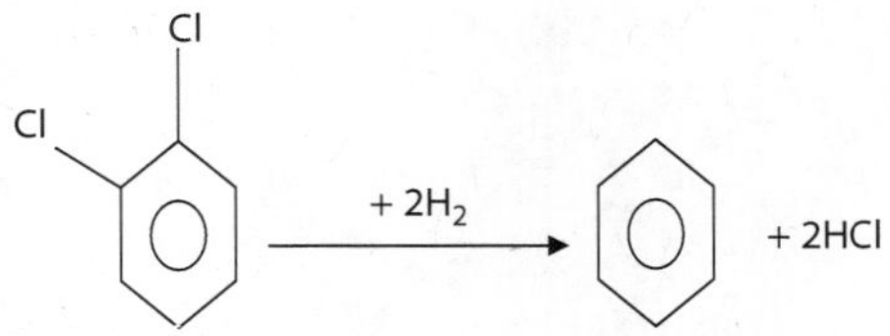

Figure 5.11 Reductive dechlorination.

After dechlorination, the aromatic ring of hydroxyl- and carboxyl-containing hydrocarbons can be opened by incorporation of oxygen as described above, or it can be opened by some anaerobic bacteria via carboxylation and decarboxylation reactions.

Products of Degradation of Aromatic Hydrocarbons

Due to the incorporation of oxygen, dechlorination, and formation of hydroxylic and carboxylic groups in different sites of the aromatic rings, there is a big diversity of the intermediate products of biodegradation (Figure 5.12), which can be accumulated if the conditions are not favorable for complete biodegradation. Some of these intermediate products of biodegradation, for example, polycyclic aromatic acids produced from polycyclic aromatic hydrocarbons, are significantly more toxic and dangerous for the environment than the initial substances themselves. Environmentally safe end products of anaerobic degradation are volatile fatty acids, alcohols, methane, hydrogen, carbon dioxide, and water. Environmentally safe end products of aerobic biodegradation are carbon dioxide and water. Therefore, during engineering biodegradation of aromatic hydrocarbons, (1) accumulation of intermediates and changes in toxicity must be monitored and (2) anaerobic and aerobic processes of biodegradation must be optimized so that only carbon dioxide and water will be the final products of biodegradation.

Nonbiodegradable and Biodegradable Polymers

Most of the plastics are produced from petrochemical products and cannot be degraded and are discharged into the environment (Figure 5.13). Even if a portion

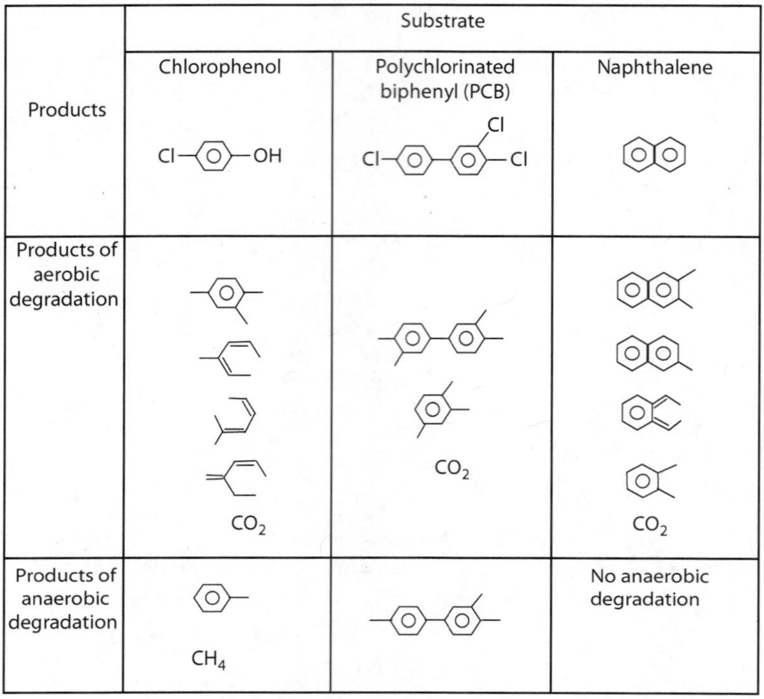

	Substrate		
Products	Chlorophenol Cl–OH	Polychlorinated biphenyl (PCB) Cl, Cl–Cl	Naphthalene
Products of aerobic degradation	CO_2	CO_2	CO_2
Products of anaerobic degradation	CH_4		No anaerobic degradation

Figure 5.12 Examples of the products of aromatic hydrocarbons degradation (radicals connected with the rings could be Cl and OH, COOH groups).

PE (polyethylene)	$[-CH_2-CH_2-]_n$
PPE (polypropylene)	$[-CH_2-CH(CH_3)-]_n$
PE (polyvinylchloride)	$[-CH_2-CHCl-]_n$

PLA (polylactic acid)	$[-O-CH(CH_3)-C(O)-]_n$
PHB (polyhydroxybutyrate)	$[-O-CH(CH_3)-CH_2-C(O)-]_n$

Figure 5.13 Examples of biodegradable and nonbiodegradable plastics.

of these plastics is reused, millions of tons of waste plastics are incinerated, landfilled, or just dumped in the environment, causing increasing pollution. New plastics, which are UV-degradable or biodegradable in landfills, are being developed. However, the cost of the biodegradable plastic production is significantly higher than that of petrochemicals-based plastics. Hence, biodegradable plastics are used currently only for some special applications like biodegradable packaging materials or biomedical devices.

Polylactic Acid and Polyhydroxyalkanoates

The most promising biodegradable plastics are polylactic acid (PLA) and polyhydroxyalkanoates (PHAs) (Figure 5.13). PLA is produced by the chemical polymerization of lactic acid, which is formed during fermentation of renewable source such as starch of corn or cassava and/or sugar.

PHAs are produced by many bacteria as an internal store of carbon and energy formed during excess of carbohydrates and organic acids in medium and shortage of electron acceptors. The content of PHAs and its properties as a plastic can be changed by the incorporation of different organic acids into this biopolymer. Many technologies have been proposed for the production of PHAs but the basic problem, the high cost of this plastic, still remains and is a major obstacle for the industrial scale production and widespread use of PHAs as a plastic. So, the ecological selections for non-aseptic cultivation of PHA-producing bacteria and using cheap raw materials and wastes can solve the problem of the high cost of biodegradable plastics.

6

Molecular Biology and Genetics

Metabolic Blocks of Biosynthesis

There are three metabolic blocks of biochemical pathways performing biosynthesis in cells:

1. Amphibolism—the block of biochemical reactions for synthesis of monomers.
2. Anabolism—the block of biochemical reactions for synthesis of polymers from monomers. Synthesized polymers are self-assembling and finally form a cell.
3. Secondary metabolism—the biochemical reactions of this metabolic block produce substances that are not essential for cell formation and reproduction, but can help cells to survive in their environment, for example, by the attachment to surfaces, production of antibiotics against neighboring microorganisms, or toxins against macroorganisms.

Energy for Biosynthesis

The sources of energy are oxidation of organic and inorganic compounds as well as energy of light, forming ATP, NADH ($+H^+$), NADPH ($+H^+$), and proton-motive force as the forms of biological energy (see Chapter 4). Biosynthetic reactions of amphibolism and secondary metabolism consume different forms of biochemical energy, mainly ATP for modification of chemical groups and NADPH ($+H^+$) for the reduction of chemical groups. Reactions of anabolism consume mainly ATP for the binding of monomers together by dehydration to form polymers.

Balance of Energy

There is no negative feedback between anabolism and catabolism in prokaryotes. Therefore, the balance between produced and consumed biological energy can be

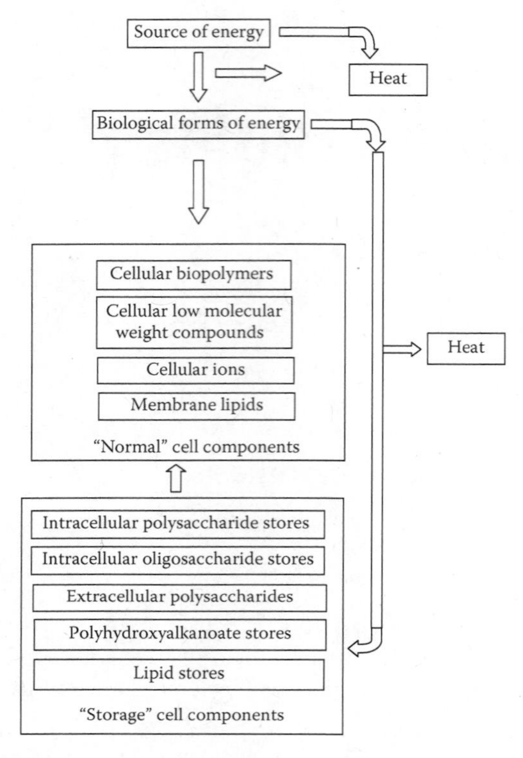

Figure 6.1 Balance of energy within the cell.

calculated only for the complete coupling between production and utilization of biological energy. It is known from experiments that under conditions of complete coupling between production and utilization of biological energy, the synthesis of 1 g of cell biomass, that is, synthesis of cellular polysaccharides, proteins, lipids, ions, and nucleic acids of bacterial cell, requires energy that is equivalent to approximately 0.1 mol of produced ATP.

Excess of Biological Energy

If there is an excess of biological energy, it is used for the production of energy, intracellular carbon, or phosphate stores like glycogen, polyhydroxybutyrate (PHB), lipids, and polyphosphates (Figure 6.1). An excess of biological energy is also transformed to heat.

Number of Enzymes

All biochemical reactions are catalyzed by enzymes. The goal of the cellular biosynthesis is to produce all necessary enzymes which will produce all necessary components for cell growth and division. For a prokaryotic cell to replicate, it must synthesize, depending on the species, from 500 to 5000 different enzymes. The number of enzymes needed for the synthesis of a eukaryotic cell is much bigger, between 5,000 and 50,000. Additionally, enzymes that control the activity of other enzymes are also needed.

Gene

Enzymes are proteins consisting of chains of amino acids joined by covalent peptide bonds. Twenty amino acids are used in protein synthesis. The sequence of these amino acids, which is specific for each enzyme, is coded by the genetic code in the sequence of nucleotides of DNA called a gene. A gene also contains noncoding sequence of nucleotides determining the activity of the gene.

Noncoding RNA Gene

Studies of genomes have revealed that only a small part of the complete genome constitutes the DNA sequences that code for proteins (enzymes). The other part of DNA constitutes the so-called noncoding RNA genes, which are the genes of rRNA, tRNA, and different small and long noncoding RNAs participating in the regulation of gene transcription.

Chromosome

Genes are arranged linearly on the chromosome, like beads on a necklace. Prokaryotic cells contain a moderate number of the genes, so almost all of their DNA is stored in one chromosome. This is a circular molecule of DNA in prokaryotic cells. Eukaryotic cells have a large number of genes, so their DNA is stored in numerous chromosomes, which are long, thread-like structures containing genetic material. The DNA from a single human cell contains about 3 billion bp, and human cells have 23 pairs of chromosomes.

Genome

Genome is the complete hereditary information of an organism. Depending on species, prokaryotic cells have from 500 to 5,000 genes and eukaryotic cells have from 5,000 to 60,000 genes, which is nearly the same as the number of enzymes. The lengths of the DNA from a human cell and a prokaryotic cell are approximately 2 m and 1 mm, respectively. The length of DNA is measured in the number of complementary base pairs (bp). DNA from human cell and *Escherichia coli* have approximately 3200 millions bp and 4.6 million bp, respectively.

Genomes of Chloroplasts and Mitochondria

Chloroplasts and mitochondria of eukaryotic cells have autonomously replicated DNA sequences, which are not considered as part of the genome of eukaryotic cells and are referred to as the plastome and the mitochondrial genome, respectively.

A Genome in Comparison with a Book

Separation of information stored in DNA in subunits such as chromosomes and genes can be understood by comparing the complete genome with a book, the chromosomes with the chapters of this book, the genes with the separate pages of the book, and nucleotides with the letters (A, G, C, T) that are used to write a book (genome). The pages (genes) contain the discrete information that can be copied (transcribed) to mRNA and the numerous copies are used (translated) on the cell-protein plants (ribosomes) for the production of specific enzymes. A book (genome) is divided into the chapters (chromosomes) to simplify the book-copying process (DNA reproduction) using bit-by-bit reproduction of genome.

Flow of Genetic Information

The central dogma of molecular biology is that flow of genetic information is directed from the sequence of nucleotides in DNA to the sequence of amino acids in protein using the genetic code. A protein-coding gene is copied into a molecule of messenger RNA (mRNA), which is complementary to the sequence of the gene. This process is called as transcription. The transcribed mRNA transfers the portion of information from DNA to the ribosome, where the sequence of mRNA of the protein-coding gene is translated into the sequence of amino acids, which is self-folding, forming protein (Figure 6.2). A noncoding gene is copied, producing short and long RNAs with different properties and cellular functions (Figure 6.2).

Genetics, Molecular Biology, Bioinformatics, and Epigenetics

Biosynthesis in cells, as well as the functions of nucleic acids and proteins in cells, are usually studied using genetics, molecular biology, and bioinformatics. Genetics is a biological science studying heredity in organisms and populations. Because the gene is the basic unit of heredity in organisms, genetics also studies the transfer of genes from ancestors to descendants and changes in genes.

The term molecular biology is usually related to the study of replication, transcription, and translation of genetic material at the molecular level. There are many molecular biology applications in environmental engineering.

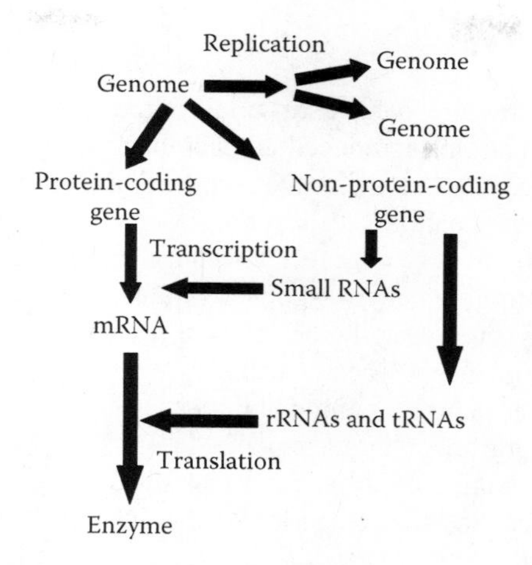

Figure 6.2 Flow of genetic information in the cell.

Bioinformatics is an application of information technology in biological sciences and bioengineering. One of the major tasks of bioinformatics is the informational analysis and comparison of DNA sequences of individual genes and complete genomes of organisms.

There are changes of cell function in cell generations, which are not caused by the changes in DNA sequence but by other factors that change the activation of genes. These changes are called epigenetics. The functions of microbial cells during the cycles of its reproduction are changed due to both genetic and epigenetic mechanisms.

Collection of DNA Sequences

The process of determination of DNA sequences is called DNA sequencing. The DNA sequences of complete genomes have been determined for many species. This information is stored in national, international, and private databases, many of which are accessible by the scientific community. One of the most informative public databases is GenBank (www.ncbi.nlm.nih.gov/Genbank), an annotated collection of all publicly available DNA sequences created by the National Center for Biotechnology Information of the United States.

Nucleus and Nucleoid

The chromosomes in eukaryotic cells are stored in the nucleus, which is separated from the cytoplasm by a membrane with pores permeable for mRNAs. The chromosome in a prokaryotic cell is a covalently closed circular molecule, which is extensively folded and twisted. This aggregate of chromosomal DNA is called as a nucleoid. In addition to the nucleoid, one or more small circular DNA molecules, called plasmids, may be present in the cell.

Plasmids

Plasmids of prokaryotes are defined as small, circular DNA molecules that are reproduced autonomously (Figure 6.3). Plasmids control their own replication, separately from the replication of the chromosome. Commonly, plasmids provide a selective advantage for cells under given environmental conditions. Therefore, they carry only a few genes, usually the genes of resistance to specific antibiotics or heavy metals, degradation of specific recalcitrant compounds, nitrogen fixation, ability to conjugate, or produce toxin. Plasmids are used in genetic engineering as a vector to transfer particular gene from cell to cell and as a tool to multiply particular gene in cell.

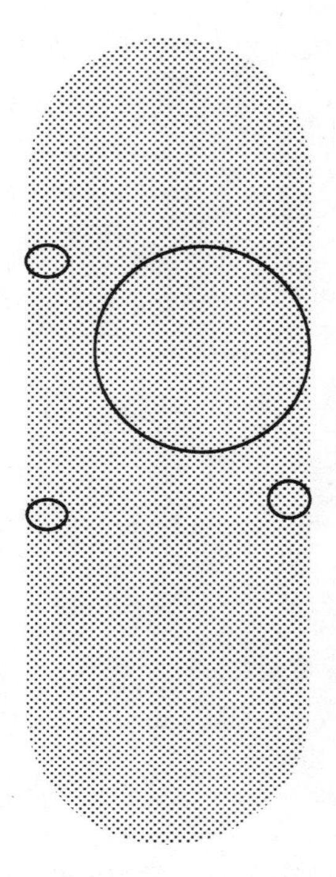

Figure 6.3 Conventional view of chromosome (big circle) and plasmids (small circles) in prokaryotic cell.

Transposons

Transposons are segments of DNA that can move around to different positions in the genome of a single cell. In the process, they may cause mutations by an increase (or decrease) in the amount of DNA in the genome. These mobile segments of DNA are sometimes called "jumping genes."

DNA Replication

Adenine on one strand of DNA pairs with thymine on the other strand, and guanine on one strand of DNA pairs with thymine on the other strand. Therefore, the two strands of DNA have a complementary sequence. This provides the cell with a convenient way of DNA replication, which is the accurate copying of a DNA sequence.

Speed of DNA Replication

DNA replication begins at a specific site of the DNA molecule called the replication origin and proceeds at the speed of about 1000 nucleotides per second. If only one point of DNA replication is present on the chromosome, the replication of the *E. coli* genome containing 4.6 millions bp is performed for about 80 min, but the replication can be performed for 40 min because several replications can be performed simultaneously in prokaryotes. The frequency of incorrect copying of nucleotides in the sequence is approximately 1 per 10^9 nucleotides copied in the sequence. The eukaryotic chromosome contains about a 100 millions nucleotide bp, which are copied at the speed of about 50 bp/s. There are many replication sites on the eukaryotic chromosome, so the total replication time could be several hours.

Transcription

Genetic information from the sequence of DNA (from a gene) is transcribed into the sequence of RNA by complementary pairing of nucleotides in the strands. G always pairs with C, A in DNA always pairs with U in RNA, and T in DNA always pairs with A in RNA (Figure 6.3). Transcription of specific genes is thoroughly controlled by different mechanisms including interactions between regulating proteins and DNA as well as between regulating small and long RNA and DNA (Figure 6.3).

Translation

Genetic information from protein-coding genes is transcribed to the sequence of mRNA, which is translated into the sequence of protein (Figure 6.3). This translation is due the correspondence between the codon and anticodon. The codon is a sequence of three nucleotides in ribosome-attached mRNA. The anticodon is the three nucleotide sequence of transfer RNA (tRNA) that is complementary to a codon sequence. There are one or several specific tRNAs for every amino acid. tRNA is a molecule 75–95 nucleotides long, with the same structure for all organisms. During translation, tRNA collects the specific amino acid, enters the ribosome, and supplies specific amino acid to the growing polypeptide chain. The formation of polypeptide chain starts from the start codon and the elongation of the protein sequence is completed by the signal of a stop codon.

Genetic Code

DNA and mRNA can be considered as sequences of trinucleotides, three-letter "words." Each specific trinucleotide of mRNA, codon, is the code for a specific amino acid. There are 64 ($=4^3$) combinations of 4 nucleotides in trinucleotides (Table 6.1). For example, the sequence of mRNA AGCAAGCGA can be read as AGC-AAG-CGA. AGC is the code for the amino acid serine (Ser), AAG is the code for the amino acid lysine (Lys), and CGA is the code for the amino acid arginine (Arg). Therefore, the sequence of mRNA AGCAAGCGA can be translated into the sequence of amino acids Ser-Lys-Arg. There are also start and stop codons for the start and finish of polypeptide chain elongation (Table 6.1).

Changes in Genome

The genome of a specific organism is relatively stable, but there are several ways in which a genome can be changed: (1) random errors in DNA replication, (2) natural genetic recombination of the genomes, (3) genetic recombination of

TABLE 6.1 mRNA Codons and Related Amino Acids

Amino Acid and Its Symbols	mRNA Codons
Ala/A	GCU, GCC, GCA, GCG
Arg/R	CGU, CGC, CGA, CGG, AGA, AGG
Asn/N	AAU, AAC
Asp/D	GAU, GAC
Cys/C	UGU, UGC
Gln/Q	CAA, CAG
Glu/E	GAA, GAG
Gly/G	GGU, GGC, GGA, GGG
His/H	CAU, CAC
Ile/I	AUU, AUC, AUA
START	AUG
Leu/L	UUA, UUG, CUU, CUC, CUA, CUG
Lys/K	AAA, AAG
Met/M	AUG
Phe/F	UUU, UUC
Pro/P	CCU, CCC, CCA, CCG
Ser/S	UCU, UCC, UCA, UCG, AGU, AGC
Thr/T	ACU, ACC, ACA, ACG
Trp/W	UGG
Tyr/Y	UAU, UAC
Val/V	GUU, GUC, GUA, GUG
STOP	UAA, UGA, UAG

the genomes in genetic engineering, and (4) mutations of DNA. The changes in the genome may inhibit the vital activity of organisms and may even prove lethal to them, but if these changes are not lethal, they could remain for generations and lead to genetic modifications of organisms, genetic diversity of populations, and better adaptations of organisms to their environment.

Natural Genetic Recombination in Prokaryotes

Genetic recombination refers to the exchange of genes between two DNA molecules to form new combinations of genes. The recombinations take place naturally and are carried out in laboratories, using specific technologies described in Chapter 10. Genetic material can be transferred between prokaryotic cells by such ways as conjugation, transformation, and transduction.

Conjugation

Conjugation is the transfer of genetic information from a donor cell that gives a portion of its total DNA to a recipient cell. The recipient cell that incorporates the donor DNA is referred to as the recombinant. Conjugation is mediated by a special plasmid. The donor develops the pilus connecting the donor with the recipient and allowing the transfer of genetic information from the donor to the recipient.

Transformation

During transformation, which occurs only in several prokaryotic groups, genes are transferred from one prokaryotic cell to another through the liquid medium. Bacterial cells in nature, in the laboratory, or in engineering systems can be donors of pieces of DNA from liquid medium.

Transduction

Transduction is the transfer of bacterial DNA from a donor cell to a recipient cell inside a virus called bacteriophage that infects a recipient bacterial cell. Transduction is often used in genetic engineering to incorporate the specific genes in DNA of cell-recipient.

Horizontal Gene Transfer

Horizontal gene transfer is the transfer of genes by transformations or transduction between organisms that have distant, evolutionarily different genomes. Horizontal gene transfer occurs at a low frequency in natural and engineering ecosystems. Horizontal gene transfer between evolutionarily distant prokaryotes created parallelism in the evolution of prokaryotes of aquatic, terrestrial, and extreme environments.

DNA Reparation and Mutations

Chemical change in a base on one strand of DNA can initiate the enzymatic reparation of DNA due to signals of noncomplementarities from the complemantary DNA strand. However, if the DNA is not repaired, the base change will be inherited. The inherited change of the base and base sequence in nucleic acid is called mutation. An organism with this change is called a mutant. Mutations can be due to random mistakes in DNA replication, but mainly due to the action of chemical, physical, and biological mutagens.

Chemical Mutagens

The following are chemical mutagens: (1) substances that alter bases by chemical reactions usually by oxidation and reduction; (2) aromatic compounds and heavy metals that react with DNA molecules, changing its replication; and (3) base analogs that incorporate into DNA but changing the correct pairing of the DNA bases. Many products and wastes of chemical and pharmaceutical industries are chemical mutagens. Therefore, the destruction and prevention of the dispersion of chemical mutagens in environment is one of the major tasks of environmental engineering.

Physical Mutagens

The main physical mutagens are UV light and ionizing radiation. UV radiation is absorbed by DNA bases, changing them chemically, thus causing errors in DNA replication. Ionizing radiation generates free radicals in water that cause chemical changes in the bases.

Biological Mutagens

Some biological agents, such as transposons and the bacteriophage Mu, cause mutations by inserting their DNA sequences into genes, disrupting the coding information of these genes.

Genetic Adaptation of Microbial Population to Changed Environment

The changes of DNA in the genome of microorganisms by genetic recombination, gene transfer, and mutations are mechanisms for genetic (inherited) adaptation of microorganisms to a changed environment. Therefore, the enhancement of these changes, usually by application of mutagens, is a way by which industry can select microorganisms with needed properties by applying the mutagens together with the medium or applying conditions suitable for the development of needed properties.

7

Bioagents of Environmental and Engineering Bioprocesses

Hierarchy of Life

Levels of life can be shown by the following sequence: cell → organism → population → ecosystem → biosphere of earth. Cells are combined into an organism, organisms are combined into a population, populations are combined into an ecosystem (Figure 7.1), and, finally, combination of all ecosystems on the earth can be considered as a global ecosystem, a biosphere, or Gaia.

Cell Aggregates

Levels of microbial systems can be shown by the following sequence: cell → cell aggregates → microbial population → microbial community → microbial ecosystem. Both suspended cells and cell aggregates are used in environmental engineering. Major types of cell aggregates are flocs, granules, and biofilms. Flocs are irregular shaped, loose aggregates of thousands or millions of microbial cells (Figure 7.2). Granules are cell aggregates with a regular spherical or ellipsoidal shape and size ranging from 0.1 to 20 mm (Figure 7.2). Biofilms are thin, 10–100 μm layers of microbial cells, or irregular shaped aggregates of microbial cells attached to a solid surface (Figure 7.2).

Microbial Communities and Ecosystems

A population of microorganisms is a group of unicellular organisms of the same species. The microbial community is a multi-species system. A microbial ecosystem includes a group of microbial communities inside a defined place, separated from other environments by physical- or chemical-gradient boundaries. Lakes are examples of microbial ecosystems, which are separated from the environment

Figure 7.1 (See color insert following page 292.) Sample ecosystem in New Zealand with different plants and animals as well as viruses and microorganisms: bacteria, fungi, and algae. The surrounding mountains form the physical boundary of the ecosystem. Component ecosystems, for example a single deer farm in New Zealand, (pictured) can be considered part of a bigger ecosystem, despite being separated from the surrounding environment by a fence.

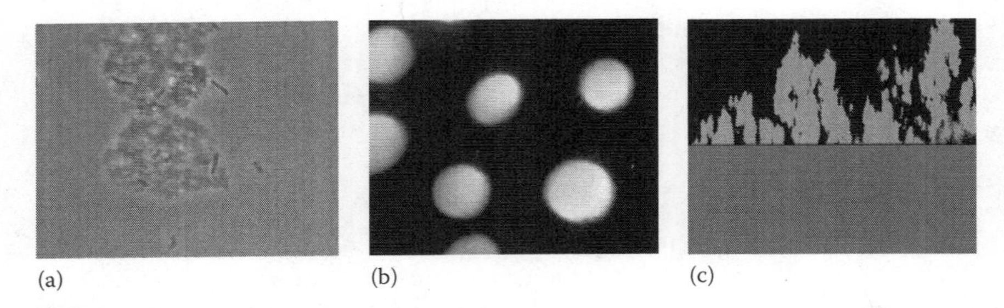

Figure 7.2 Cell aggregates used in environmental engineering: (a) flocs, (b) granules, and (c) biofilm attached to surface.

by the air–water interface, bottom and soil–water interface. There are different microbial communities living at the bottom, in the bulk of water, on water–air, and soil–water interfaces.

Shapes of Prokaryotic Individual Cell and Connected Cells

The most common shapes are spherical cell (coccus), rod-shaped cell (bacterium), slightly curved cell (vibrion), and helix-like cell (spirillum). The cells cannot be disconnected after cell division. These connected cells form a sheath-covered or uncovered filament of cells, chain of cells, 2D- or 3D-structures (Figure 7.3).

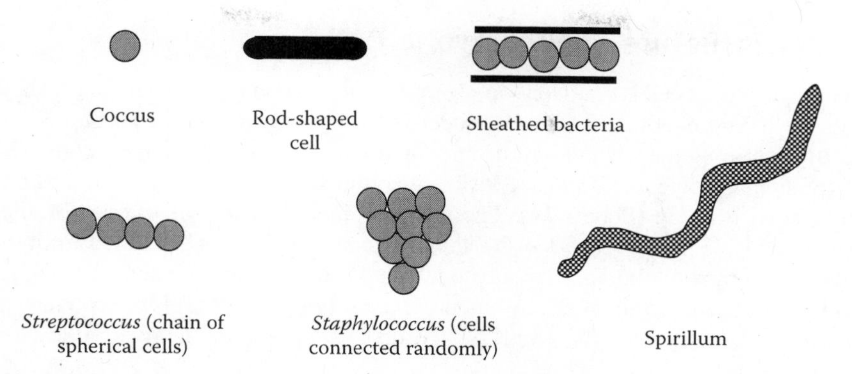

Figure 7.3 Most common shapes of prokaryotic cells and cell aggregates.

Shapes of Prokaryotic Cells as Evolutionary Adaptation to Environment

The shape of the prokaryotic cell can be considered as an adaptation to environment.

The spherical cell (coccus) is an adaptation to a homogenous environment, without gradients of nutrients. Examples of this environment include freshwater and seawater.

The elongated rod-shaped cell (bacterium) is an adaptation to a heterogeneous nonviscous environment with gradients of nutrients. It is the most abundant shape of prokaryotes. The elongated shape increases the vector of directional movement (called chemotaxis or phototaxis) of cells toward the source of nutrients. An example of this environment with gradients of nutrients is the aquatic microenvironment, which is close to the surface of solid matter of soil particles, suspended particles, bottom sediments, as well as to the surface of animal or plant tissue.

The curved (vibrio) or spiral shape (spirillum) of prokaryotic cells is an adaptation to a viscous environment. This shape ensures a spiral rotating movement of cell through viscous environments, thus decreasing the resistance of directional movement of cell toward the source of nutrients. Examples of this environment are viscous bottom sediments of aquatic ecosystems and mucosal surfaces of animal and plant tissues.

Filaments of cells are an adaptation to a heterogeneous environment with particles, where the nutrients are concentrated on the surface of these particles. So, the spread of filaments onto the surface of the particles is an optimal way to obtain these nutrients. Examples of this environment are soil particles, particles of dead organic matter, and particles suspended in water.

This explains the dominance of cocci (spherical prokaryotic cells) in oligotrophic (not polluted) freshwater and seawater, dominance of rod-shaped cells in eutrophic (polluted) water, vibrioforms (curved cells) and spirilla (spiral cells) in bottom sediments and surfaces of tissues, and filamentous cells on the surface of soil particles and particles suspended in water.

Inner Structure of Prokaryotic Cell

The prokaryotic cell is relatively simple in structure in comparison with the eukaryotic cell. The major cellular component is the cytoplasm (liquid portion inside cell) containing the nucleoid, ribosomes, and intracellular storage inclusions. The nucleoid and some inclusions are visible in cytoplasm under light microscope due to special staining (Figure 7.4). The nucleoid contains the prokaryotic chromosome, which is a circular DNA molecule. The nucleoid is not separated from the cytoplasm by a membrane like a true nucleus is in a eukaryotic cell. Ribosomes are particles consisting of rRNA and proteins, serving as the tools of protein synthesis. Several hundred thousands of ribosomes are present in a prokaryotic cell.

Intracellular Inclusions of Prokaryotic Cell

Granules of storage polymers, glycogen or poly-β-hydroxybutyrate (PHB), are present in the cytoplasm. Granules of elemental sulfur are found in the cytoplasm of phototrophic or sulfur-oxidizing bacteria. Gas vesicles are intracellular hollow spaces in cells providing cell buoyancy. Magnetosomes are the magnetite granules of specific shape. The accumulation of polyphosphate granules in some bacterial cells is the main approach for the bioremoval of phosphorus from wastewater.

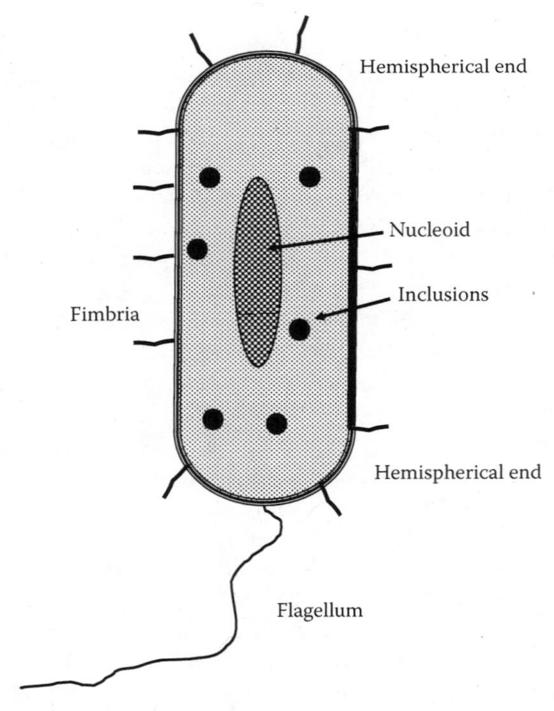

Figure 7.4 Typical structure of a prokaryotic cell.

The presence of different granules inside bacterial cells is an important parameter used in environmental engineering for cell identification or monitoring of physiological status of microbial communities. For example, the accumulation of glycogen is a sign of the shortage of nutrients other than being a source of carbon and energy, the accumulation of poly-β-hydroxybutyrate (PHB) granules is a sign of oxygen limitation, and the accumulation of polyphosphate granules is a parameter for evaluation of phosphate removal efficiency. The presence of shining granules of liquid sulfur in cells is an identification feature of anoxic phototrophic bacteria or microaerophilic bacteria oxidizing H_2S.

Outer Components of Prokaryotic Cell

The cytoplasm is enclosed by the following structures (Figure 7.5):

- Cytoplasmic membrane
- Cell wall
- Outer membrane (in Gram-negative bacteria)
- Capsule (not in all cells)

with attached fibrillar structures (not in all cells) such as

- Flagella(s)
- Fimbriae
- Pili

Cytoplasmic Membrane

The cytoplasmic membrane is an elastic cover of the cytoplasm of prokaryotes. It consists of two lipid layers with inserted protein gates for molecules. The following are the functions of the cellular cytoplasmic membrane:

- Control of entry and exit of charged molecules into the cell
- Production of biological forms of energy

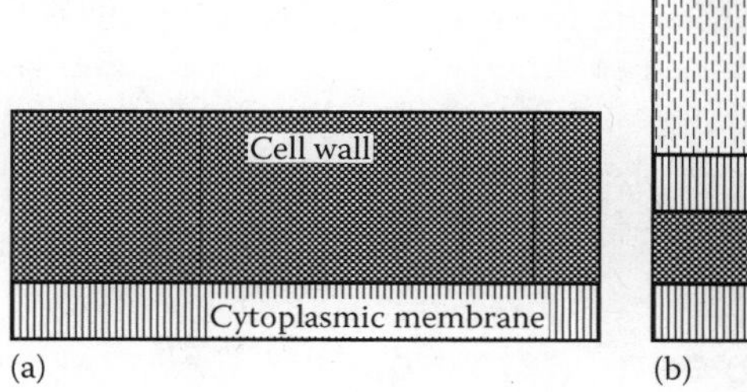

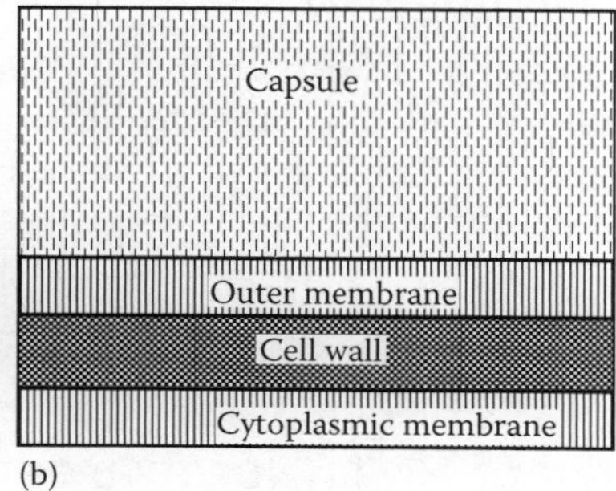

Figure 7.5 Outer structures of (a) Gram-positive and (b) Gram-negative prokaryotic cell.

- Attachment of prokaryotic chromosome and plasmids
- Attachment of flagella, fimbriae, and pili

Charged molecules and ions cannot penetrate through the layers of hydrophobic lipids of the cytoplasmic membrane. They can enter or exit only through cell-controlled protein gates in the membrane.

Biological forms of energy such as ATP, HADH ($+H^+$), and proton-motive force (transmembrane electrochemical gradient of protons) are also produced using enzymes located inside the cytoplasmic membrane.

Vulnerability of Cytoplasmic Membrane

The cytoplasmic membrane is the most vulnerable cell component because it is thin, elastic, and its integrity and the activity of membrane molecular gates are major conditions for generation of biological forms of energy. In environmental engineering, the integrity of the cytoplasmic membrane determines cell survival during the following processes:

1. The activity of oxidants during disinfection due to the inactivation of molecular gates present in the cytoplasmic membrane—used in many water and wastewater disinfection technologies.
2. The activity of surfactants due to the solubilization of lipids by the surfactants—washing hands with soap, a routine activity, is an example of this action on the cytoplasmic membrane of microorganisms.
3. The action of organic solvents, which dissolve lipid cytoplasmic membrane.
4. The action of low or high temperature—low temperature makes the membrane rigid and fragile and inactivates membrane gates, freezing temperature initiates formation of ice crystals destroying integrity of membrane, and high temperature melts lipid membrane and inactivates membrane gates.

Prokaryotic Cell Wall

Almost all prokaryotes require a cell wall to protect themselves from the effects of changes in osmotic pressure. There are two major types of cell walls, Gram-positive and Gram-negative, which can be differentiated using the staining procedure described by Gram. Prokaryotes with a thick cell wall stain positively and are called Gram-positive bacteria. Cells that have a thin cell wall and outer membrane as a part of cell wall stain negatively (Figure 7.5).

Origin of Gram-Negative and Gram-Positive Cells

The difference in cell wall structure could be considered as an adaptation to the osmotic pressure of the environment. If the cell wall is stable, for example, in seawater, freshwater, or in the tissue fluids of macroorganisms, there is no need for a cell to have a strong and rigid cell wall preventing it from swelling and disruption due to the entry of water present in the environment with osmotic pressure lower than that inside the cell. Therefore, Gram-negative cells with elastic and thin cell walls are mainly cells of aquatic origin. Gram-negative bacteria are mostly applicable in environmental engineering for water and wastewater treatment and bioremediation of aquatic systems.

However, osmotic pressure in soil can vary significantly due to drying under the sun or rain. Thick and rigid cell walls in these conditions can prevent cells from swelling and disruption due to entry of water in the environment that has an osmotic pressure lower than that inside the cell. Therefore, Gram-positive cells with rigid and thick cell walls are mainly cells of terrestrial origin. The stability of Gram-positive bacteria to the changes in osmotic pressure ensures their applicability in environmental engineering for soil bioremediation and production of dry bacterial compositions as starter cultures.

Flagella, Fimbria, and Pilus

A prokaryotic cell contains either one flagellum or several flagella (Figure 7.4). The flagellum is a rotating fibrillar structure ensuring cell movement toward the source of nutrients or away from a place with unfavorable conditions. The fimbria is a fibrillar structure that is shorter and thinner than a flagellum. Its major function is the specific attachment of a cell to a tissue surface. The pilus is a fibrillar structure, which is used by cells to attach to another cell and to transfer plasmid or chromosomal DNA from one cell to another. From the environmental engineering point of view, these external fibrillar structures can be considered as a valuable by-product of wastewater treatment because they can be used in nanotechnology as biodegradable nanofibrills and nanowires.

Outer Membrane, Lipopolysaccharides, and S-Layer

Gram-negative bacteria are covered outside the cell wall with an outer lipid membrane, which also contains lipids covalently linked to polysaccharides. These lipopolysaccharides are important in environmental engineering because they can be released into water after cell death and are often toxic or allergenic for humans and animals. Some prokaryotic cells are also covered with an S-layer, which is a rigid protein envelope functioning as a barrier for macromolecules.

Glycocalyx and Capsule

The glycocalyx is the extracellular slime where cells are embedded. Some prokaryotic cells can be covered with a structured slime layer (capsule) or a nonstructured accumulation of slime outside cell. The glycocalyx is important for environmental engineering from the following points of view:

1. It contributes to the pathogenicity of bacteria.
2. It helps bacterial cells to attach to surfaces.
3. It protects cells against desiccation.
4. It binds heavy metals and protects cells from their negative effects.
5. It binds chlorine and protects cell from disinfection by chlorine.
6. It aggregates bacterial cells, forming flocs and granules and improving the sedimentation of these aggregates in the activated sludge process.
7. It aggregates bacterial cells, forming a biofilm in fixed biofilm reactors.

Anabiotic Prokaryotic Cells

When nutrients become exhausted, some bacterial cells are transformed into a dormant endospore, which is anabiotic (temporarily out of life) cell. The inner dry matter of an endospore is covered by a thick envelope; therefore, endospores are very resistant to high temperature, sterilization, and disinfection. Endospores can remain dormant even for thousands of years, but germinate to form a vegetative cell when nutrients and optimal conditions for growth are provided.

Some groups of prokaryotes under starvation are also able to transform vegetative cells into a cyst, which is a spherical anabiotic dry cell, resistant to desiccation but not to high temperature. The cysts germinate when nutrients and optimal conditions for growth are provided. Some cyanobacteria, oxygenic phototrophic prokaryotes, form big cysts, which undergo multiple fissions, forming numerous small cells during germination.

Actinomycetes, which are filamentous Gram-positive prokaryotes, form, under conditions unfavorable for growth, numerous abiotic cells, called exospores, whose function is not only the survival of organisms during starvation and desiccation but also the dispersion of organisms in the environment.

Structure of Eukaryotic Cell

The eukaryotic cell is bigger and more complex, in that they contain organelles that are compartments for special metabolic functions and that are isolated from the cytoplasm by membranes. The major functions of eukaryotic cell components are the same as in prokaryotic cells:

1. The cell often has external filaments (cilia, flagella) for locomotion and attachment.
2. The cell is covered with a cell wall from cellulose (plants), chitin (fungi), silica (some algae), or other polymers to keep cell volume stable under

varying osmotic pressures; however, there may be no cell wall (animals) if osmotic pressure is stable.

3. The cytoplasm is isolated from the environment by a cytoplasmic (plasma) membrane, which is selectively permeable to nutrients and metabolites.

The major difference between eukaryotic and prokaryotic cells is the presence of organelles.

Organelles

Three important organelles of eukaryotic cells are mitochondria, chloroplasts, and the nucleus.

These organelles are bound by lipid membranes, which are more specifically permeable to molecules than the cytoplasmic membrane.

The nucleus in eukaryotes contains the genetic material. The genome is contained in a number of separated DNA molecules called chromosomes. The nucleus is a place where DNA for new cells is synthesized and transcribed into mRNA.

Mitochondria are organelles that generate biological forms of energy from the oxidation of low-molecular-weight organic compounds.

Chloroplasts are organelles assimilating light energy. The energy and reducing power generated by photosynthesis is used to convert carbon dioxide into organic carbon according to the equation:

$$CO_2 + H_2O + \text{energy of light} \rightarrow CH_2O + O_2 + \text{biological forms of energy}$$

where CH_2O is the conventional empirical formula for carbohydrates, primary products of photosynthesis.

8

Reproduction, Proliferation, and Growth

Reproduction is defined as the increase in the number of organisms or viral particles. Growth is defined as an increase of biomass.

Reproduction of Viruses

There are six basic stages in viral reproduction:

1. Attachment: a specific binding between virus and host cell. This binding determines the host range of a virus.
2. Penetration: the attached virus enters the host cell.
3. Uncoating: the viral envelope is degraded, thus releasing the viral nucleic acid.
4. Replication: the synthesis of viral nucleic acid and protein.
5. The self-assembly of virus particles.
6. The lysis of host cell and the release of viruses.

Reproduction of Prokaryotes

The most common method of cell reproduction in bacteria and archaea groups is an increase of cell volume, followed by the binary fission of the adult cell into two daughter cells (Figure 8.1). Other less common methods of cell reproduction include multiple fissions and budding (Figure 8.2).

Vegetative Reproduction of Microscopic Eukaryotes

Reproduction in microscopic eukaryotes is either by vegetative reproduction, which starts with the growth of cell biomass following the separation of mother and daughter cells, or by other methods of asexual and sexual reproduction.

For example, vegetative reproduction of mycelial fungi takes place by the elongation and separation of cells at the tip of the hyphae. The network of hyphae,

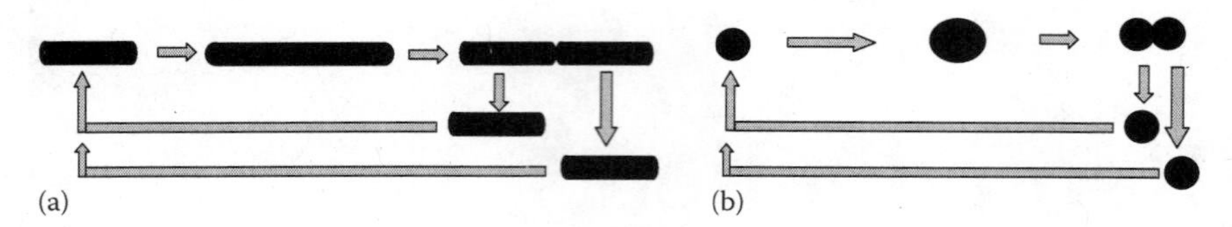

Figure 8.1 Reproductive cycles of (a) rod-shaped and (b) spherical prokaryotic cells by binary fission.

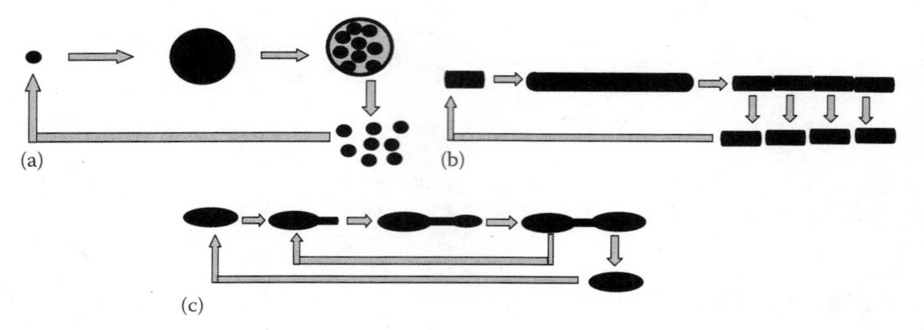

Figure 8.2 Reproductive cycles of prokaryotic cells with multiple fission (a and b) or budding (c). The bud is formed on the mother cell. The bud then grows and is transformed into a daughter cell that is separated from the mother cell.

mycelium, is an effective structure to penetrate soil and adsorb nutrients from the terrestrial environment. Mycelial fungi grow by the elongation of hyphae at their tips. Nutrients adsorbed by mycelia are distributed between the cells of mycelium.

Vegetative reproduction of unicellular fungi, yeasts, takes place usually by budding. This process includes the increase of mother cell volume, formation of bud, increase of bud volume, and separation of mother and daughter cells.

Asexual and Sexual Reproduction of Microscopic Eukaryotes

Eukaryotic cells can be reproduced after replication of DNA of the same organism (asexual reproduction) or after replication of the recombinant DNA, i.e. DNA from organisms of same species but different mating types ("sexes") (sexual reproduction).

Asexual reproduction often takes place through the formation of numerous spores (seeds). The major function of this reproduction is the dispersion of the organism in the environment; it is similar to the function of plant seeds. Spores of molds, produced by aerial mycelium, are easily dispersed in air and can cause allergy and other diseases in humans, especially indoors.

Sexual reproduction of microscopic eukaryotes includes mating between individual organisms of opposite mating type; mixing (recombination) of genetic material of organisms follows with the formation of spores. The function of sexual reproduction is the exchange of genetic material between organisms and dispersion of spores in the environment.

Cell Growth and Cell Division Cycle

Cell growth is the increase of cell biomass. It can be measured as an increase in cell mass, volume, or cell polymers. The time required to form two new cells from one is termed generation time, t_g.

In prokaryotic cells, there is a certain coordination between the cell division cycle (period between consecutive cell divisions) and the DNA replication cycle (period between initiation and termination of chromosomal DNA replication). However, this coordination is not as strong as in eukaryotes. Depending on growth or proliferation rates, there may be some cycles of DNA replication within a cell division cycle or even a cell division cycle without a DNA replication cycle accompanied by formation of DNA-free daughter cells.

In eukaryotic cells, there is strict coordination of a cycle of individual cell growth and division, with the DNA replication cycle. An eukaryotic cell cycle (mitotic cycle) has the following phases:

1. G_1-phase is a period between cell division and initiation of DNA replication; the duration of mitotic cycle is usually proportional to the duration of G_1-phase, and differentiation of cells starts from G_1-phase.
2. S-phase is a period of chromosomal DNA replication.
3. G_2-phase is a period between termination of DNA replication and mitosis (splitting of nucleus).
4. M-phase is a period of mitosis and splitting of nucleus.

Coordination of Cell Cycle Events

There are many levels of coordination between biochemical and physiological cell activities during a cell cycle:

1. Individual RNAs and enzyme synthesis and degradation
2. Regulation of enzyme activity by metabolites and cofactors
3. Regulation of catabolism and energy storage
4. Regulation of whole-cell activity by different cell regulators

Periods of Exotrophy and Endotrophy in Cell Cycle

A theory of the author explains the coordination of cell cycle events as an alternation of the periods of exotrophy and endotrophy. It was demonstrated

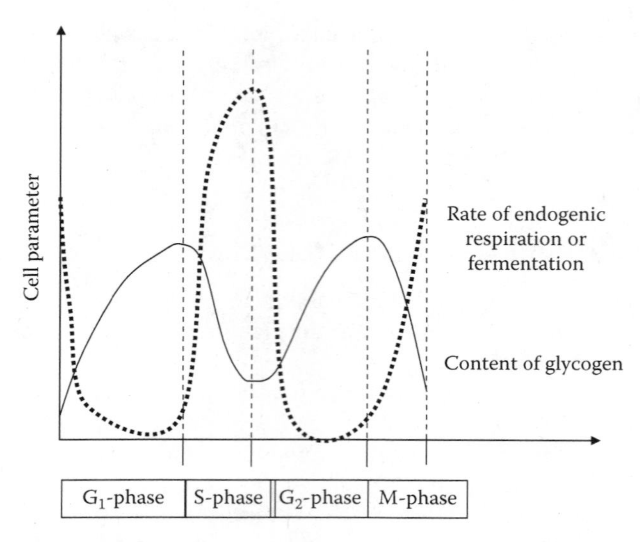

Figure 8.3 Periods of exotrophy and endotrophy in cell cycle.

experimentally that G_1- and G_2-phases of the eukaryotic cell cycle are mainly the periods of exotrophy. Exotrophy is a cell physiological state when the external sources of carbon and energy are extensively transformed into energy and carbon stores (glycogen, starch, lipids). Alternatively, S- and M-phases, which are the most sensitive periods of a cell cycle, are mostly the periods of endotrophy. During endotrophy, the accumulated stores of energy and carbon are utilized for DNA replication and mitosis. External sources of energy and carbon are not assimilated during these periods (Figure 8.3). The alternations between the periods of exotrophy and endotrophy take place due to the increase or decrease of intracellular concentration of cyclic AMP and are accompanied by alternation of the charge of membrane potential.

Cell Differentiation

Cell differentiation is the transformation of microbial cells into specialized cells. Examples of such cells are as follows:

1. An endospore is an anabiotic (i.e., temporarily not active) cell with low water content, covered by a thick envelope, serving for survival under unfavorable conditions for growth, e.g., starvation, dry environment, and high temperature.
2. An exospore is similar to an endospore in its properties but it does not form in the mother cell; the main functions of these cells are to increase survival and dispersion of cells in the environment.
3. An anabiotic cyst is an enlarged cell whose main function is increasing the cell survival rate.

4. Nitrogen-fixing cysts and bacteroides are enlarged cells whose main function is the transformation of atmospheric nitrogen into amino groups of organic substances.

Growth and Proliferation of Cell Population

Growth is defined as an increase of the biomass of a cell population. Proliferation is an increase of cell number in a population. Balanced growth is a proportional increase of both biomass and cell number in a population. There may be "unbalanced growth" without proliferation or proliferation without growth under unfavorable conditions for either growth or proliferation.

Generation Time and Number of Generations

The rate of proliferation can be determined as the increase of cell concentration (N). The time required to double the cell number in a population can be considered as the average cell generation time, t_g.

The number of cells as well as the biomass can increase linearly in case of formation of new cells on the tips of hyphae, which can be defined by the equation

$$N = kN_0$$

where

N_0 is the initial cell concentration
k is growth rate

It can change exponentially in case of cell reproduction by fission:

$$N = N_0 2^n$$

where n is the number of generations.

Specific Growth Rate

Specific growth rate, μ, is the quantity of new biomass produced by unit of biomass per unit of time:

$$m = \frac{dX}{dt} X = \frac{\ln X - \ln X_0}{t}$$

Therefore, exponential growth can be described by the following equation:

$$X = X_0 e^{\mu t}$$

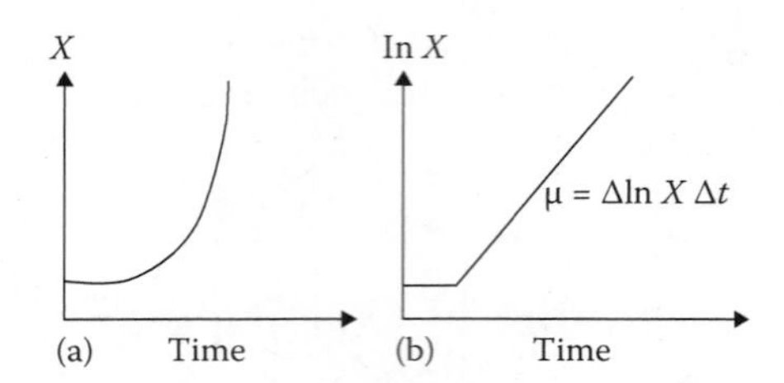

Figure 8.4 Exponential growth (a) can be represented linearly on the semilogarithmic graph (b).

Exponential growth can be represented linearly on semilogarithmic graph, where logarithm of cell concentration is plotted versus time (Figure 8.4).

Typical specific growth rates are

–1.0 to 0.2 h^{-1} for aerobic heterotrophic bacteria under optimal conditions in the medium with carbohydrates
0.2 to 0.05 h^{-1} for fermenting bacteria and methanogens under optimal conditions for growth
0.2 to 0.05 h^{-1} for fungi under optimal conditions for growth

Stoichiometry of Microbial Growth

Stoichiometry shows the ratios between substrates and products of the reactions. Growth yield ($Y_{X/S}$) is the ratio between quantity of produced biomass (dX) and consumed substrate (dS):

$$Y_{X/S} = \frac{\Delta X}{\Delta S}$$

for the batch system, the growth yield is determined by the equation

$$Y_{X/S} = \frac{X_T - X_0}{S_0 - S_t},$$

where

S_t is substrate concentration in the system at the end of period t
S_0 is the initial substrate concentration

For the continuous system without recycling of the biomass, the growth yield is determined by the following equation:

$$Y_{X/S} = \frac{X}{S_\mathrm{I} - S_\mathrm{E}},$$

where
X is the biomass concentration
S_I and S_E are the concentration of substrate in the influent and effluent, respectively

Typical growth yield of bacteria in medium with glucose is 0.5 g of dry biomass/g glucose for aerobic growth and 0.1 g of dry biomass/g glucose for anaerobic growth.

The average water content in cells is 75%. The average contents of major cell components are as follows:

- Proteins, 60%
- RNA, 15%
- Polysaccharides, 10%
- DNA, 5%
- Ions, 10%

Major elements in the composition of microbial cells are C, H, O, and N. The average content of elements in the biomass may be shown by the empirical formula $CH_{1.8}O_{0.5}N_{0.2}$.

Cell Age and Cell-Trophic-State Distributions in Microbial Population

Due to asynchronous cell cycles of individual microbial cells, there is a distribution of cells with different ages and cell-trophic states in a population. Cell size, DNA content, and percentage of exotrophic or endotrophic cells can be used to monitor cell population by flow cytometry. For example, duration of exotrophy (Δt_{ex}) and duration of G_1 phase (Δt_{G_1}) of yeasts are linearly related to the duration of cell cycle (T):

$$\Delta t_{ex} = 0.5T - 1.0 \quad \text{and}$$

$$\Delta t_{G_1} = 0.7T - 0.9$$

Using these or similar equations, the specific growth rate in a population (μ) can be determined from the microscopic image of cells or flow cytometry distribution, taking into account that $T = \ln 2/\mu$.

G_1-cells of the budding yeasts can be determined as cells without buds. Exotrophic and endotrophic cells of bacteria can be also distinguished after adding a small quantity of co-oxidizing substrate, producing toxic products of oxidation. For example, a solution of allyl or amyl alcohol can be added to cells that utilize ethanol. As a result, cells produce allyl or amyl aldehyde, which cannot be oxidized further and, therefore, kills cells. Exotrophic cells die after this incubation but endotrophic cells remain alive because they do not consume and oxidize external sources of carbon and energy. The share of exotrophic cells increases during starvation and other unfavorable conditions because the S phase cannot be started until intracellular accumulation of sufficient quantity of carbon and energy sources occurs.

Quantification of Microbial Biomass

Determination of cell number (enumeration) or quantity of cell biomass can be performed by the following methods:

1. Microscopic or flow cytometric enumeration of cells
2. Physical measurement of microbial cells and biomass concentration
3. Chemical measurement of microbial cells and biomass concentration
4. Biological methods of cells and viruses enumeration
5. Physiological measurement of biomass
6. Molecular-biological methods of cells and viruses enumeration

Factors that affect the choice of method include

1. Cost and length of time required for analysis
2. Sensitivity and specificity of the method
3. Availability of the equipment
4. Characteristics of the interest

Microscopic Enumeration

Light microscopes (bright field, phase contrast, fluorescence), confocal laser scanning microscopes (CLSM), transmitted electron microscopes (TEM), scanning electron microscopes (SEM), and other kinds of microscopes can be used to visualize a number of microbial cells or virus particles on a defined area.

The particular cell structure or virus surface may be labeled by a specific stain, and DNA or RNA of cells can be hybridized with oligonucleotide probes labeled by fluorescence, radioactive, or other labels. This specific staining can ensure cell enumeration together with cell identification and with the analysis of cell physiological state.

Flow Cytometry Enumeration

Flow cytometry is used to quantify cells that are stained with specific fluorescent stains that are excited by lasers in the flow of a small diameter path (Figure 8.5). The fluorescence of individual cells is then measured by photomultipliers and the signals are collected and processed by a computer. In addition to the cell number, three to six other parameters of thousands of individual microbial cells or viral particles can be analyzed in seconds.

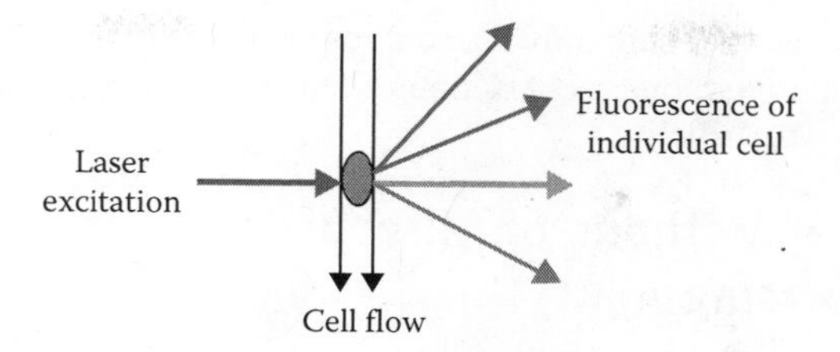

Figure 8.5 (See color insert following page 292.) Flow cytometry of individual cells.

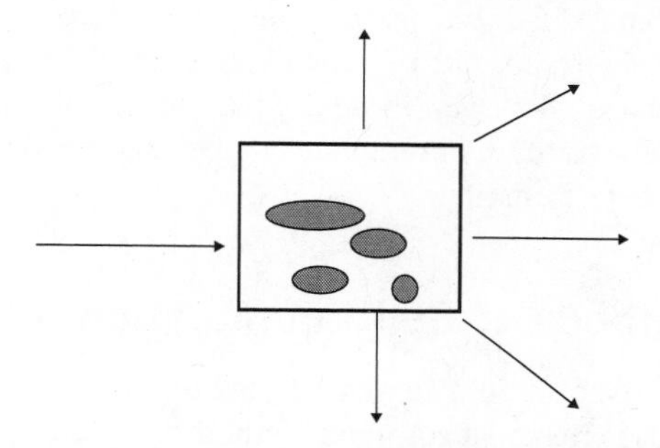

Figure 8.6 Light scattering is proportional to the cell sizes and cell concentration in the sample.

Physical Methods for Microbial Biomass Measurement

Biomass can be measured as the weight, optical density, turbidity, fluorescence, or radioactivity of microbial suspensions and solid matter. A convenient method of suspended biomass estimation is turbidity measurement. The amount of scattered light is proportional to the cell sizes and cell concentration in the sample (Figure 8.6).

Autofluorescence of microbial cell components (chlorophyll of algae, bacteriochlorophyll and carotenoids of cyanobacteria, F_{420} of methanogens) or fluorescence of stained cells also can be used for the measurement of biomass. Fluorescence spectrometry can be used to quantify microorganisms in environmental engineering systems using determination of the binding of specific oligonucleotide probes.

Chemical Methods for Microbial Biomass Measurement

Measurement of protein, DNA, the components of cell wall, ATP, photopigments, cytochromes, coenzymes $NADH_2$ or F_{420} could be useful in determination of biomass. ATP levels are a sensitive indicator of a small quantity of viable microorganisms.

Chemical changes in the medium caused by microbial growth and those which change in proportion to growth can be monitored using electrochemical sensors and fiber optic sensors. There may be changes of pH, oxidation–reduction potential (ORP), conductivity, or concentration of ions in medium.

Physiological Methods of Microbial Biomass Measurement

The measurements of physiological activity of cells, for example, respiration rate, specific biochemical transformation rate, and ATP concentration could be used for evaluation of biomass. It could be, for example, the rate of production of lactic acid for lactic acid bacteria, or rate of ethanol production for yeasts, or respiration rate for aerobic bacteria. However, these are parameters of physiological activity and not biomass itself, so these parameters can be used for the evaluation of biomass but not for its determination.

Biological Methods of Cell Enumeration: Plate Count

Plate count, i.e., cultivation on a semisolid medium and enumeration of colony-forming units (CFU) is the most common method in environmental microbiology (Figure 8.7). It is assumed in this method that one cell placed on a plate with semisolid medium produces one colony on this medium but cells are often aggregated in clamps or attached to the particles (Figure 8.8).

Usually, 0.1 mL of bacterial suspension is spread on the surface of a semisolid medium. However, there may be 10^3–10^{12} cells in 1 mL of the sample. Therefore, it should be diluted in a sterile medium before being spread onto a Petri dish to produce not more than 100–300 colonies per plate (Figure 8.8).

Some bacterial groups cannot be cultivated in the laboratory because the medium or growth conditions for them have not yet been defined.

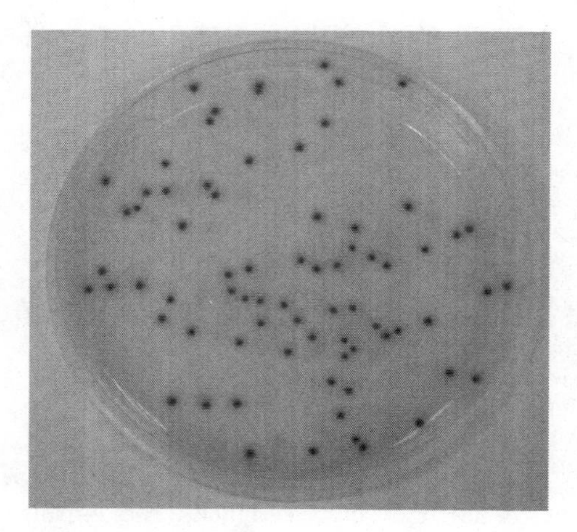

Figure 8.7 (See color insert following page 292.) Colonies on Petri dish.

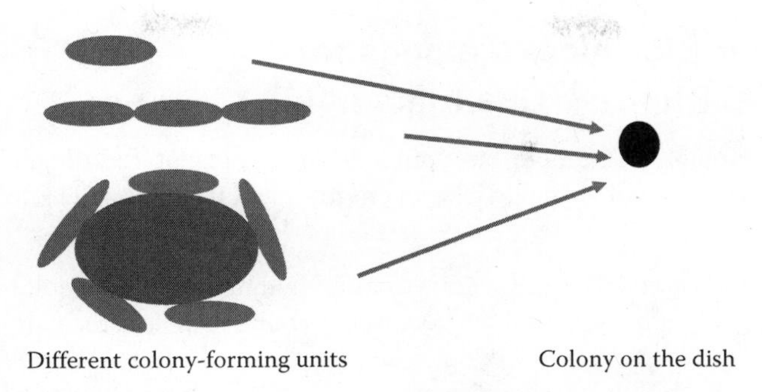

Figure 8.8 Colony-forming units.

Biological Methods of Cell Enumeration: Most Probable Number Count

The most probable number (MPN) count, i.e., identification of the maximum dilution at which the growth or microbial activity can be easily detected by color change, precipitation, or formation of gas bubbles. For example, if the maximum dilution to detect microbial activity in 1 mL of specific medium is 5×10^{-4}, the most probable number of cells in the sample is 2×10^{3} cells/mL. It is assumed that multiplication of one cell in the tube with maximum dilution can produce the detectable result (color change, gas bubbles), but this assumption is often incorrect. MPN count is often used in analysis of water quality for the determination of water pollution by cells of an indicator bacterial group of fecal coliforms.

Enumeration of Cells and Viral Particles at Low Concentration

If the cells or viral particles being studied are present at a low concentration, they must be concentrated using the following methods:

1. Filtration of the sample through a sterile membrane filter having a pore size <0.45 mm to retain bacterial cells and <2–5 mm to retain eukaryotic cells
2. Precipitation or centrifugation of cells or viral particles of the sample
3. Chromatography of the sample
4. Adsorption of cells and viral particles in the column with a specific adsorbent

Viruses are enumerated biologically by spreading a diluted suspension on the surface of a lawn of actively growing cells susceptible to the virus. As virus particles infect and reproduce, the newly produced viruses kill surrounding cells, forming a zone of clearing in the cell layer.

Molecular-Biological Methods for Microbial Biomass Quantification

Molecular-biological methods are often used at present for quantification of microbial biomass and viral particles in environmental samples. The major methods are

1. Immunochemical quantification of microbial biomass using color change produced by the reaction between specific antibody and the cell or viral particle surface.
2. Molecular-biological quantification of microbial biomass using fluorescence change in the in situ reaction between a specific oligonucleotide probe and cell RNAs or DNA.
3. Quantitative polymerase chain reaction (PCR) called real-time PCR. It involves the extraction of DNA from the sample and amplification of specific genes with its quantification after every cycle of DNA amplification. This method is especially important for bacterial groups that cannot be cultivated in the laboratory because the medium or growth conditions for them are not yet defined, or which are symbiotic or parasitic species.

9

Microbial Ecology

Ecosystem

Microbial ecology is a science studying the interactions of microorganisms with the environment, other organisms, and between themselves. An ecosystem comprises biotic (biological) and abiotic (physical, chemical) components interacting with each other and isolated from the environment by a boundary (Figure 9.1). An interaction between two elements can also affect the interaction between two other elements.

Level of Ecosystems

The hierarchy of life units in microbial ecosystems can be represented in the sequence of increasing spatial and biological complexity:

1. Suspended cells (unicellular organisms) of one species
2. Suspended cells (unicellular organisms) of a microbial community
3. Aggregated cells and multicellular microorganisms
4. Ecosystems of located biotope
5. Ecosystems of whole biosphere

Boundary of Ecosystem

The boundary between an ecosystem and its surrounding environment is a steep gradient of physical and/or chemical properties. The physical boundary is formed by an interface between solid and liquid phases, solid and gas phases, and liquid and gas phases. For example, the microbial ecosystem of an aerobic tank for wastewater treatment is separated from the environment by the reactor walls and an air–water interface. The steep gradient of chemical substances, for example, oxygen, ferrous ions, hydrogen sulfide, etc., forms a chemical barrier. Such barriers separate, for example, aerobic and anaerobic ecosystems, in a lake. The steep gradient of conditions can be also created by cell aggregation in flocs, granules, or biofilms. The main function of the boundary is to maintain integrity of an

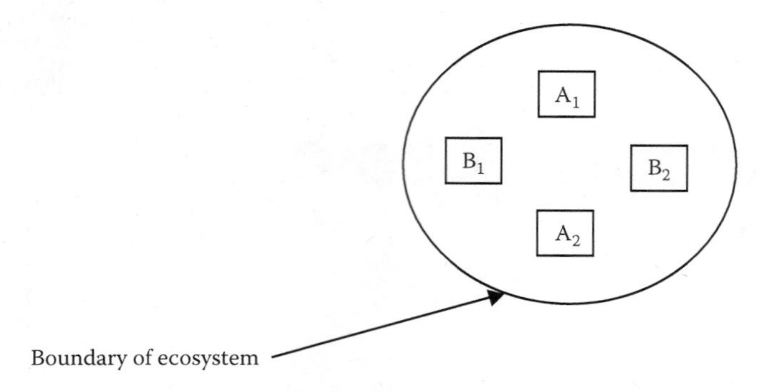

Figure 9.1 Example of the simplest ecosystem consisting of two biotic (B_1 and B_2) and two abiotic elements (A_1 and A_2).

ecosystem by controlled isolation from the environment and to protect an ecosystem from the destructive effects of the environment.

Boundaries of Unicellular Organisms

The following are boundaries of unicellular organisms:

1. The cell membrane (cytoplasmic membrane) performs a selective and controlled exchange of molecules between cell and environment. It is the most sensitive boundary, because even a small break in the cell membrane will destroy isolating and energy-generating properties of a cell membrane. Surface-active substances, organic solvents, oxidants, and high temperature destroy the integrity of a cell membrane.
2. The cell wall protects a cell from changes in osmotic pressure and mechanical impulses. Bacteria with a thick cell wall are stained as Gram-positive cells.
3. Bacteria that are stained as Gram-negative cells have a thin cell wall covered by an outer membrane. Lipopolysaccharides of the outer membrane of Gram-negative bacteria are very specific. These molecules interact with the human immune system and are often toxic or allergenic. Gram staining is just one and not always a reliable method to differentiate bacteria with Gram-positive and Gram-negative types of cell walls.
4. Some prokaryotes, for example, mycoplasmas, have no cell wall.
5. Fungi and algae often have cell walls containing polysaccharides such as cellulose or chitin. Some algae have inorganic compounds such as calcium carbonate or silica in their rigid walls. Animal cells often have no cell walls.
6. The glycocalyx (capsule) is an extracellular polysaccharide covering microbial cells of some species. Its functions include attachment of the cells to the surface, aggregation of cells, and protection of cells against drying, oxidants, heavy metals, and antibiotics.

Boundary of Multicellular Aggregate

A multicellular aggregate is formed and separated from its surrounding environment due to:

1. Aggregation by hydrophobic force, electrostatic interactions, or salt bridges
2. Loose polysaccharide or inorganic matrix (iron hydroxide as example); combining the cells altogether by mechanical embedding, chemical bonds, hydrogen bonds, electrostatic forces, or hydrophobic interactions
3. Formation of mycelia, which is a net of branched cell filaments
4. Polysaccharide matrix with a filamentous frame
5. Structured matrix with layers parallel to the boundary or sub-aggregates, which are perpendicular to the boundary
6. Coverage by a common sheath of organic (polysaccharides, proteins) or inorganic origin (iron hydroxide, silica, calcium carbonate)
7. Coverage by a common sheath ("skin" of microbial aggregate) consisting of dead cells

A microbial aggregate can be considered as a multicellular organism if its parts have different coordinated or synchronized physiological functions, that is, growth, motility, sexual interactions, assimilation of atmospheric nitrogen, production of extracellular polysaccharides, transport and distribution of nutrients, and reduction of oxygen.

Boundaries of Microbial Communities in Environmental Engineering

Microbial communities of environmental engineering systems are usually suspended or adhered to surface cells and microbial aggregates such as fixed biofilms and suspended flocs or granules. The boundaries of these ecosystems are as follows:

1. Side walls of the equipment with a fixed microbial biofilm
2. Bottom of the equipment with the sediment of microbial aggregates
3. Gas–liquid interface with accumulated hydrophobic substances (lipids, hydrocarbons, aromatic amino acids) and cells or aggregates with high hydrophobicity of their surface or cells and aggregates containing gas vesicles

Diversity of a Microbial Ecosystem

Diversity of ecosystem refers to the heterogeneity in genotypes (diversity of strains, species, physiological groups), in space (different zones, layers, aggregates, and chemical or physical gradients), and in time (temporal changes in diversity of genotypes and spatial structure of ecosystem). Succession refers to the typical sequence of temporal changes in an ecosystem. Stagnation or climax is a state of ecosystem characterized by weak changes caused by poor environment, degeneration, or aging of the system.

Numerous mathematical expressions that quantify diversity are known. For example, the Shannon–Weaver index (H) is

$$H = \sum_{i=0}^{i=S} [p_i - \ln(p_i)] \tag{9.1}$$

where
p_i is the proportion of the ith group in the community
S is the number of groups in the community

Evenness index (E) is a measure of how similar the abundances of different groups are:

$$E = \frac{H}{\text{In}\, S} \tag{9.2}$$

When there are similar proportions of all groups, then the evenness index is 1. The evenness index is larger than 1 when the abundances are very dissimilar. An example of quantitative characterization of microbial diversity in an anaerobic digester of activated sludge is given below.

Diversity in an Anaerobic Digester

There are at least five microbial groups involved in anaerobic digestion:

1. Hydrolytic bacteria degrading polymers (polysaccharides, proteins, nucleic acids) to monomers (glucose, amino acids, nucleosides)
2. Acidogenic bacteria fermenting monomers to organic acids and alcohols
3. Acetogenic bacteria producing acetate from other organic acids and alcohols
4. Acetotrophic methanogens producing methane from acetate
5. Hydrogenotrophic methanogens producing methane from hydrogen and carbon dioxide

If the cell concentration of these organisms per 1 mL is 4×10^7, 7×10^8, 2×10^7, 5×10^8, and 1×10^8, respectively, the Shannon–Weaver index (H) of microbial diversity by physiological functions will be 13, and the evenness index (E) will be 8.1. The diversity indices are related to the process efficiency and stability and can be used in environmental engineering to compare processes with different operational parameters.

Types of Interactions in Microbial Ecosystems

The types of interactions between the biotic elements of a microbial ecosystem (microbial cells, microbial populations, microorganisms, and macroorganisms) are positive and negative. The positive interactions are as follows:

1. Commensalism (only one biotic element has benefits)
2. Cooperation, mutualism (both elements have benefits)
3. Essential mutualism, symbiosis (both elements cannot live separately)

The negative interactions are as follows:

1. Neutral competition (organisms compete in the rate and efficiency of nutrient consumption, growth rate, or in the resistance to unfavorable factors for growth)
2. Antagonism (both abiotic elements suffer from interaction because they produce specific factors that negatively affect the growth rate or other physiological or biochemical properties of competitors)
3. Amensalism (only one element suffers from the interaction)
4. Predation and parasitism (it is interactions when one element [prey] suffers and the other element [predator] benefits)

There may also be neutralism, that is, absence of positive or negative interactions between biotic elements.

Population Density Determines the Type of Interaction

The population density or average distance between biotic elements determines the type of interaction (Figure 9.2).

When the population density is low, organisms have neither positive nor negative interactions. When the population density is medium, organisms compete among themselves for the availability of resources, by rate or efficiency of growth, and by the production of metabolites, which negatively affect the growth of competitors. When the population density is high, cells usually aggregate and cooperate between themselves. Both competition and cooperation are carried out mainly due to the changes of chemical factors of environment such as concentration of nutrients, pH, and redox potential of the medium, excretion of antibiotics, extracellular digestive enzymes, or heavy metals binding exopolysaccharides, and simultaneous biodegradation of substances.

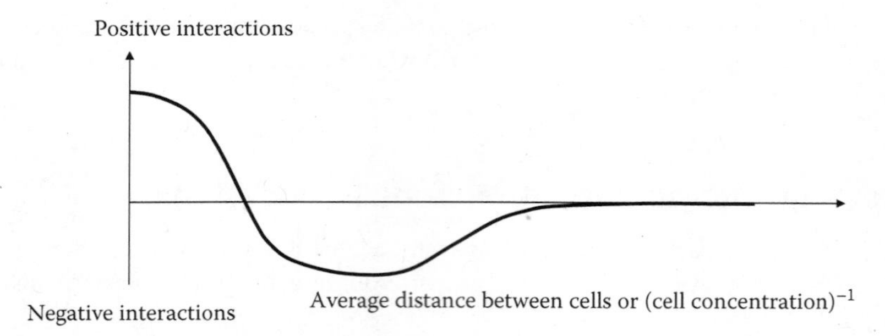

Figure 9.2 Microbial interactions depend on cell concentration in the ecosystem and the distance between cells within the cellular community.

Commensalism

Commensalism is a relationship between biotic elements in which only one biotic element benefits. It is realized by different ways (Figure 9.3).

There are thousands of examples of this interaction in environmental biotechnological systems. Some of them are as follows:

1. Facultative anaerobes use oxygen and create the conditions for the growth of obligate anaerobes; this interaction is important in the formation of an anaerobic layer in microbial aggregates existing under aerobic conditions.
2. One group of microorganisms produces a growth factor essential for another group; this interaction is an obvious condition for the outbreak of pathogenic *Legionella pneumophila* originating from such engineering systems as air conditioners, cooling towers, and fountains.
3. Sequential biodegradation of xenobiotics is carried out by different groups of microorganisms; the microbial group performing biodegradation does not depend on the activity of the groups degrading its product of metabolism.

Mutualism

Mutualism is a type of interaction in which both biotic elements (microbial groups) derive advantages from their interaction (Figure 9.4).

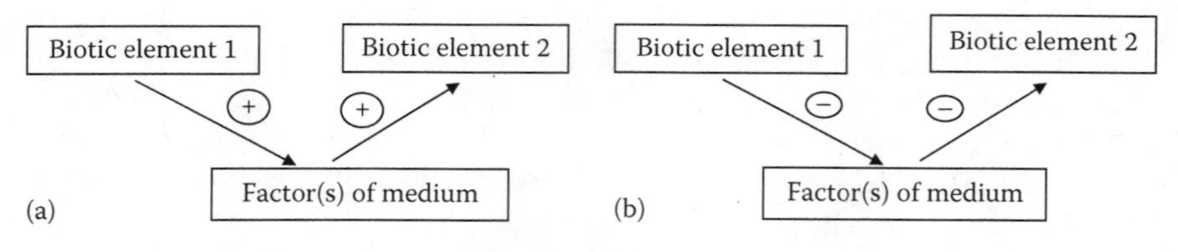

Figure 9.3 Examples of microbial commensalism: (a) one group of microorganisms produces growth factor(s) essential for another group, (b) facultative anaerobes use oxygen and create anaerobic habitat suitable for the growth of obligate anaerobes.

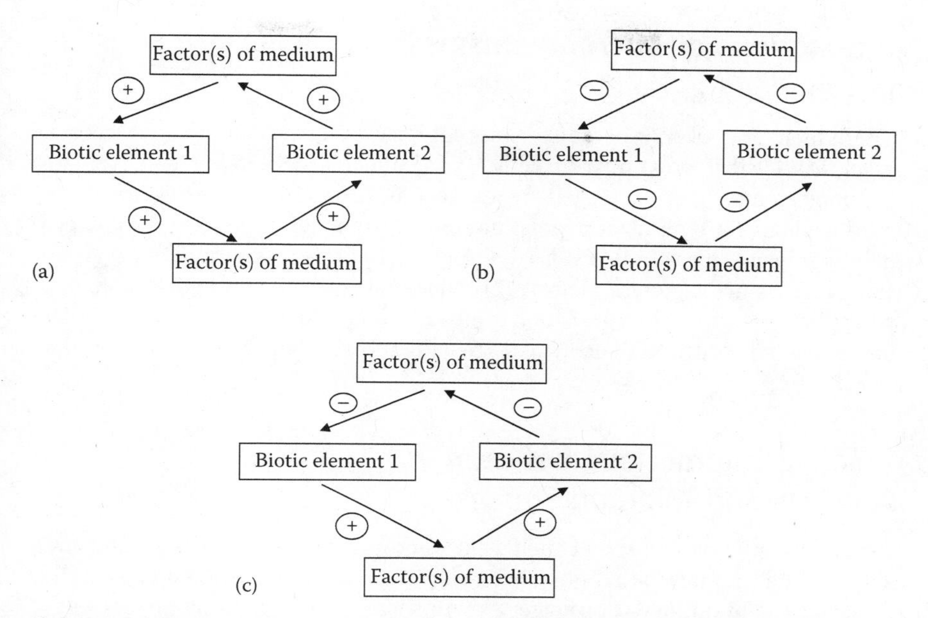

Figure 9.4 Examples of mutually beneficial microbial interactions: (a) both microbial groups exchange growth factors or nutrients, (b) both microbial groups diminish negative factors in the medium, and (c) removal of oxygen by group 2 creates anaerobic conditions for fixation of nitrogen by group 1, which supplies nitrogen compounds for group 2.

Mutual interactions are facilitated by close physical proximity in microcolonies, biofilms, and flocs. Physiological cooperation in biofilm or aggregates is supplemented and supported by its spatial structure, that is, formation of microhabitats for individual populations. Some examples of mutualism are as follows:

1. Syntrophy (co-eating)—both microbial groups supply growth factors or nutrients.
2. Sequential biodegradation of xenobiotics—when the product of biodegradation inhibits biodegradation.
3. Cycling of elements by two microbial groups:

   a. $$\text{Phototrophic bacteria} + \text{light} + H_2S + CO_2 \rightarrow S + \text{organics} \tag{9.3}$$

   b. $$\text{Facultative anaerobic bacteria} + S + \text{organics} \rightarrow H_2S + CO_2 \tag{9.4}$$

4. Removal of oxygen by heterotrophic bacteria creates anaerobic conditions for fixation of nitrogen by phototrophic cyanobacteria, which supply nitrogen compounds for heterotrophic bacteria.

Interactions of Microorganisms in Cellular Aggregates

The interactions of microorganisms in aggregates are usually positive because of close physical proximity of cells. Physiological cooperation in aggregates is supplemented and supported by its spatial structure, that is, formation of microhabitats for individual populations. Some examples of mutualism in microbial aggregates are as follows: (1) syntrophy (co-eating), both microbial groups supply nutrients or growth factors; (2) sequential biodegradation of xenobiotics, when the product of biodegradation inhibits biodegradation; and (3) biochemical oxidation and reduction of element by two microbial groups.

Positive Interactions between Animals and Microorganisms

Positive interactions between animals and microorganisms are common and often essential for animals. Microorganisms can improve the digestion and assimilation of food by animals, produce growth factors like vitamins and essential amino acids for animals, and keep out the pathogenic microorganisms from the surface and cavities of macroorganisms. For example, ants and termites cultivate cellulose-degrading fungi in their chambers to enhance the feeding value of plant material. Insects provide cellulose-degrading fungi with favorable conditions and a supply of cellulose and mineral components. Another example is interactions between microorganisms and humans. The human body contains complex and stable microbial communities on the skin, hairs, body cavities, and within the gastrointestinal tract. Macroorganisms provide favorable conditions and supply nutrients to microorganisms, which produce some vitamins and keep out the pathogenic microorganisms from the skin and surface of cavities.

Symbiotic Mutualism

Symbiotic mutualism, or simply symbiosis, means that two groups cannot live separately. This is commonly the case for the interaction between macroorganisms and microorganisms. Many protozoa have symbiotic relations with bacteria and algae, often including them into the cell as endosymbionts. Bacterial endosymbionts supply the growth factors to the protozoan partner. A well-known example is the symbiosis of ruminant animals (cow, deer, sheep) and anaerobic, cellulose-degrading microorganisms in their rumen. Ruminant animals ensure crushed organics and mineral components, optimal pH and temperature, and microorganisms hydrolyze and transform cellulose to assimilated fatty acids.

Positive Interactions between Plants and Microorganisms

These interactions are common and often essential for plants. Epiphytic microorganisms live on aerial plant structures such as stems, leaves, and fruits. The habitat and microorganisms on the plant leaves is called phyllosphere. The yeasts and lactic acid bacteria, for example, dominate in the phyllosphere. They receive carbohydrates and vitamins from the plant. High microbial activity occurs also in the soil surrounding the roots, called the rhizosphere. Organic compounds that stimulate heterotrophic microbes are excreted through the roots. Some fungi are integrated into the roots and contribute to plant mineral nutrition. This type of symbiotic interaction is called mycorrhizae. An example of mycorrhizae is the interaction between pine and fungi. Fungi integrated into the roots of pine contribute to plant mineral nutrition in exchange for a supply of organic nutrition from the plant.

Symbiotic Mutualism of Plants and Microorganisms

A well-known example is the symbiotic fixation of atmospheric nitrogen, which is a major reservoir of nitrogen for life. The roots of some plants are invaded by nitrogen-fixing bacteria, mainly from the genus *Rhizobium*, which form tumor-like aggregates (nodules), where the bacteria are transformed into large cells (bacteroids) capable of fixing N_2 from air. A plant supplies the bacteroids with organic and mineral feed and the bacteroids supply organic nitrogen to the plant. Symbiotic relations ensure the existence of lichens where photosynthetic algae or the cyanobacterial component of lichens produce organic matter, and microscopic fungi provide mineral nutrient transport and the mechanical frame for the photosynthetic organisms. Most lichens are resistant to extreme temperatures and drying and are capable of fixing nitrogen and occupying hostile environments.

Neutral Competition

Neutral competition between the biotic elements (organisms/populations/groups) means competition by the rate of nutrients consumption or growth rate. There may also be neutral competition by affinity with the nutrients or by resistance to environmental factors unfavorable for growth. It is the most typical interaction for aquatic natural ecosystems and wastewater treatment engineering systems.

Amensalism

It is an active competition in which one biotic element produces a substance that inhibits the growth of another biotic element. There may be, for example, changes in pH caused by the production of inorganic and organic acids by one population. Neutral competition and amensalism are the main mechanisms for forming an enrichment culture where one or some species dominate after cultivation of an environmental sample. The production of antibiotics is a specific example of amensalism because antibiotics are substances that are able to inhibit growth of sensitive cells at a low concentration. Antibiotic-producing microorganisms dominate in rich environments with optimal conditions for growth, that is, in the biotopes where neutral competition is not sufficient to ensure domination of one biotic element. These biotopes are soil, phyllosphere, skin, or cavities of animals, but not aquatic biotopes with a low concentration of nutrients.

Antagonism

Antagonism is the active competition between two biotic elements, that is, competition enhanced by specific tools such as excretion of chemical substances including antibiotics, by two competing biotic elements.

Predation

Predation occurs when one organism engulfs and digests another organism. A typical predator–prey relationship exists between predator protozoa and bacteria. Therefore, the predator protozoa improves the bacteriological quality of the effluent after aerobic wastewater treatment because it helps to reduce the number of free-living bacteria.

Parasitism

Parasitism is a very common interaction between microorganisms and macroorganisms and between different microorganisms. The benefiting parasite derives its nutritional requirements from the host, which is the harmed organism. All viruses are parasites of bacteria, fungi, algae, plants, and animals. Some prokaryotes are parasites of prokaryotes. For example, *Bdellovibrio* spp., small curved cells, are parasites of Gram-negative bacteria, and *Vampirococcus* spp. sucks the cytoplasm out of another bacterium. Enumeration of microbial parasites in the environmental sample by measuring the zones of lysis on a Petri dish with a layer of specific bacteria is the simplest non-direct way to evaluate the pollution of environment with these bacteria. Growth of bacterial viruses (bacteriophages) can deteriorate the industrial cultivation process because of spontaneous lyses of bacterial cells. Bacteriophages are widely used in the genetic engineering of bacterial strains as the vector for transfer of defined genes into bacterial cells.

Plant Parasites

Phytopathogenic viruses, prokaryotes, and fungi cause plant diseases. Typical stages of disease are as follows:

1. Contact of the microorganism with the plant
2. Entry of the pathogen into the plant
3. Growth of the infecting microorganism
4. Development of plant disease symptoms

Microbial pathogens disrupt normal plant functions by producing enzymes, toxins, and growth regulators. Some plant pathogens such as white-rot fungi or bacteria from genus *Pseudomonas* can degrade xenobiotics and are widely used in environmental engineering. Therefore, the risk of plant infection must be accounted for in environmental biotechnology operations, especially during soil bioremediation.

Parasites of Humans and Animals

Pathogenic viruses, prokaryotes, fungi, or protozoa attach and then grow on/in human or animal tissue and can cause diseases in macroorganisms. Saprophytic microorganisms feed on dead organic matter. Opportunistic pathogens are normally harmless but have the potential to be pathogens for debilitated or immunocompromised organisms. More details about infectious microorganisms and aspects of public health for civil and environmental engineers can be found in Chapter 13.

Effects of Nutrients on Biotic Elements

Kinetic limitation means that the specific growth rate depends on the concentration of limiting nutrient. Usually, there is one limiting nutrient but there may be simultaneous limitation by some nutrients. If one nutrient limits the specific growth rate, this dependency is often expressed by Monod's equation:

$$\mu = \mu_{max}\left[\frac{S}{(S + K_s)}\right] \qquad (9.5)$$

where

μ_{max} is the maximum of specific growth rate
S is the concentration of the nutrient (substrate) limiting growth rate
K_s is a constant

However, there are hundreds of other known models describing μ, as $f(S_i)$. A double limitation of μ by donor of electrons and oxygen is typical for the cases when the initial step of catabolism is catalyzed by oxidase or oxygenase incorporating atom(s) of oxygen into a carbon molecule or energy source.

Effects of Nutrients on Yield

The stoichiometric limitation of growth means that the dosage of the nutrient in the medium linearly determines the yield of biomass. For some groups of prokaryotes, the sources of carbon and energy are separated. Growth efficiency depends on energy extracted during catabolism of the energy source. Growth yield reflects the balance of energy produced in catabolism and energy assimilated in biosynthesis. Typically in microbial growth, there is no feedback regulation between the rates of biosynthesis and catabolism. Therefore, limitation of biosynthesis due to nutrient limitation or unfavorable physical factors of environment diminishes growth yield. However, there may be a paradoxical increase of growth yield under unbalanced biosynthesis and catabolism, and excess of carbon source. This can be caused by the redirection of carbon flow under excess of carbon source to the synthesis of storage carbohydrates or polyhydroxybutyrate, which require less energy for their synthesis than cell biomass. Another portion of energy in the case of unbalanced catabolism and biosynthesis can be used for the synthesis of extracellular polysaccharides or intracellular accumulation of polyphosphates, polypeptides, or low-molecular-weight osmoprotectors.

Effect of Starvation on Microorganisms

There are three typical responses of microorganisms to starvation, that is, shortage of some nutrients in a medium. The bacteria known as R-tactics are fast growing in a rich medium but can quickly die under a shortage of nutrients. Typical representatives of this group are *Pseudomonas* spp. The L-tactics bacteria are fast growing in a rich medium but under starvation they form dormant spores and cysts. Typical representatives of this group are *Bacillus* spp. K-tactics bacteria are adapted to grow slowly in the medium with a low concentration of nutrients. Typical representatives are the oligotrophs *Hyphomicrobium* spp.

Effect of Oxygen on Biotic Elements

Relation to oxygen is one of the main features of microorganisms. The generation of biologically available energy in a cell is due to oxidation–reduction reactions. Oxygen is the most effective acceptor of electrons in the generation of energy from the oxidation of substances, but not all microorganisms can use it. The following groups of microorganisms differ in their relation to oxygen:

1. Obligate anaerobic prokaryotes producing energy by fermentation (it is intramolecular oxidation–reduction without an external acceptor of electrons) or anoxic respiration (electron acceptors are not oxygen); they can die after contact with oxygen because they have no protection against such toxic products of oxygen reduction as hydrogen peroxide (H_2O_2), superoxide radical (O_2^-), and hydroxyl radical ($OH^\bullet$).
2. Tolerant anaerobes producing energy by fermentation or anoxic respiration but that survive after contact with oxygen due to a protective mechanism against oxygen radicals.
3. Facultative anaerobic bacteria, which are capable of producing energy either anaerobically if oxygen is absent or by aerobic respiration if oxygen is present.
4. Microaerophilic bacteria, which prefer a low concentration of dissolved oxygen in a medium.
5. Obligate aerobes, which produce energy by aerobic respiration only.

The relationship to oxygen can be easily determined in the laboratory and is one of the most important identification properties of microorganisms because it was created in microbial evolution in parallel with the planet's evolution from an anaerobic to an aerobic atmosphere. The concentration of dissolved oxygen for the specific growth rate of aerobic microorganisms is usually limited to below 0.1 mg/L but in the cases where the initial step of catabolism is catalyzed by oxidase or oxygenase, it can be significantly higher, up to 1 mg/L.

Anoxic Microorganisms

Anoxic (anaerobic) respiration is typical for prokaryotes only and is the oxidation of organic or inorganic substances by electron acceptors other than oxygen. Different electron acceptors are used for energy generation by specific physiological groups of prokaryotes, including

1. Nitrate (NO_3^-) and nitrite (NO_2^-) are used by denitrifying bacteria (denitrifiers).
2. Sulfate (SO_4^{2-}) is used by sulfate-reducing bacteria.
3. Ferric ions (Fe^{3+}) are used by iron-reducing bacteria.
4. Ions of different oxidized metals also are used as acceptor of electrons.
5. Carbon dioxide (CO_2) is used by methanogens.

Effect of Temperature on Growth

The maximum temperature for growth depends on the thermal sensitivity of secondary and tertiary structures of proteins and nucleic acids. The minimum temperature depends mainly on the freezing temperature of the lipid membrane. The optimal temperature is close to the maximum temperature. Different physiological groups of microorganisms adapt to different temperatures. Psychrophiles have optimal temperatures for growth below 15°C. Mesophiles have optimal growth temperatures in the range between 20°C and 40°C. Thermophiles grow best between 50°C and 70°C. There are known thermoextremophyles growing at temperatures higher than 70°C.

Effect of pH on Growth

Natural biotopes have differing pH values: pH 1–3 (gastric juice, volcanic soil, mine drainage); pH 3–5 (plant juices, acid soils); pH 7–8 (freshwater and seawater); and pH 9–11 (alkaline soils and lakes). Acidophiles grow at pH lower than 5, neutrophiles grow within the pH range from 5.5 to 8.5, and alkalophiles grow at pH higher than 9. Intracellular pH is an approximately neutral pH. Extracellular pH affects the dissociation of carboxylic, phosphate, and amino groups of a cell's surface, thus changing its charge and adhesive properties. This feature is important for the sedimentation of activated sludge, cell aggregation, and formation of microbial biofilm.

Effect of Osmotic Pressure on Growth

The majority of microorganisms can live with a concentration of salts in medium up to 30 g/L. A higher concentration of salts or organic substances can cause water to diffuse out of the cell by osmosis. However, some groups of microorganisms adapt to high osmotic pressure or low activity of water. These halophiles require the addition of NaCl in the medium during their isolation and cultivation. Extreme halophiles require a high concentration of NaCl (15%–30%) in the medium. Xerophiles are able to live in a dry environment. The main adaptation characteristic of halophiles and xerophiles is their ability for intracellular accumulation of such low-molecular-weight hydrophilic osmoprotectors as polyoles, oligosaccharides, and amino acids.

Natural Death of Microorganisms

The equally splitting cells of bacteria are considered almost immortal creatures. However, some bacterial cells die even in pure culture at optimal growth conditions. Hypothetically, the death of microorganisms at optimal conditions for growth can be caused by a small asymmetry always determined in cell division and accumulation or depletion of asymmetrically separated cell components. The asymmetry of the division of eukaryotic cells is visible so that mother and daughter cells can be often distinguished. There may be accumulation or depletion of inert or essential cell components in this asymmetrical division, budding, or splitting of eukaryotic cells. Therefore, the percentage of dead cells in the population of pure eukaryotic culture under optimal conditions for growth can be from 1% to 5%. Other reasons for natural cell death may include the production of toxic oxygen radicals, shortage of essential cell components due to starvation, changes in structure of cell biopolymers, and lipid components due to unfavorable physical parameters of the microenvironment.

Fate of Released Microorganisms in Environment

The natural death of pathogenic microorganisms released into the environment is the most important factor in the termination of infectious disease outbreaks. Active biodegraders, which are used in environmental engineering, are often opportunistic pathogens or genetically modified strains. Therefore, the study of environmental fate, death rate, and survivability of microorganisms, which are used in bioremediation of polluted soil or spills in marine environment and released into the environment, is essential for determining process feasibility. A rule of thumb is that applied microorganisms must have some reasonable limits of lifetime in the treated and surrounding areas. This short lifetime can prevent accumulation or spread of unwanted microorganisms in the environment.

10

Classification of Viruses and Microorganisms

Biological Classification

Biological classification is an abstract grouping of biological objects by some conventionally defined criteria, for example, shape, size, chemical content, and biological properties. Classification is used for understanding the mechanisms of diversity of the studied groups and for identification of new objects.

Units of Biological Classification

The elementary unit is a strain (organisms isolated from one colony) or a clone (organisms originated from one organism). Similar strains are combined in a species. Higher classification units are *genus* (pl. *genera*), family, order, and kingdom.

The names of species are conventionally given and read in *Latin* and include the genus name and species name, for example, *Bacillus subtilis* (from genus *Bacillus*), *Escherichia coli* (from genus *Escherichia*), *Pseudomonas aeruginosa* (from genus *Pseudomonas*). After the first mention of a full name in the text, a researcher may write *B. subtilis*. Name *Bacillus* sp. means that the species is not defined.

Classification of Viruses

Viruses are particles, which are self-assembled from the biopolymers, capable of multiplying inside living prokaryotic or eukaryotic cells. Extracellular virus particles are metabolically inert. The typical virus size ranges from 0.02 to 0.2 μm. Viruses contain a single type of nucleic acid, either DNA or RNA. Virus-like agents called prions, which are infectious proteins, have also been identified.

A simple method of classifying viruses is grouping them into bacterial, plant, and animal viruses. Viruses are not included in biological classifications because they are obligate intracellular parasites of cells and thus cannot self-reproduce.

However, their classification is similar to biological one. Classification and naming of viruses is set out by the International Committee on Taxonomy of Viruses.

A more detailed classification of viruses is based on their type of nucleic acid (RNA, DNA), type of their replication, and shape of envelope.

Groups of Viruses

The established classification includes seven groups of viruses. DNA viruses (groups 1 and 2) include double-stranded and single-stranded DNA viruses. Examples of these are viruses causing lysis of bacteria (bacteriophages) as well as oral herpes, chickenpox, and smallpox in humans. RNA viruses (groups 3, 4, and 5) include double-stranded and single-stranded RNA viruses. Examples of these are the rotaviruses and enteroviruses, which are major agents of water-borne infections, and viruses causing Hepatitis A and C, polio, foot-and-mouth disease, SARS, yellow fever, and influenza. Reverse transcribing viruses (groups 6 and 7) are single-stranded RNA viruses (HIV is an example) and double-stranded DNA viruses (hepatitis B virus is an example) replicating using reverse transcription, which is the formation of DNA from an RNA template.

Importance of Viruses for Environmental Engineering

Bacterial viruses (bacteriophages) are especially important for environmental engineering because they are used for the monitoring of environmental quality and affect the activity of bacterial communities that are used in waste biotreatment.

Viruses are important for environmental engineering because of the following reasons:

1. Pathogenic viruses must be removed, retained, or destroyed during water and wastewater treatment.
2. Viruses of bacteria (bacteriophages) can infect and degrade the bacterial cultures.
3. Bacteriophages can be used for the detection of specific microbial pollution of the environment.
4. Viruses may be a vector (carrier) of the genes in artificial or natural genetic recombinations.

Isolation and Collection of Microbial Strains

Unicellular microorganisms can be isolated from nature as a strain (cells of one colony) or a clone (cells originated from one cell). Isolation of pure culture (microbial strain) is usually performed by spreading a diluted microbial suspension on a Petri dish with a semisolid medium to produce 10–50 colonies on the dish after several days of cultivation. Cells of one colony are picked up for the next round of cultivation on a semisolid or liquid medium. However, the following methods can also be effectively used for the isolation of a pure microbial culture:

1. Mechanical separation of cells by micromanipulator
2. Sorting of cells or microbeads with immobilized cell using flow cytometer
3. Magnetic or immunomagnetic separation
4. Cell chromatography

Microbial Collections

Every organization, which studies microorganisms, has its own collection of microbial strains. The main functions of culture collection are the deposition and storage of strains for their further use or protection of strain property rights. The biggest university and national collections contain type strains to serve as the standards of the characteristics attributed to a particular species. The best known culture collections are Deutsche Sammlung von Mikroorganismen und Zellkulturen (DSMZ) and American Type Culture Collection (ATCC). ATCC, for example, is a nonprofit bioresource center that provides biological products, technical services, and educational programs to private industry, government, and academic organizations around the world. Its functions are to acquire, authenticate, preserve, develop, and distribute biological materials, information, technology, intellectual property, and standards for the advancement, validation, and application of scientific knowledge.

A strain is identified by its assigned number in a microbial collection and the name of the species. For example, *B. subtilis* ATCC6633 refers to a strain of species *B. subtilis* stored under number 6333 in the American Type Culture Collection (ATCC).

Classification of Microorganisms

There are existing and used phenotypic and genotypic classifications of prokaryotes. Phenotypic classification is based on the phenotypic characteristics, that is, characteristics, which are determined under real conditions of vital activity. Genotypic classification is based on the analyses of genetic characteristics of the organisms, which are stored in the sequences of DNA. The goals of microbial classification are (1) studying microbial diversity and (2) identification of microorganisms in nature and engineering systems.

Bergey's Manual of Systematic Bacteriology (2nd edition), issued during 2001–2007, presents both phenotypic and genotypic characters that are used to classify prokaryotes with special emphasis on phylogenetic classification based on the comparison of 16S rRNA gene sequences. Groups of fungi, protozoa, and algae are described in specific manuals approved by related international scientific associations.

Phenotypic Characteristics

The strains collected in the collections are combined into groups using phenotypic or genotypic characteristics.

Phenotypic characteristics, which are used in classification, are as follows:

1. Cell structure characteristics: size, shape, typical cell aggregates, membrane structures, etc. One of the most important cytological characteristics of prokaryotes is the Gram-positive or Gram-negative type of cell wall. A Gram-positive cell wall is a thick and rigid 3-D layer of polymer. A Gram-negative cell wall is a thin and more elastic layer of polymer, which is covered by an outer membrane and a lipopolysaccharide layer.
2. Physiological characteristics: type of energy production, relation to oxygen, pH, temperature, chemical content of cell wall and membranes, enzyme profiles, etc.
3. Ecological characteristics: habitats, econiches, colonial structures, and interrelationships with other organisms.
4. Chemical characteristics: content of cell wall, membrane, presence of specific pigments, content of membrane lipids, etc.

Genotypic Classification

Genotypic classification (phylogenetic taxonomy) is based on the analyses of genetic characteristics of the organisms, which are stored in the sequences of DNA. Genotypic characteristics include

1. G+C content in DNA
2. Sequences of genes (the sequences of DNA, which store information on the biopolymers of homologous, similar function in different species, are compared)
3. Sequences of homologous (similar) proteins
4. Level of hybridization between the sequences of DNA and RNA of compared strains

The sequences of not only DNA but also RNA and proteins with homologous (similar) function in different species are compared in genotypic classification. It is considered in phylogenetic classification that the number of differences between the sequences of DNA, RNA, or protein with similar function and originated from common ancestor reflects the evolutionary distance between compared sequences.

For the prokaryotic classification, the genes of ribosomal 16S RNA and 23S RNA have been exceptionally useful because of the low rate of evolutionary changes and the sufficient number of sites for comparison (≈1500 for 16S rRNA and ≈3000 for 23S rRNA). The evolutionary-distance tree, after calculation of the percentage of nonidentical sequences between the genes of rRNAs of any two organisms, is generated by computer.

Guanine/Cytosine (G+C) Content in DNA and Genotypic Classification

The Guanine and Cytosine (G+C) content of prokaryotic chromosomal DNA ranges from 25 to 80 mol% and can be used to classify two different evolutionary branches of bacteria. The author believes that the proportion of G+C content in prokaryotic DNA evolved from low to high in one branch of bacteria and from high to low in another branch. This means that one branch and its modern representatives can be called GC+, while another branch and its modern representative can be called GC–. This means that the absolute value of the difference between G+C contents of GC+ and GC– prokaryotes (GC distance) can be used to calculate the evolutionary time that has passed since the branches split. The deviation from the G+C content of 50 mol% is proportional to the length of time that has passed since species origin (Figure 10.1).

Comparison of the Steps in Phenotypic and Genotypic Identification of Strains

The steps of identification using phenotypic or genotypic properties are similar by the function (Table 10.1). The object of phenotypic classification is organism of strain. The object of genotypic identification is a sequence of DNA.

Phylogenetic Groups of Prokaryotes

Traditional Phyologenetic classification divides prokaryotes into two distinct groups: Bacteria with about 15,000 known sequences of 16S rRNA and Archaea with about 1,200 known sequences of 16S rRNA. The Bacterial domain can be subdivided into groups based on rRNA sequence. One of the most important of these groups for environmental engineers is the Proteobacteria (about 7000 known sequences of 16S rRNA). The Proteobacteria are a major group of Gram-negative bacteria and are further subdivided into the following subgroups

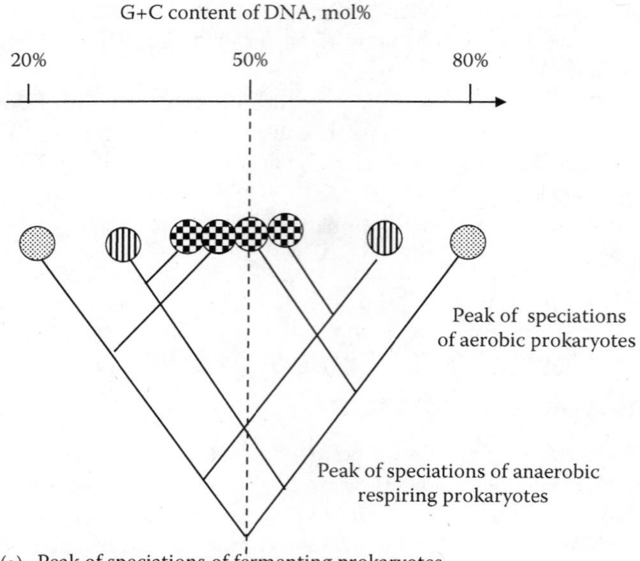

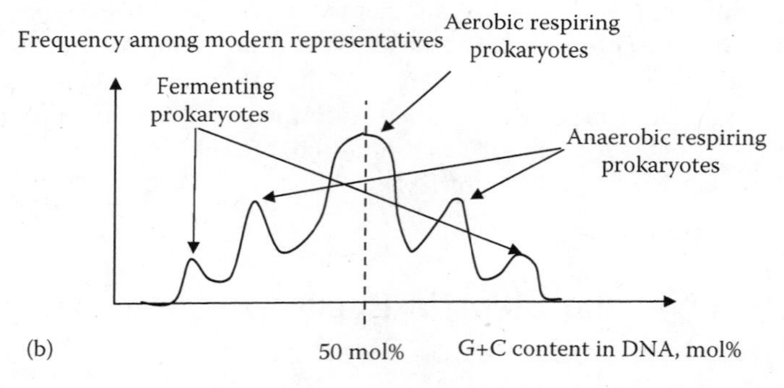

Figure 10.1 (a) Waves of speciation and (b) symmetry of the distribution of G+C (Guanine/Cytosine) content in DNA within groups of prokaryotes engaged in fermentation or anaerobic, and aerobic respiration. The line indicating deviation from G+C content of 50 mol% also reflects the two branches of prokaryote evolution (GC+ and GC–).

according to the differences in rRNA documented by the ribosomal database project (RDP-II):

- Alpha subdivision
- Beta subdivision
- Gamma subdivision
- Delta subdivision
- Epsilon subdivision and some other small groups

TABLE 10.1 The Steps of Strain Identification

Phenotypic Taxonomy	Genotypic Taxonomy
1. Isolation of strains from environment	1. Isolation of genes from environment
2. Amplification of cells	2. Amplification of gene
3. Study of characteristics	3. Study of sequence
4. Collection of strains in cell culture collection	4. Collection of sequences in digital database
5. Identification of new strain by comparison with reference strains	5. Identification of new sequence by comparison with reference sequences

Major rRNA phylogenetic divisions of Archaea are as follows:

1. *Euryarchaeota*
   a. *Methanococcales*
   b. *Methanobacteriales*
   c. *Thermococcales*
   d. *Methanopyrales Crenarchaeota*
      i. Thermophilic *Crenarchaeota*
      ii. Non-thermophilic *Crenarchaeota*

Classification of Microscopic Fungi

Fungi are eukaryotic microorganisms, mostly multicellular, which assimilate organic substances and absorb nutrients through the cell surface. The typical cell size is between 5 and 20 μm. Cells are often combined in the branched filaments called hyphae, which are combined in a web known as mycelium. Fungi are important degraders of polymers and used in the composting and biodegradation of toxic organic substances. Fungi are used in environmental engineering in composting, soil bioremediation, and biodegradation of xenobiotics. Mycelia effectively penetrate solid wastes and soil.

The classification of fungi is based on the diversity of their sexual reproduction. There are five major groups of fungi:

1. *Oomycetes* (water molds)
2. *Zygomycetes* (molds, produce numerous asexual spores)
3. *Ascomycetes* (reproduced by spores stored in a sac called ascus or by spores called conidia)
4. *Basidiomycetes* (club fungi and mushrooms)
5. *Deuteromycetes* (or *Fungi imperfecti*) have no known sexual stage

Molds are filamentous fungi (from *Zygomycetes* and *Ascomycetes*) that have widespread occurrence in nature. They have a surface mycelium and aerial hyphae that contain asexual spores (conidia). These spores are airborne allergens in damp or poorly constructed buildings.

Yeasts (from *Ascomycetes*) are fungi that grow as single cells, producing daughter cells either by budding (the budding yeasts) or by binary fission (the fission yeasts). Mushrooms are filamentous fungi that form large above-ground fruiting bodies, although the major portion of the biomass consists of hyphae below ground.

White-rot fungi (from *Basiodiomycetes*) are able to degrade lignin and many recalcitrant organic compounds due to the production of oxygen radicals.

Classification of Microscopic Algae

Microscopic algae are floating eukaryotic microorganisms that assimilate energy from light. The typical size of a cell is 10–20 µm. Algae live primarily in aquatic habitats and on the soil surface. Algae should not be confused with cyanobacteria, which are bacteria but referred to occasionally as "blue-green algae," which is an obsolete and incorrect term.

The classification of algae is based on the type of chlorophyll and other pigments, cell wall structure, and the nature of carbon reserve material:

1. *Chlorophyta* (green algae)
2. *Chrysophita* (golden-brown algae)
3. *Euglenophyta* have no cell
4. *Pyrrophyta* (dinoflagellates)
5. *Rhodophyta* (red algae)
6. *Phaeophyta* (brown algae)

Algae are important for environmental engineering for the following reasons:

1. They remove nutrients from water and are active microorganisms in waste stabilization ponds.
2. Some algae are fast growing in polluted water and produce toxic compounds; these cause the "red tides" in polluted coastal areas.
3. Selected species of microscopic algae in natural waters are used as indicators of water quality.
4. There may be value-added products, for example, oil for biodiesel production, pigments, and unsaturated fatty acids, developed from algae grown during wastewater management.

Classification of Protozoa

Protozoa are unicellular organisms that absorb and digest organic food inside a cell. The typical cell size ranges from 10 to 50 µm. Some protozoa are pathogenic and must be removed from water and wastewater. Four major groups of protozoa are distinguished by their mechanism of motility:

1. Amoebas move by means of false feet
2. Flagellates move by means of flagella
3. Ciliates use cilia for locomotion
4. Some protozoa have no means of locomotion

Protozoa obtain nutrients by ingesting other microbes or by ingesting macromolecules. The cells form cysts under adverse environmental conditions and are resistant to desiccation, starvation, high temperature, and disinfection. Changes in the protozoan community reflect the operating conditions of aerobic wastewater treatment:

1. Amoebas can be found in high concentrations of organic matter (at high values of biochemical oxygen demand—BOD).
2. Flagellated protozoa and free-swimming ciliates are associated with high bacterial concentrations in activated sludge and medium concentration of values of BOD.
3. Protozoa contribute significantly to the reduction of bacteria including pathogens in activated sludge.
4. Stalked ciliates occur at low bacterial and BOD concentrations in water.

11

Physiological Classification of Prokaryotes

Physiology of Prokaryotes

Prokaryotes are microorganisms with prokaryotic type of cell. They consist of two phylogenetic groups: bacteria and archaea. Prokaryotes are most active in the degradation of organic matter and are used in wastewater treatment and soil bioremediation. Physiology (meaning in Greek "study of nature") of prokaryotes studies the functioning of cell and cell population. One of the most important physiological features of prokaryotic cell is the method of biological energy generation.

Three Types of Energy Generation by Chemotrophy

There are three different mechanisms of generating biologically available energy by chemotrophs and three related physiological groups:

1. Fermenting microorganisms produce biologically available energy under anaerobic conditions using intramolecular oxidation/reduction.
2. Anaerobically respiring (or anoxic) microorganisms produce biologically available energy under anoxic (no oxygen) conditions using oxidation of organic matter by electron acceptors other than oxygen, for example, Fe^{3+}, SO_4^{2-}, and CO_2.
3. Aerobically respiring organisms produce biologically available energy under aerobic conditions by aerobic respiration, using reduction of oxygen.

Three Types of Energy Generation by Phototrophy

There are also three different ways of generating biologically available energy by phototrophs and three related physiological groups, which

1. Use products of fermentation (organic acids, alcohols, hydrogen) as donors of electrons and light as a source of energy to reduce CO_2 (anoxygenic photosynthesis with organics as electron donor)

$$CH_3COOH + CO_2 + \text{light energy} \rightarrow \langle 2CH_2O \rangle + CO_2$$

2. Use products of anoxic respiration (H_2S, Fe^{2+}) as donor of electrons and light as a source of energy to reduce CO_2 (anoxygenic photosynthesis with organics as electron donor)

$$2H_2S + CO_2 + \text{light energy} \rightarrow \langle CH_2O \rangle + 2S + H_2O$$

3. Use the product of aerobic respiration, water, as a donor of electrons and light as a source of energy to reduce CO_2 (oxygenic photosynthesis)

$$H_2O + CO_2 + \text{light energy} \rightarrow \langle CH_2O \rangle + O_2$$

Evolution of Atmosphere and Prokaryotes

According to geological data, the age of the earth is about 4.6×10^9 years old. The first organisms, prokaryotes, appeared about $3.5–3.8 \times 10^9$ years ago. There was no oxygen in that atmosphere, so the first organisms could be anaerobes. The next step was an accumulation of atmospheric oxygen by oxygenic (oxygen-producing) phototrophic prokaryotes on the boundary $2.2–2.0 \times 10^9$ years ago. Eukaryotes appeared about $1.8–1.5 \times 10^9$ years ago, probably due to the intracellular symbiosis of smaller and bigger cells. The aerobic atmosphere led to the formation of the ozone barrier for intensive UV radiation on the earth's surface. It was the primary condition for the creation of terrestrial life and multicellular organisms $0.6–0.5 \times 10^9$ years ago.

The relationship to oxygen and the related types of biological energy generation are the most important identification properties of microorganisms because these properties were created in microbial evolution in parallel with the planet's evolution from an anaerobic to an aerobic atmosphere.

Contradictions between rRNA-Based Phylogenetic Classification and Physiological Classification of Prokaryotes

The main contradiction between physiological classification and phylogenetic classification, which is currently based on comparison of 16S rRNA sequences, is that physiological groups often do not correspond to rRNA-based phylogenetic groups. Almost all phylogenetic divisions and subdivisions consist of species with different physiological features. It can be explained by the evolutionary stability of such basic physiological properties of cells as the mechanism of biological energy generation. At the same time, rRNA and other polynucleotide and polyamino acid sequences were changed significantly during evolution. Therefore, the distance between 16S rRNA gene sequences, which is used in phylogenetic classification, reflects time after divergency (branching) of phylogenetic lines but not the physiological discrepancies between the representatives of these lines. Future classification methods, which will be based on the comparison of complete genome sequences, can help to eliminate this contradiction between the genotypic and physiological classifications.

If the main and additional evolutionary branches separated a long time ago but their representatives were developed under the same conditions, they can have similar basic physiological features. These representatives will have a large phylogenetic difference between 16S rRNA sequences but very similar physiological properties. Alternatively, if the evolutionary branches separated during transition from anaerobic to anoxic or aerobic atmosphere, their modern representatives will have different mechanisms of energy generation but small differences between 16S rRNA sequences.

Absence of Predictive Power in rRNA-Based Phylogenetic Classification

The existing phylogenetic classification of prokaryotes, which is based on the comparison of 16S rRNA gene sequences, is a very useful tool for the experimental identification of microbial species. Identification can be currently performed by PCR of 16S rRNA gene and sequencing this gene, which requires several hours. Physiological identification requires a significantly longer and more laborious procedure performed by experienced researchers. Therefore, rRNA-based classification and identification is more popular in experimental research. However, this classification has no logical basis and scientific predictive power. As a result of the widespread use of 16S rRNA-based phylogenetic classification in experimental study, the diversity of prokaryotes is often perceived by new users and students as a random mixture of microbial species and groups.

Parallelism in Evolution of Genes

Another reason for the differences between physiological and rRNA-based classifications is the parallelism in the evolution of genes in different evolutionary lines due to the transfer of genes between the representatives of these lines. In rRNA-based phylogenetic classification, it is thought that the number of differences in sequence of rRNA reflects the evolutionary distance of the origin of compared sequences from a common ancestral sequence. However, it was proved experimentally that there may be lateral transfers of genes in environment, that is, transfers of genes not only from ancestor to descendant but also between neighboring organisms.

Periodic Table of Prokaryotes

The periodic table of prokaryotic classification (Table 11.1) was proposed to give predictive power to prokaryotic classification, clarify the physiological and evolutionary connections between microbial groups, and give a logical basis for students to understand microbial diversity.

The three major types of biological energy generation are the results of evolution of the earth's atmosphere from an anaerobic to aerobic one. Therefore, microbial physiological diversity can be shown as created in three evolutionary periods related to fermenting, anaerobically respiring, and aerobically respiring microorganisms. There are also intermediate groups, for example, microaerophilic or facultative anaerobic microorganisms between these groups.

These groups exist in three parallel, semi-independent but parallel (evolutionary coordinated) phylogenetic lines:

1. Line of aquatic organisms
2. Line of terrestrial organisms
3. Line of organisms in extreme environment

TABLE 11.1 Evolutionary Lines and Periods of the Periodic Table of Chemotrophic Prokaryotes

	Evolutionary Period		
Evolutionary Line	**Fermenting Prokaryotes**	**Anoxic Prokaryotes**	**Aerobic Prokaryotes**
Prokaryotes of aquatic origin (Gram-negative type of cell wall, *Gracilicutes*)	*Bacteroides* *Prevotella* *Ruminobacter*	*Desulfobacter* *Geobacter* *Wolinella*	*Pseudomonas* *Acinetobacter* *Nitrosomonas*
Prokaryotes of terrestrial origin (Gram-positive type of cell wall, *Firmicutes*)	*Clostridium* *Peptococcus* *Eubacterium*	*Desulfotomaculum* *Desulfitobacterium*	*Bacillus* *Arthrobacter* *Streptomyces*
Prokaryotes of extreme environment origin (*Archaea*)	*Desulfurococcus* *Thermosphaera* *Pyrodictium*	*Methanobacterium* *Thermococcus* *Haloarcula*	*Picrophilus* *Ferroplasma*

Note: Selected examples of conventional genera are shown in the groups.

These lines are semi-separated because of the low frequency of genetic exchanges between organisms in aquatic, terrestrial, and extreme environments.

Prokaryotes of aquatic, terrestrial, and extreme environments are in the following lines: (1) Gram-negative bacteria (*Gracilicutes*)—cells with thin wall, originated from environments with stable osmotic pressure (e.g., water, tissues of macroorganisms); (2) Gram-positive bacteria (*Firmicutes*)—cells with rigid cell wall, originated from environments with changeable osmotic pressure (e.g., soil); and (3) Line of Archaea (*Mendosicutes*)—cells without conventional peptidoglycan, originated from environment with some extreme conditions, usually temperature or oxidation–reduction potential.

So, the physiological diversity of chemotrophic prokaryotes can be shown in the periodic table of prokaryotes as three evolutionary periods of three parallel lines of aquatic, terrestrial, and extreme environment evolution (Table 11.1).

This periodic table is a logical basement to understand microbial physiological diversity. It gives the predictive power of yet to be discovered groups of prokaryotes, and clarifies the evolutionary connection between microbial groups.

Phototrophic Prokaryotes in the Periodic Table

Phototrophic organisms are also included in this periodic table as three sublines of aquatic, terrestrial, and extreme environment origin. Three periods of phototrophs can be integrated with three periods of chemotrophs:

1. The first period is related to phototrophs using products of fermentation (organic acids, alcohols, hydrogen) as donors of electrons and light as a source of energy to reduce CO_2.
2. The second period is related to phototrophs using products of anoxic respiration (H_2S, Fe^{2+}) as donors of electrons and light as a source of energy to reduce CO_2.
3. The third period is related to phototrophs using the product of aerobic respiration, water, as a donor of electrons, and light, as a source of energy to reduce CO_2.

Integrated periodic table of chemotrophic and phototrophic prokaryotes is shown in Table 11.2.

Some groups of phototrophic prokaryotes have not yet been discovered, like phototrophic Gram-positive bacteria and Archaea, which hypothetically use products of fermentation or anaerobic respiration as electron donors for CO_2 reduction. The existence of these groups in nature could be predicted following the periodic table of prokaryotes.

TABLE 11.2 Evolutionary Lines and Periods of the Periodic Table of Chemotrophic and Phototrophic Prokaryotes

		Periods of Evolution		
Evolutionary Line	**Subline**	**Fermenting Prokaryotes or Prokaryotes Using Products of Fermentation as Electron Donors**	**Anoxic Prokaryotes or Prokaryotes Using Products of Anoxic Respiration as Electron Donors**	**Aerobic Prokaryotes or Prokaryotes Using Products of Aerobic Respiration as Electron Donors**
Prokaryotes of aquatic origin (Gram-negative type of cell wall, *Gracilicutes*)	Chemotrophs	*Bacteroides* *Prevotella* *Ruminobacter*	*Desulfobacter* *Geobacter* *Wolinella*	*Pseudomonas* Acinetobacter *Nitrosomonas*
	Phototrophs	*Rhodopseudomonas*	*Chlorobium* *Rhodocyclus* *Chromatium*	Cyanobacteria *Prochloron*
Prokaryotes of terrestrial origin (Gram-positive type of cell wall, *Firmicutes*)	Chemotrophs	*Clostridium* *Peptococcus* *Eubacterium*	*Desulfotomaculum* *Desulfitobacterium*	*Bacillus* *Arthrobacter* *Streptomyces*
	Phototrophs	*Heliobacterium* *Heliobacillus*	Not known yet	Not known yet
Prokaryotes of extreme environment origin (*Archaea*)	Chemotrophs	*Desulfurococcus* *Thermosphaera* *Pyrodictium*	*Methanobacterium* *Thermococcus* *Haloarcula*	*Picrophilus* *Ferroplasma*
	Phototrophs	Not known yet	Not known yet	*Halobacteria*

Note: Selected examples of conventional genera are shown in the cells of periodic table.

Reasons of Parallelism and Periods in the Periodic Table of Prokaryotes

The existence of three evolutionary lines can be explained by parallel synchronized evolution in three different habitats:

1. Gram-negative bacteria (*Gracilicutes*, bacteria of aquatic origin) were adapted to life in biotopes with constant osmotic pressure such as seawater, freshwater, or animal fluids.
2. Gram-positive bacteria (*Firmicutes*, bacteria of terrestrial origin) were adapted to life in biotopes with changeable osmotic pressure such as soil.
3. *Archaea* (prokaryotes-extremophiles) were adapted to life in hyperextreme environments with high temperature, negative oxidation–reduction potential, salinity, or low pH.

According to the hypothesis of parallelism of prokaryotic evolution, the frequency of the lateral transfer of the genes between major phylogenetic lines of *Gracilicutes*, *Firmicutes*, and *Archaea* can by synchronized by evolutionary changes of atmosphere and waves of organic matter accumulation due to waves of glaciations on the planet.

Geological Synchronization of the Periods in the Periodic Table of Prokaryotes

The evolutionary parallelism in phylogenetic lines can be explained by the geological synchronization of the frequency of speciations in three phylogenetic lines. Hypothetically, the synchronization could have been caused by the waves of glaciations in the Pre-Cambrian era. Every period of glaciation decreased the microbial population in the biosphere, decreased the concentration of CO_2 in the atmosphere, and increased the accumulation of dead organic matter and products of fermentation, anoxic respiration, or aerobic respiration, thus creating conditions for a new wave of speciations in a warm period of geological evolution on the earth.

Another factor in the synchronization of the evolutionary periods in phylogenetic lines could be the evolution of the anaerobic atmosphere to an aerobic one. Therefore, the periods of prokaryotic evolution and related groups of anaerobic fermenting, anaerobic respiring, microaerophilic, and aerobic prokaryotes could be synchronized by the changes of oxygen concentration in the atmosphere. The creation of new groups of chemotrophs likely produced the conditions for the creation of a new group of phototrophs because the final products of this new group were used as electron acceptors and carbon sources for a related new group of phototrophs. Also likely, the horizontal gene transfers between three habitats enhanced the synchronization of parallel evolution in three phylogenetic lines.

Practical Importance of the Periodic Table of Prokaryotes

The periodic table of prokaryotic phylogeny, shown in Table 11.2, provides a theoretical understanding of prokaryotic diversity. Due to the logical connection between the earth's evolution and the parallel evolution in three lines of prokaryotes, the table possesses predictive power. Some groups of prokaryotes have not been discovered yet. Using the periodic table of prokaryotic phylogeny, it would be possible to forecast the discovery of aerobic and microaerophilic phototrophic *Firmicutes* (Gram-positive bacteria).

The periodic table of prokaryotic evolution gives a general overview and is not too suitable for practical classification and identification. However, it would be possible to produce more detailed identification tables, taking into account the parallelism of not only physiological but also additional cytological, biochemical, ecological, and molecular biological features of prokaryotes. Future developments, especially future classification based on the comparison of complete genomes of microorganisms, will remove the current contradictions between the physiological and phylogenetic classifications.

12

Groups of Prokaryotes

Twenty-Four Major Physiological Groups of Prokaryotes

There are nine groups of chemotrophs and nine groups of phototrophs in the periodic table of prokaryotes. However, the groups of facultative anaerobic and microaerophilic prokaryotes can also be considered in all phylogenetic lines of the periodic table. So, the number of major physiological groups covering all prokaryotic diversity is 24.

There are three chemotrophic and three phototrophic groups in four periods of the periodic table, that is, among

- Fermenting prokaryotes
- Anoxic prokaryotes
- Facultative anaerobic and microaerophilic prokaryotes
- Aerobic prokaryotes

A brief description of these 24 groups is given in this chapter.

Functions of Anaerobic Prokaryotes

These microorganisms are related to the first period of the periodic table of prokaryotes. Their natural biotopes are the sediments of aquatic ecosystems, tissues of macroorganisms, anaerobic microzones of soil, and hot springs. The periodic table with the groups of prokaryotes mentioned in this section is shown in Table 12.1.

Gram-Negative (Aquatic), Chemotrophic, Fermenting Bacteria

The main functions of Gram-negative, chemotrophic, fermenting bacteria in engineering systems are as follows:

TABLE 12.1 Evolutionary Lines and Periods of the Periodic Table of Chemotrophic and Phototrophic Prokaryotes

Evolutionary Line	Subline	Periods of Evolution		
		Fermenting prokaryotes or prokaryotes using products of fermentation as electron donors	Anoxic prokaryotes or prokaryotes using products of anoxic respiration as electron donors	Aerobic prokaryotes or prokaryotes using products of aerobic respiration as electron donors
Prokaryotes of aquatic origin (Gram-negative type of cell wall, *Gracilicutes*)	Chemotrophs	*Bacteroides* spp. *Prevotella* spp. *Ruminobacter* spp. *Acetogenium* spp. *Syntrophococcus* spp. *Syntrophobacter wolinii* *Veillonella*		
	Phototrophs	*Rhodopseudomonas*		
Prokaryotes of terrestrial origin (Gram-positive type of cell wall, *Firmicutes*)	Chemotrophs	*Clostridium botulinum* *Clostridium tetani* *Clostridium perfringens* *Peptococcus* *Peptostreptococcus* *Eubacterium* *Acetobacterium*		
	Phototrophs	*Heliobacterium* *Heliobacillus* *Heliophilum* *Heliorestis*		
Prokaryotes of extreme environment origin (*Archaea*)	Chemotrophs	*Desulfurococcus* *Thermosphaera* *Pyrodictium*		
	Phototrophs	Not known yet		

Note: Selected examples of conventional genera of fermenting prokaryotes, mentioned in this section, are shown in the cells of periodic table.

1. Fermentation of saccharides to organic acids, alcohols, and hydrogen
2. Syntrophic formation of acetate from other organic acids during anaerobic digestion of organic matter
3. Indication of fecal pollution of water

Examples of the functions of selected genera are given below.

Bacteroides spp. are the predominant organisms in the human colon and are generally isolated from the gastrointestinal tract of humans and animals; some species are pathogenic. Their functions in environmental biotechnology include: (a) anaerobic degradation of polysaccharides in an anaerobic digester and the anaerobic zones of microbial biofilms and (b) indication of fecal pollution in water.

Prevotella spp. are mainly pathogenic organisms, but some species can be used for the biodegradation of collagen-containing wastes. *Ruminobacter* spp. facilitate anaerobic fermentation in the rumen and can be used for anaerobic fermentation of organic wastes.

Acetogenium spp. and *Syntrophococcus* spp. are so-called acetogens, capable of producing acetate from other organic acids. This function ensures a supply of acetate to acetotrophic methanogens in the anaerobic digestion of organic wastes.

Syntrophococcus spp. can also metabolize some C_1-compounds (methanol, formaldehyde, formic acid, carbon monoxide) and remove metoxilic groups from lignin. *Syntrophobacter wolinii* also produces acetate by degradation of propionate only in coculture (syntrophically) with hydrogen-utilizing prokaryotes. The acetate produced is sequentially used by acetotrophic methanogens during anaerobic digestion of organic wastes.

Veillonella spp. are parasitic microorganisms. Some species can be used for the reduction of nitroaromatic compounds, including 2,4,6-trinitrotoluene (TNT), as the first step in their biodegradation.

Gram-Positive (Terrestrial), Chemotrophic, Fermenting Bacteria

The main functions of Gram-positive, chemotrophic, fermenting bacteria in engineering systems are as follows:

1. Hydrolysis of biopolymers
2. Fermentation of saccharides and amino acids to organic acids, alcohols, and hydrogen
3. Formation of acetate from hydrogen and carbon dioxide during the anaerobic digestion of organic matter
4. Indication of fecal pollution in water

Examples of the functions of selected genera are shown below.

Species from the genus *Clostridium* are able to form endospores; usually they have no tolerance toward oxygen but the spores can survive in an aerobic environment. Both pathogenic and nonpathogenic species exist. The pathogenic species, for example, *Clostridium botulinum* and *Clostridium tetani*, are agents of severe diseases and produce strong toxins that can be considered as bioweapons. Clostridia can hydrolyze biopolymers, ferment monomers and amino acids, and produce alcohols, organic acids, and hydrogen. It is the first rate-determining step in the anaerobic digestion of organic waste. Some thermophilic clostridia can be used for the hydrolysis of cellulose and production of fuel ethanol. Some clostridial species may be used for the reductive dechlorination of xenobiotics, for example, pesticides and herbicides. *Clostridium perfringens* is an indicator species in water quality evaluation. Spores of *Clostridium* spp. are used as test cultures in disinfection studies.

Species from the genera *Clostridium*, *Peptococcus*, *Peptostreptococcus*, and *Eubacterium* ferment saccharides to form fatty acids (butyric, propionic, lactic, succinic, and acetic acids), ethanol, hydrogen, and carbon dioxide. It is the second step in the anaerobic digestion of organic waste.

Species from genera *Acetobacterium*, together with some species from the genus *Clostridium* are homoacetogenic bacteria; they are able to reduce carbon dioxide and produce acetate that serves as a substrate for acetotrophic methanogens. It is the third important step in the anaerobic digestion of organic wastes. The production of organic acids by anaerobic fermenting bacteria can contribute to microbially induced corrosion of metal engineering systems.

Chemotrophic, Fermenting *Archaea*

Obligate anaerobic species from genera *Desulfurococcus*, *Thermosphaera*, *Pyrodictium*, and some others from the phylum *Creanarchaeota* are able to ferment organic substances but some of them can also generate energy using sulfur as an electron acceptor. Those species capable of sulfur respiration can be classified also in the group of anoxic *Archaea*. These species are not currently used in engineering systems but they may be potentially effective in the anaerobic biodegradation of organic wastes.

Gram-Negative (Aquatic), Phototrophic Bacteria Using Products of Fermentation as Electron Donors

This group has not yet been discovered, but in future such organisms could be found among filamentous anoxygenic phototrophs close to the representatives of the family Chloroflexaceae. Anaerobic Gram-negative phototrophic bacteria using reduced products of fermentation (organic acids and alcohols) could be important in anaerobic photobiodegradation of organics in stabilization ponds and microbial mats of shallow rivers or springs.

Gram-Positive (Terrestrial), Phototrophic Bacteria Using Products of Fermentation as Electron Donors

There is one family Heliobacteriaceae in this group of obligate anaerobic Gram-positive phototrophic bacteria. The genera of heliobacteria are *Heliobacterium*, *Heliobacillus*, *Heliophilum*, and *Heliorestis*. Heliobacteria use such fermentation products such as pyruvate, lactate, acetate, butyrate, and ethanol as electron donors. Heliobacteria reside in soil, especially in paddy fields, mainly in plant rhizosphere, and can form spores for survival under unfavorable growth conditions. Symbiotic relationships may exist between heliobacteria performing strong nitrogen fixation for rice plants and rice plants supplying organic substances for Heliobacteria.

Heliobacteria could be useful in the biodegradation of organics in soil or microbial mats of springs and in nitrogen fixation in paddy soils and soils under bioremediation.

Phototrophic *Archaea* Using Products of Fermentation as Electron Donors

This group of extremophilic microorganisms has not been discovered yet. It could be possible that there is no species of phototrophic *Archaea* in nature because of the sensitivity of phototrophic energy production to extreme temperature.

Functions of Anoxic Prokaryotes

These microorganisms are related to the second period of the periodic table. Their natural biotopes are anoxic zones of aquifers, aquatic sediments, hot springs, and anoxic microzones of soil. The periodic table with the groups of prokaryotes mentioned in this section is shown in Table 12.2.

Gram-Negative (Aquatic), Chemotrophic, Anoxic Bacteria

There are many important functions of Gram-negative, chemotrophic, anoxic bacteria in engineering systems using the following biotechnological methods:

1. Biotechnological methods coupled with the reduction of sulfate for removal of heavy metals or sulfate
2. Biotechnological methods coupled with the reduction of nitrate and nitrite for denitrification of wastewater and anoxic biodegradation of organic substances
3. Biotechnological methods coupled with the reduction of Fe^{3+} for removal of phosphate and anoxic biodegradation of organic substances

Examples of the functions of selected genera are shown below.

Dissimilatory sulfate-reducing bacteria are obligate anaerobes that use organic acids, alcohols, and hydrogen as donor of electrons and sulfate or other oxoanions of sulfur as acceptors of electrons, for example:

$$CH_3COOH + SO_4^{2-} \rightarrow 2CO_2 + H_2S + 2OH^-$$

$$4H_2 + SO_4^{2-} \rightarrow H_2S + 2H_2O + 2OH^-$$

There is a large diversity of morphological forms (spherical, ovoid, rod-shaped, spiral, vibrioid-shaped cells, etc.), physiological properties, and related genera in phenotypic taxonomy, for example, *Desulfococcus*, *Desulfobacter*, *Desulfobacterium*, *Desulfobulbus*, *Desulfosarcina*, *Desulfovibrio*, etc. Sulfur-reducing bacteria from genera *Desulfurella* and *Desulfuromonas* are unable to reduce sulfate or other oxoanions of sulfur.

TABLE 12.2 Evolutionary Lines and Periods of the Periodic Table of Chemotrophic and Phototrophic Prokaryotes

Evolutionary Line	**Subline**	**Periods of Evolution**		
		Fermenting prokaryotes or prokaryotes using products of fermentation as electron donors	Anoxic prokaryotes or prokaryotes using products of anoxic respiration as electron donors	Aerobic prokaryotes or prokaryotes using products of aerobic respiration as electron donors
Prokaryotes of aquatic origin (Gram-negative type of cell wall, *Gracilicutes*)	Chemotrophs		*Desulfobacter* *Desulfococcus* *Desulfobacter* *Desulfobacterium* *Desulfobulbus* *Desulfosarcina* *Desulfovibrio* *Desulfurella* *Desulfuromonas* *Geobacter* *Wolinella* *Dehalobacter*	
	Phototrophs		*Chlorobium* *Rhodocyclus* *Chromatium*	
Prokaryotes of terrestrial origin (Gram-positive type of cell wall, *Firmicutes*)	Chemotrophs		*Desulfotomaculum* *Desulfitobacterium*	
	Phototrophs		Not known yet	
Prokaryotes of extreme environment origin (*Archaea*)	Chemotrophs		*Methanobacterium* *Methanobrevibacter* *Methanothermus* *Methanococcus* *Methanolobus* *Methanothrix* *Methanomicrobium* *Methanogenium* *Methanospirillum* *Methanoplanus*	

TABLE 12.2 (continued) Evolutionary Lines and Periods of the Periodic Table of Chemotrophic and Phototrophic Prokaryotes

Evolutionary Line	Subline	Periods of Evolution
		Methanocorpusculum
		Methanoculleus
		Methanohalobium
		Methanohalophilus
		Methanosarcina
		Methanosphaera
		Methanobacterium
		Haloarcula
		Desulfurolobus
		Metallosphaera
		Pyrobaculum
		Thermofilum
		Thermoproteus
		Hyperthermus
		Staphylothermus
		Thermodiscus
		Desulfurococcus
		Pyrodictium
		Thermococcus
		Pyrococcus
	Phototrophs	Not known yet

Note: Selected examples of conventional genera of anoxic prokaryotes are shown in the cells of periodic table.

Typical habitats of sulfate-reducing bacteria are anoxic sediments of freshwater, marine, or hypersaline aquatic environments. Thermophilic species occur in hot springs and submarine hydrothermal vents.

Sulfate-reducing bacteria can be used in environmental engineering to precipitate undissolved heavy metal sulfides from solutions or for the removal of sulfate from wastewater with a high concentration of sulfate. However, dihydrogen sulfide is the toxic, bad-smelling and corrosive product of this anaerobic respiration. So, sulfate reduction is usually an unwanted process in environmental and other engineering systems. Sulfate-reducing bacteria are the agents of corrosion of steel and concrete constructions, as well as engineering equipment in oil and gas industry and in wastewater treatment.

Close phylogenetic relatives of sulfate-reducing bacteria are in the genus *Syntrophus*, benzoate-reducing bacteria oxidizing fatty acids with benzoate as an electron acceptor in syntrophic association with hydrogen-using microorganisms such as methanogen *Methanospirillum hungatei*.

Iron-reducing bacteria can reduce different Fe(III) compounds using organic substances. These bacteria are important in the anaerobic biodegradation of organic matter in aquifers because they can reduce different Fe(III) natural compounds such as iron-containing clay minerals:

$$4Fe^{3+} + CH_2O + H_2O \rightarrow 4Fe^{2+} + CO_2 + 4H^+$$

This group of bacteria includes many genera. Species of the genus *Geobacter* are phylogenetically similar to sulfate-reducing bacteria and compete with them for electron donors in anaerobic zones. *Geobacter metallireducens* and *Geobacter sulfurreducens* are able to reduce not only Fe(III) but also Mn(VI), U(VI), Tc(VII), Co(III), Cr(IV), Au(III), Hg(II), As(V), and Se(VII) using aliphatic and some aromatic acids and alcohols as electron donors. It is the dominant group of iron-reducing bacteria detected in aquifers and subsurface environments. Therefore, they can be used for the bioremediation of these biotopes. Other biotechnological methods involving iron-reducing bacteria are the removal of phosphate, sulfide, and ammonia from return liquor of municipal wastewater-treatment plants. Two new cultures of iron-reducing bacteria, *Stenotrophomonas maltophilia* and *Brachymonas denitrificans*, have been recently isolated. These cultures were able to remove phosphate and degrade xenobiotics using iron hydroxide as an oxidant and branched fatty acids of liquid after anaerobic digestion of biomass.

Halobacteria from *Dehalobacter* genus oxidize such electron donors as formate, acetate, pyruvate, lactate, and H_2 due to anaerobic reductive dechlorination and can be used for degradation of chlorinated ethenes in soil or wastewater. The ability to reduce Fe(III), Mn(VI), Se(VI), and As(V) and anaerobic reductive dechlorination is often a common property among Gram-negative, chemotrophic, anoxic bacteria.

The species *Wolinella succinogenes* can use fumarate, nitrate, nitrite, nitrous oxide (N_2O), and polysulfide as terminal electron acceptors with formate, molecular hydrogen, or sulfide as electron donors. The species has been proven to be a bioagent for the treatment of hazardous industrial wastewater containing ammonium perchlorate (AP) and rocket motor components on the sites of demilitarization and disposal of solid rocket motors.

Denitrifying bacteria, which are capable of oxidizing organic substances, hydrogen, Fe^{3+}, H_2S, or S using nitrite or nitrate as electron acceptors, are usually not only anaerobic respiring bacteria but facultative anaerobic prokaryotes.

Gram-Positive (Terrestrial), Chemotrophic, Anoxic Bacteria

The functions of anaerobic, chemotrophic, Gram-positive anaerobic respiring bacteria in engineering systems are similar to the functions of Gram-negative, chemotrophic, anoxic bacteria, that is, to perform reductions of sulfate, nitrate, Fe(III), and other metals. Sulfate-reducing bacteria in this group are classified in the genus *Desulfotomaculum*. They form heat-resistant endospores and use organic acids and alcohols as electron donors. The *Desulfitobacterium* genus is an important group of bacteria used for anaerobic dechlorination of such xenobiotics as chlorinated phenols, chlorinated ethenes, and polychlorinated biphenyls (PCBs). The species of *Desulfosporosinus* genus may be important in the bioremediation of groundwater contaminated with benzene, toluene, ethylbenzene, and xylene (BTEX compounds). *Bacillus infernus* is an anaerobic species that is able to reduce Fe(III) and Mn(VI) using formate or lactate.

Chemotrophic, Anoxic *Archaea*

The majority of *Archaea*, the methanogens and extreme thermophiles, are representatives of this group.

Methanogens are obligate anaerobes, which convert CO_2, molecular hydrogen, methyl compounds, or acetate to methane by anaerobic respiration:

$$4H_2 + CO_2 \rightarrow CH_4 + 2H_2O$$

$$CH_3COOH \rightarrow CO_2 + CH_4$$

Methane is produced by methanogens in the rumen (forestomach) of ruminant animals, paddy fields, and wetlands. It is also produced and utilized as a fuel during industrial anaerobic digestion of organic wastes of municipal wastewater-treatment plants and on landfills. Representatives of microbial genera *Methanobacterium*, *Methanobrevibacter*, *Methanothermus*, *Methanococcus*, *Methanolobus*, *Methanothrix*, *Methanomicrobium*, *Methanogenium*, *Methanospirillum*, *Methanoplanus*, *Methanocorpusculum*, *Methanoculleus*, *Methanohalobium*, *Methanohalophilus*, *Methanosarcina*, and *Methanosphaera* can be easily distinguished under a transmission electron microscope (TEM) by the shape of cell or cell arrangement or under confocal laser scanning microscope (CLSM) using specific oligonucleotide probes and fluorescence in situ hybridization with cells of methanogens.

Sulfate reducers (*Archaeoglobus*) and extremely thermophilic and hyperthermophilic S^0-metabolizers of *Archaea* (*Desulfurolobus*, *Metallosphaera*, *Pyrobaculum*, *Thermofilum*, *Thermoproteus*, *Hyperthermus*, *Staphylothermus*, *Thermodiscus*, *Desulfurococcus*, *Pyrodictium*, *Thermococcus*, *Pyrococcus*) require temperatures from 70°C to 105°C for growth. Some organisms use sulfur as an electron acceptor. Hyperthermophiles are inhabitants of hot and sulfur-rich volcanic springs on the surface or on the ocean floor. They are not used in environmental biotechnology currently, but they may be useful in thermophilic biodegradation of organic wastes, production of environmentally useful enzymes, recovery of metals at a temperature close to the boiling point of water, and probably for the removal of sulfur from coal and oil.

Gram-Negative (Aquatic), Phototrophic, Anoxic Bacteria

The majority of anoxygenic phototrophic bacteria are in this group. Their main functions in engineering systems are as follows:

1. Anaerobic photoremoval of bad-smelling and toxic H_2S during anaerobic treatment of waste
2. Removal of sulfate from water by a bacterial system consisting of sulfate-reducing bacteria and anoxygenic phototrophic bacteria with the formation of elemental sulfur
3. Anaerobic removal of nutrients
4. Removal of Fe^{2+} from water by photooxidation

The stoichiometry of anoxygenic photosynthesis can be shown by the following equation:

$$2H_2S + CO_2 + \text{energy of light} \rightarrow \langle CH_2O \rangle + 2S + H_2O$$

where $\langle CH_2O \rangle$ is a conventional formula showing synthesized carbohydrates.

The following are groups of anoxygenic phototrophic Gram-negative bacteria classified by conventional taxonomy:

1. Purple sulfur bacteria with internal or external sulfur granules, for example, the genus *Chromatium*. By the phylogenetic classification of 16S rRNAs phototrophic purple bacteria belong to the α-, β-, and γ-proteobacteria.
2. Purple nonsulfur bacteria, for example, the *Rhodocyclus* genus. According to the phylogenetic classification of 16S rRNAs phototrophic purple nonsulfur bacteria can be found in the α- and β-proteobacteria.
3. Green sulfur bacteria, for example, the genus *Chlorobium*. By the phylogenetic classification of 16S rRNAs phototrophic green sulfur bacteria can be found in the group 2.16 of bacteria—they are typical aquatic microorganisms and grow where light reaches the anaerobic water layer of a lake or sediment.
4. Multicellular filamentous green nonsulfur bacteria, for example, the genus *Oscillochloris*, live in hot and cold springs, freshwater lakes, river water and sediments, in both marine and hypersaline environments. It is a group phylogenetically distant from green sulfur bacteria.

Some green nonsulfur bacteria, such as the species of genus *Chloroflexus*, are able to perform aerobic respiration and may be included in the groups of microaerophilic or aerobic phototrophs. Some species of purple bacteria, green sulfur bacteria, and *Chloroflexus* oxidize ferrous iron as an electron donor for photosynthesis instead of H_2S.

Gram-Positive (Terrestrial), Phototrophic, Anoxic Bacteria

Representatives of this group of anoxygenic Gram-positive phototrophic bacteria, using products of anaerobic respiration (H_2S, Fe^{2+}) have not yet been discovered. Such species could likely have properties close to Heliobacteria and could be similar in 16S rRNA genes to the subgroups of Heliobacteria, *Desulfotomaculum* and *Desulfitobacterium*.

Phototrophic *Archaea* Using Products of Anoxic Respiration as Electron Donors

This group of extremophilic microorganisms has not yet been discovered. It could be possible that there is no such species of phototrophic *Archaea* in nature because of the sensitivity of phototrophic energy production to extreme temperatures.

Functions of Facultative Anaerobic and Microaerophilic Prokaryotes

Microorganisms originated from the boundary between the periods of anaerobic and aerobic atmosphere on the earth selected two strategies of adaptation to environment:

1. The ability to switch methods of energy production between fermentation, anaerobic, and aerobic respirations depending on redox conditions (concentration of oxygen) in their habitat. This ability is characteristic of diverse groups of facultative anaerobic (or facultative aerobic) microorganisms; modern representatives of this group dominate in biotopes where aerobic and anaerobic conditions frequently change.
2. The ability for aerobic respiration only at low oxygen concentration, for example, conventionally lower than 1 mg/L. The natural habit of microaerophilic modern representatives of this group is the interface between aerobic and anaerobic zones of ecosystem or hot aquatic biotopes where the concentration of dissolved oxygen is low.

The periodic table with the groups of prokaryotes mentioned in this section is shown in Table 12.3.

Gram-Negative (Aquatic), Chemotrophic, Facultative Anaerobic and Microaerophilic Bacteria

These bacteria can produce energy by aerobic respiration and either anaerobic respiration (usually by denitrification) or fermentation. Most of these bacteria are active destructors of organic substances and are used in engineering systems, where aerobic and anaerobic conditions frequently change.

Many facultative anaerobic species are enterobacteria, that is, their typical habit is the human or animal intestine. There are many agents of waterborne diseases in the genera of *Salmonella*, *Shigella*, and *Vibrio*. The cell numbers of indicator species of *Esherichia coli*, group of physiologically similar coliforms, and enterococci are common indicators of water pollution by feces or sewage.

The species of the genus *Shewanella* from the gamma-subdivision of Proteobacteria are facultative anaerobic bacteria that are able to perform anaerobic respiration using thiosulfate, elemental sulfur, nitrate, iron (III), and manganese (VI) as electron acceptors. The growth yield of these processes is low. However, the

TABLE 12.3 Evolutionary Lines and Periods of the Periodic Table of Chemotrophic and Phototrophic Prokaryotes

Evolutionary Line	Subline	Periods of Evolution			
		Fermenting prokaryotes or prokaryotes using products of fermentation as electron donors	Anoxic prokaryotes or prokaryotes using products of anoxic respiration as electron donors	Microaerophilic and facultative anaerobic prokaryotes	Aerobic prokaryotes or prokaryotes using products of aerobic respiration as electron donors
Prokaryotes of aquatic origin (Gram-negative type of cell wall, *Gracilicutes*)	Chemotrophs			*Escherichia coli* *Helicobacter* *Salmonella Shigella* *Vibrio* *Shewanella* *Beggiatoa Thiothrix* *Siderocapsa* *Naumanniella* *Siderococcus* *Siderocystis* *Gallionella* *Leptothrix* *Sphaerotilus* *Magnetospirillum*	

	Phototrophs	*Cyanobacteria*, for example, *Microcoleus* *Oscillatoria*
Prokaryotes of terrestrial origin (Gram-positive type of cell wall, *Firmicutes*)	Chemotrophs	*Propionibacterium* *Staphylococcus* *Streptococcus* *Enterococcus* *Actinomycetes* *Frankia* *Microthrix* *Nocardia Trichococcus*
	Phototrophs	Not known yet
Prokaryotes of extreme environment origin (*Archaea*)	Chemotrophs	*Sulfolobus* *Acidianus Metallosphaera* *Thermoplasma* *Picrophilus* *Ferroplasma*
	Phototrophs	Not known yet

Note: Selected examples of conventional genera of microaerophilic and facultative anaerobic prokaryotes are shown in the cells of periodic table.

biomass of these species can be grown aerobically with high yield on fermentation end products, that is, lactate, formate, and some amino acids. *Shewanella* spp. live naturally in association with fermentative prokaryotes that supply them the needed nutrients. Some species have been isolated from the deep sites and are tolerant of high pressures and low temperatures. The main application of *Shewanella* spp. in environmental biotechnology may be the aerobic growth of biomass with a further application for anoxic remediation of polluted sites using Fe(III) of ferric oxides, hydroxides, or iron-containing clay minerals as electron acceptors. Reduction of iron and manganese makes these metals dissolved but reduction of dissolved U(VI) and Cr(VI) makes U(IV) and Cr(III) insoluble, respectively.

The ability to reduce nitrate or nitrite is widespread among facultative anaerobic prokaryotes. Denitrifying bacteria are capable of oxidizing organic substances, hydrogen, Fe^{3+}, H_2S, or S using nitrite or nitrate as electron acceptors:

$$5\langle CH_2O\rangle + 4NO_3^- + 4H^+ \rightarrow 2N_2 + 5CO_2 + 7H_2O$$

Active denitrifiers, which are used for the removal of nitrate from groundwater and wastewater, are, for example, *Pseudomonas denitrificans* and *Paracoccus denitrificans*. Electron donors for industrial scale denitrification can be methanol or ethanol. Hydrogen or sulfur can be used for industrial scale autotrophic denitrification of drinking water and seawater, respectively:

$$10H_2 + 4NO_3^- \rightarrow 2N_2 + 8H_2O + 4OH^-$$

$$5S + 6NO_3^- + 2H_2O \rightarrow 5SO_4^{2-} + 3N_2 + 4H^+$$

Denitrifiers can also be used for the anoxic biodegradation of toxic organic substances in the case of fast bioremediation of anoxic clay soil.

Filamentous microaerophilic H_2S-oxidizing bacteria from genera *Beggiatoa* and *Thiothrix* cause a problem of wastewater treatment called bulking. Bulking or bulking foaming in poorly aerated or overloaded aerobic tanks may be due to the excessive growth of filamentous bacteria forming loose and poor settled flocs. Growth of *Beggiatoa* spp. in sulfide rich, microaerophilic environment leads to the formation of sulfur-containing slime. Similar hydrogen and sulfide-oxidizing thermophilic filamentous microaerophilic bacteria include species from *Aquifex* and *Hydrogenobacter* genera.

Neutrophilic iron-oxidizing and manganese-oxidizing bacteria are used in environmental biotechnology for the removal of iron and manganese from water. These have also been proposed to be used for the removal or even recovery of ammonia from wastewater instead of nitrification. Iron-oxidizing bacteria clog drains, pipes, and wells with iron hydroxide deposits. Their natural habitats are springs from iron-rich soil, rocks, and swamps. Species from *Siderocapsa* genus are usually suspended or attached to the soil, rock, and plant surfaces in springs or lakes with an input of Fe(II) from iron-rich soil, deposits, or swamps. Cells are covered by a slimy, iron hydroxide–containing capsule. Species from *Naumanniella* genus are usually psychrophilic and attached to the walls of the pipes and wells. Cells are slim rods with a thin iron-containing capsule. *Siderococcus* spp. are cocci with a capsule-like iron hydroxide deposition. *Siderocystis* spp. form chain-connected spherical ferric hydroxide particles. There are also stalk-forming bacteria and thread-forming sheathed bacteria of *Gallionella*, *Leptothrix*, and *Sphaerotilus* genera capable of oxidizing iron (II) and precipitating iron (III) hydroxides in the stalk or sheath. Sheaths of neutrophilic iron-oxidizing bacteria can adsorb heavy metals and radionuclides from hazardous streams. The microaerophilic filamentous species, *Sphaerotilus natans*, has false branches of sheathed cells with a mycelium-like appearance and is called "sewage fungus." They are also responsible for bulking in poorly aerated or overloaded aerobic tanks. Some microaerophilic bacteria participate in the transformation of metals. Microaerophilic spirilla, for example, *Magnetospirillum* spp., can produce magnetite from Fe^{2+}.

Microaerophiles can form H_2O_2 and other reactive oxygen species (ROS), such as superoxide radical and hydroxyl radical as the final products of oxygen detoxication. ROS can oxidize nonspecific xenobiotics. Microaerophilic bacteria are known to cause the biodegradation of benzene, phenol, toluene, and naphthalene.

Together with the species useful for engineering, there are many pathogenic microaerophilic organisms, for example, some strains of *E. coli* that cause intestinal infections, *Salmonella* spp., and *Campylobacter* spp. that cause the life-threatening infections salmonellosis and campylobacteriosis, respectively. Some strains of bacteria from the species *Vibrio cholera* are agents of waterborne infectious disease. The bacteria *Helicobacter pylori* cause stomach ulcer, and *Treponema pallidum* cause syphilis. Some facultative anaerobic bacteria, for example, *Stenotrophomonas maltophila*, are active degraders of xenobiotics but are opportunistic pathogens. These strains are used in environmental engineering, but strict biosafety rules must be heeded in their applications.

Gram-Positive (Terrestrial), Chemotrophic, Facultative Anaerobic and Microaerophilic Bacteria

There are facultative anaerobic Gram-positive bacteria belonging to the genera *Propionibacterium*, *Staphylococcus*, *Streptococcus*, and *Enterococcus*, which are closely associated with the surfaces of the human body and animals. Some are agents of infectious diseases. These bacteria, for example, *Enteroccoccus*, are used in environmental engineering as indicators of bacteriological quality of the environment.

Facultative anaerobic species among *Actinomycetes*, a group of Gram-positive bacteria with high G+C content in DNA, are important in the degradation of organic compounds in soil. The microaerophilic representatives of genus *Frankia* are able to fix molecular nitrogen in symbiosis with nonleguminous plants.

Microaerophic filamentous Gram-positive bacteria, for example, *Microthrix parvicella*, *Nocardia* spp., and *Trichococcus*, are common to wastewater-activated sludge; however, the abundance of these organisms is associated with bulking, foaming and scum formation, and finally with wastewater-treatment failure.

Chemotrophic, Facultative Anaerobic and Microaerophilic *Archaea*

The species from genera *Sulfolobus*, *Acidianus*, and *Metallosphaera* are microaerophilic or facultative anaerobic, grow in sulfur-rich, hot acid biotopes, are capable of oxidizing organic substances, S, $S_4O_6^{2-}$, S^{2-}, and Fe^{2+} using oxygen or Fe^{3+} as electron acceptors. Another group of facultatively anaerobic, thermoacidophilic *Archaea*, are the species of the genera *Thermoplasma*, *Picrophilus*, and *Ferroplasma*. Applications of acidophilic *Archaea* in environmental engineering may include bioextraction (bioleaching) of heavy metals from sewage sludge at high temperatures. Another potential application is the bioremoval of inorganic and organic sulfur from coal and oil to diminish the emissions of sulfur oxides in the atmosphere.

Gram-Negative (Aquatic), Phototrophic, Facultative Anaerobic and Microaerophilic Bacteria

The oxygenic photosynthetic Gram-negative bacteria comprise the cyanobacteria and the prochlorophytes that are distinguished by their photosynthetic pigments. Some representatives from the *Microcoleus* and *Oscillatoria* genera are facultative anaerobic organisms capable of anoxygenic photosynthesis with hydrogen or sulfide as electron donors for the reduction of CO_2 or for oxygenic photosynthesis:

$$H_2O + CO_2 + \text{energy of light} \rightarrow \langle CH_2O \rangle + O_2$$

where $\langle CH_2O \rangle$ is the conventional formula for synthesized carbohydrates.

The ability of cyanobacteria to grow in both aerobic and anaerobic environments is related to the life of some cyanobacteria in a dense microbial mat where the conditions are aerobic during the day and become anaerobic at night. During anoxygenic photosynthesis under CO_2 limitation, the electrons from sulfide may be also used for fixation of molecular nitrogen or for the production of molecular hydrogen. This feature can be used in the biogeneration of hydrogen in fuel cells.

Gram-Positive (Terrestrial), Phototrophic, Facultative Anaerobic and Microaerophilic Bacteria

Currently, Gram-positive, phototrophic, facultative anaerobic and microaerophilic bacteria using products of anaerobic respiration, H_2S, Fe^{2+}, or the product of aerobic respiration, H_2O, as electron donor, have not yet been discovered. Such species could likely have properties close to Heliobacteria and could be similar by 16S rRNA genes to the subgroups of Heliobacteria, *Desulfotomaculum* and *Desulfitobacterium*.

Phototrophic, Facultative Anaerobic and Microaerophilic *Archaea*

This group of extremophilic microorganisms has not been discovered yet. It could be possible that there is no such species of phototrophic *Archaea* in nature because of the sensitivity of phototrophic energy production to extreme temperature.

Functions of Aerobic Prokaryotes

Gram-Negative (Aquatic), Chemotrophic, Aerobic Bacteria

Strictly aerobic chemotrophic Gram-negative bacteria are the most active organisms in the biodegradation of xenobiotics, aerobic wastewater treatment, and soil bioremediation. Examples of the functions of selected genera are shown below. The periodic table with the groups of prokaryotes mentioned in this section is shown in Table 12.4.

Members of the genus *Pseudomonas* (the pseudomonads), for example, *P. putida*, *P. fluorescence*, and *P. aeruginosa*, are used in environmental biotechnology as active degraders of xenobiotics in wastewater treatment and soil bioremediation. These organisms oxidize aliphatic hydrocarbons, monocyclic and polycyclic aromatic hydrocarbons, halogenated aliphatic and aromatic compounds, different pesticides, and oxidize or cometabolize halogenated ethanes and methanes. Biodegradation often depends on the presence of specific plasmids. Therefore, both native and genetically engineered strains with amplified and diverse degradation ability are used in environmental engineering. Some selected strains of *Pseudomonas* genus are used instead of chemical biocides to control plant diseases.

Other active biodegraders are the species from genera of *Alcaligenes*, *Acinetobacter*, *Burkholderia*, *Comamonas*, and *Flavobacterium*. The majority of active biodegraders are opportunistic bacteria, that is, they can cause diseases in immunosuppressive, young or old people. Therefore, all experiments and treatments of water and soil by these bacteria must be performed with precautions against the dispersion of these bacteria in the environment and reasonable biosafety rules must be used in their applications.

There are many pathogens among the above mentioned genera, for example, *Pseudomonas aeruginosa*, *Burkholderia cepacia*, and *Burkholderia pseudomallei*. Therefore, the test of acute toxicity and other pathogenicity tests of all microbial strains, isolated as the active biodegraders of xenobiotics for environmental engineering applications, must be made after selection and before pilot scale research.

Representatives of genus *Xantomonas* are also active biodegraders but there are many phytopathogenic species. Therefore, they cannot be used for soil bioremediation but used as test cultures to test new biocides for agriculture.

The formation of activated sludge flocs is enhanced by the production of extracellular slime. *Zoogloea ramigera* is considered an important organism in floc

TABLE 12.4 Evolutionary Lines and Periods of the Periodic Table of Chemotrophic and Phototrophic Prokaryotes

Evolutionary Line	Subline	Periods of Evolution		
		Fermenting prokaryotes or prokaryotes using products of fermentation as electron donors	Anoxic prokaryotes or prokaryotes using products of anoxic respiration as electron donors	Aerobic prokaryotes or prokaryotes using products of aerobic respiration as electron donors
Prokaryotes of aquatic origin (Gram-negative type of cell wall, *Gracilicutes*)	Chemotrophs			*Pseudomonas* *Acinetobacter* *Alcaligenes* *Acinetobacter* *Burkholderia* *Comamonas* *Flavobacterium* *Nitrosomonas* *Methylococcus* *Methylomonas* *Yersinia pestis* *Bdellovibrio* *Nitrobacter*
	Phototrophs			Cyanobacteria, for example, *Chroococcales* *Pleurocapsales* *Oscillatoriales* *Nostocales* *Stigonematales* *Prochloron*
Prokaryotes of terrestrial origin (Gram-positive type of cell wall, *Firmicutes*)	Chemotrophs			*Bacillus* *Mycobacterium* *Nocardia* *Rhodococcus* *Gordona* *Arthrobacter* *Streptomyces*
	Phototrophs			Not known yet
Prokaryotes of extreme environment origin (*Archaea*)	Chemotrophs			*Picrophilus* *Ferroplasma*
	Phototrophs			*Halobacteria*

Note: Selected examples of conventional genera of aerobic prokaryotes are shown in the cells of periodic table.

formation because of its strong self-aggregation. Probably, the most important role in the formation of activated sludge floc belongs to gliding bacteria from the genera *Flavobacterium*, *Cytophaga*, *Myxobacterium*, *Flexibacterium*, and *Comamonas*. They are called gliding bacteria because of cell translocation on a solid surface due to the interaction of cell surface and solid surface. They produce extracellular polysaccharides and have the ability for strong aggregation of their cells.

Species of the genus *Acinetobacter* are capable of accumulating polyphosphate granules and are used for biological removal of phosphate from wastewater. This removal diminishes the supply of phosphate ions and dissolved phosphates of heavy metals from the environment. Another important environmental biotechnology feature of these bacteria is the ability to produce extracellular polyanionic heteropolysaccharides that can emulsify hydrocarbons and thus enhance their degradation in an aqueous environment. Together with these useful properties, *Acinetobacter* spp. are opportunistic pathogens that cause different infections in immunocompromised people. These infections are difficult to treat successfully because the clinical strains of *Acinetobacter* spp. acquire resistance to antibiotics.

Azotobacter and *Azomonas* are the genera of free-living, nitrogen-fixing soil bacteria that accumulate organic nitrogen and improve soil fertility. Selected strains of genus *Azotobacter* are used for the industrial biosynthesis of polyhydroxybutyrate and its derivates for biodegradable plastics. Biodegradability of plastic materials is very important for environmental sustainability.

Methanotrophs do not grow on organic compounds and oxidize only such single-carbon compounds as methane, methanol, formaldehyde, and formate. Methane-oxidizing species, for example, from the *Methylococcus* and *Methylomonas* genera, are important for the removal of methane from the atmosphere, thus diminish the greenhouse effect due to the accumulation of carbon dioxide and methane:

$$CH_4 + 0.5O_2 \rightarrow CH_3OH$$

A well-known application of methylotrophs in environmental biotechnology is the bioremoval of halogenated methanes and ethanes from the polluted groundwater by cometabolism. For example, cometabolism of trichloroethylene (TCE) by methylotrophic bacteria is considered an effective approach for the remediation of a polluted aquifer. The main reaction of cometabolism catalyzed by the enzyme methane monooxygenase of methanotrophs is described by the following equation:

$$Cl_2C(O){=}CHCl + NADH_2 + O_2 \rightarrow Cl_2C{-}CHCl + H_2O + NAD$$

There are many oligotrophs in the group of Gram-negative, chemotrophic, aerobic bacteria.

These organisms are adapted to live in environment with a low concentration of nutrients including carbon and energy sources. Their adaptation is so stable that many oligotrophs are obligate ones and cannot grow in a medium with a high concentration of carbon and energy sources. Oligotrophic microorganisms are important for the treatment of groundwater, seawater, and freshwater with a low concentration of carbon source. For example, *Hyphomicrobium* spp. are budding oligotrophic bacteria capable of oxidizing single-carbon compounds by oxygen or nitrate. They are used in environmental engineering for the removal of nitrate from water using methanol as an electron donor. Stalked oligotrophic bacteria from genus *Caulobacter* spp. are able to survive during long-term starvation. It is thought that they may perform gene transfer between different bacteria participating in water and wastewater treatment because they are often adhered to the cells of other bacteria.

Bdellovibrio spp. are Gram-negative bacteria, which are parasites of other Gram-negative bacteria. Therefore, their presence in water is an indication of water pollution by Gram-negative bacteria. *Bdellovibrio* spp. can be applied in environmental engineering for the biological control of human, animal, and plant pathogens in water and soil and for the control of excessive growth of microbial biofilm in fixed biofilm reactors, which are used for water and wastewater treatment.

Rhizobia are bacteria from the genera *Rhizobium*, *Bradyrhizobium*, and *Azorhizobium*, which are able to grow and fix atmospheric molecular nitrogen symbiotically with leguminous plants, for example, peas, beans, and clover. It is considered that this fixation supplies about a half of the nitrogen used in agriculture. The application of selected strains of rhizobia for the inoculation of soil, where leguminous plants are planted for the first time, can double the yield of these plants. Therefore, these bacteria are used in environmental engineering for the enhancement of soil fertility after bioremediation of polluted soil.

There are many pathogenic species in the group of Gram-negative, chemotrophic, aerobic bacteria. *Yersinia pestis* was responsible for the devastating outbreaks of plague in Asia and Europe in the sixth, fourteenth, seventeenth, eighteenth, and beginning of the twentieth centuries. Another example is *B. pseudomallei*, the agent of melioidosis. This disease is common in Southeast Asia. It affects people exposed to soil and soil aerosols: farmers on rice paddies, construction workers, or people living close to the soil excavation area. The disease may be misdiagnosed as syphilis, typhoid fever, or tuberculosis. Symptoms of pulmonary melioidosis can range from bronchitis to a severe pneumonia. During the period from 1989 to 1996, a total of 372 melioidosis cases, with 147 deaths, were reported in Singapore. *Legionella* spp., is an agent of Legionnaires disease, which is a lung infection caused by inhalation of water droplets from poorly maintained cooling towers, air conditioners, fountains, and artificial waterfalls.

Gram-Negative (Aquatic), Chemolithotrophic, Aerobic Bacteria

Chemolithotrophs are aerobic prokaryotes, which can use a reduced inorganic compound as an energy source. The species grow as autotrophs but some can also grow as organotrophs.

Nitrifying bacteria comprise two groups of aerobic bacteria: ammonia-oxidizers (*Nitrosomonas* spp., *Nitrosococcus* spp., *Nitrosovibrio* spp., *Nitrosospira* spp., and *Nitrosolobus* spp.) performing the reaction

$$NH_4^+ + 1.5O_2 \rightarrow NO_2^- + 2H^+ + H_2O$$

and nitrite-oxidizers (*Nitrobacter* spp., *Nitrococcus* spp., and *Nitrospira* spp. from different subdivisions of 16S rRNA classification) that use the nitrite to form nitrate:

$$NO_2^- + 0.5O_2 \rightarrow NO_3^-$$

Nitrifying bacteria are widely used in environmental biotechnology to transform toxic ammonium to a less toxic nitrate. The nitrate produced can be transformed further to molecular nitrogen by denitrifying bacteria. The problems of large-scale nitrification in wastewater treatment include

1. Washing out of these microorganisms from a bioreactor of continuous cultivation due to their slow growth rate in comparison with the growth rate of heterotrophs
2. High sensitivity of nitrifiers to toxic substances, surfactants, or organic solvents due to the large folded surface of a cell membrane

The activity of nitrifying bacteria was the basis of ancient Chinese environmental biotechnology to convert organic wastes and household ashes into fertilizer and gun powder as a value-added by-product. The transformation of waste to gun powder can be described by the following sequence of the reactions:

1. Formation of ammonia from amino acids of proteins by bacteria—ammonifiers in anaerobic microzones of waste

$$RNH_2 \rightarrow NH_4^+$$

2. Production of nitrate from ammonium by bacteria—nitrifiers in aerobic microzones of waste

$$NH_4^+ + 2O_2 \rightarrow HNO_3 + H_2O + H^+$$

3. Neutralization of acid solution of nitrate by potassium oxide from ash

$$K_2O + 2HNO_3 \rightarrow 2KNO_3 + H_2O$$

4. Formation of hydrogen sulfide by sulfate-reducing bacteria in anaerobic microzones of waste

$$SO_4^{2-} + CH_2O + 6H^+ \rightarrow H_2S + CO_2 + 3H_2O$$

5. Oxidation of hydrogen sulfide by microaerophilic bacteria

$$H_2S + 0.5O_2 \rightarrow S + H_2O$$

These reactions result in a suspension containing carbon particles from ash, particles of elemental sulfur, and potassium nitrate. Drainage of this suspension from the waste pile and drying it under the sun in a drainage collector produces a mixture of carbon, sulfur, and potassium nitrate, which is gun powder.

Aerobic sulfur-oxidizing chemolithotrophic bacteria oxidize reduced sulfur compounds producing sulfuric acid:

$$S + 1.5O_2 + H_2O \rightarrow H_2SO_4$$

$$H_2S + 4O_2 \rightarrow H_2SO_4$$

These bacteria are used or can be used in environmental biotechnology for the following purposes:

1. Bioremoval of toxic H_2S from gas, water, and wastewater
2. Industrial bioleaching of metals, for example, copper, zinc, or uranium from the ores
3. Bioleaching of heavy metals from anaerobic sewage sludge before its utilization as an organic fertilizer
4. Bioleaching of heavy metals and radionuclides from polluted soil
5. Acidification of alkaline soil

Sulfur oxidation can cause corrosion of steel and concrete constructions due to the production of sulfuric acid. Some species, for example, *Thiobacillus ferrooxidans*, (*Acidithiobacillus ferrooxidans*) can grow at an extremely low pH and be isolated from acid mine drainage.

Oxidization of Fe(II) is performed by two groups of aerobic bacteria: acidophilic and neutrophilic iron-oxidizers. Fe(II) is stable in acid solutions, if the pH is lower than 3. Its chemical oxidation by oxygen under a low pH is slow. Some bacteria, however, for example, *T. ferroooxidans* (*A. ferrooxidans*), can oxidize Fe(II) some thousand times faster than that in a chemical process.

Neutrophilic iron- and manganese-oxidizers are usually microaerophilic. The main point for neutrophilic oxidation of iron is that Fe(II) is not stable under aerobic conditions and neutral pH and must be immediately oxidized by oxygen. However, atoms of Fe(II), surrounded by chelated organic acids, are protected from chemical oxidation by oxygen. Therefore, the functions of neutrophilic microaerophilic "iron-oxidizers" are production of H_2O_2 and chemical degradation of the organic "envelope" of the Fe^{2+} atom by H_2O_2. Precipitation of iron hydroxide by these bacteria can clog pipelines and wells. They are used in environmental engineering for the removal of iron and manganese from drinking water treated in a slow sand filter. Another important application is the removal of ammonia from wastewater by co-precipitation with fine particles of positively charged iron hydroxide produced by neutrophilic "iron-oxidizers."

Gram-Positive (Terrestrial), Chemotrophic, Aerobic Bacteria

These bacteria have diverse functions in environmental engineering. Representatives of the genus *Bacillus* dominate in the aerobic treatment of wastewater or solid waste, which is rich in such polymers as starch or protein. The species of genus *Bacillus* can degrade different xenobiotics. They produce endospores providing cell survival in variable soil environment and after drying. Therefore, the endospores can be used in commercial compositions to start up soil bioremediation or biodegradation of certain substances during wastewater treatment. The shape and cellular location of endospores are used as identifying characteristics in the differentiation of species. Some species, for example, *Bacillus subtilis*, live in the human intestine; viable cells are used as probiotics, a medical application to normalize intestinal microflora. However, there are also pathogenic species in this genus. For example, *Bacillus anthracis* cause the deadly infection anthrax due to the production of strong toxins and are even considered as biological weapons.

Bacteria of the genus *Arthrobacter* are commonly found in soil and are active in the biodegradation of xenobiotics. Their specificity is a rod–coccus growth cycle; the cells are rod-shaped during active growth and cocci-shaped during starvation. Saprophytic species of mycobacteria are efficient degraders of xenobiotics but there are also many pathogenic species, for example, *Mycobacterium tuberculosis* (causative agent of tuberculosis) and *Mycobacterium leprae* (causative agent of leprosy). *Nocardia* spp. and *Rhodococcus* spp. degrade hydrocarbons and waxes; they form a mycelium that breaks into rods and cocci. Excessive growth of *Nocardia asteroides*, *Rhodococcus* spp., and *Gordona amarae* in aerobic tanks causes foaming (brown scum) of activated sludge due to the high hydrophobicity of the cell surface of these species, the production of biosurfactants, or hydrolysis of lipids. The foaming produces obnoxious odors and can increase the risk of infection of wastewater workers by opportunistic pathogenic actinomycetes such as *N. asteroides*.

Actinomycetes are aerobic, heterotrophic Gram-positive prokaryotes with a high G+C content in DNA, growing with aerial mycelia. They are used in the aerobic treatment of wastewater, soil bioremediation, and composting of solid wastes because many of them are active degraders of natural biopolymers and xenobiotics. The genus *Streptomyces*, containing half a thousand species, is extremely important for medical biotechnology because many antibiotics are produced by the strains from this genus. A significant part of soil microbial biomass is the biomass of streptomycetes. Therefore, the characteristic odor of moist earth is due to the volatile substances such as geosmin produced by these microorganisms. Thermophilic actinomycetes play an important role in the composting of organic wastes.

Chemotrophic, Aerobic *Archaea*

Some halophilic *Archaea* are aerobic microorganisms. Hypothetically, they can be used for the biotreatment of polluted industrial brines, but there are no applications of these prokaryotes in environmental biotechnology at present.

Gram-Negative (Aquatic), Phototrophic, Aerobic Bacteria

Cyanobacteria carry out oxygenic photosynthesis with water as the electron donor:

$$CO_2 + H_2O + \text{energy of light} \rightarrow \langle CH_2O \rangle + O_2$$

where ⟨CH_2O⟩ is the conventional formula of photosynthesized carbohydrates. Cyanobacteria are used in environmental biotechnology for the light-dependent removal of nitrogen and phosphorus from wastewater. However, a main problem of environmental engineering, related to cyanobacteria, is the control of their blooming in surface water polluted with ammonia or phosphate. This bloom causes obnoxious odors, bad taste of water, and accumulation of allergens and toxins in water. The cyanobacteria are morphologically diverse. There are unicellular organisms that reproduce by binary fission, budding (*Chroococcales*), or by multiple fission (*Pleurocapsales*). There are also filamentous forms without cell differentiation (*Oscillatoriales*) or with cell differentiation (*Nostocales* and *Stigonematales*).

Prochlorophytes are a group of aerobic oxygenic phototrophic Gram-negative bacteria that differ from cyanobacteria by their set of photosynthetic pigments. Group *Prochloron* is close to cyanobacteria by 16S rRNA phylogeny. This group has no importance for environmental engineering because they have been found only as extracellular symbionts of ascidians, marine animals, in the tropical areas of the Pacific and Indian oceans.

Gram-Positive (Terrestrial), Phototrophic, Aerobic Bacteria

There are currently no known obligate aerobic Gram-positive phototrophic bacteria using H_2O as an electron donor. Such species could likely have properties close to Heliobacteria and could be similar by 16S rRNA genes to the subgroups of Heliobacteria, *Desulfotomaculum* and *Desulfitobacterium*.

Phototrophic, Aerobic *Archaea*

Some halophilic *Archaea* can produce biological energy using light and the pigment bacteriorhodopsin, similar in structure to the light-sensitive pigment of human eye. This pigment could be used for biosensors and optical computers.

13

Public Health and Water Disinfection

Saprophytic, Pathogenic, and Opportunistic Microbes

Saprophytic microorganisms feed on dead organic matter because their main habitat is soil or water. Pathogenic microorganisms grow in or on animal tissue and can cause diseases in macroorganisms. Opportunistic pathogens are normally harmless but have the potential to be pathogenic in immunocompromised individuals or by penetrating the skin of the macroorganism.

Parasites of Humans and Animals

Pathogenic viruses, prokaryotes, fungi, or protozoa are parasites of macroorganisms from the ecological point of view. They attach and then grow on/in human or animal tissue and can cause diseases in macroorganisms such as humans, animals, or plants.

Stages of Infectious Disease

An infection is a disease caused by the transfer of pathogenic microorganisms from the environment or from one macroorganism to another. Typical stages of disease are as follows:

1. Contact of the microorganism with humans
2. Entry of the pathogen into the organism
3. Growth of the infecting microorganism
4. Development of disease symptoms

Microbial pathogens disrupt normal body functions by producing enzymes and toxins.

Transmission of Infection

Infectious microorganisms can enter a human through direct contact between individuals or through reservoir-to-person contact. The diseases may be conventionally distinguished as airborne, waterborne, soilborne, vectorborne (through insects), and foodborne infectious diseases.

Transmission of waterborne diseases is directly related to the bacteriological quality of water and effluents of wastewater-treatment plants. Sources of pathogens other than sewage outlets are wildlife watersheds, farms, and landfills. The prevention of outbreaks of waterborne and airborne diseases is one of the main goals of environmental biotechnology.

Environmental engineers and epidemiologists must work closely to identify the reason of outbreak, find its source (reservoir), define the major means of transmission of infectious microorganisms, and to develop a way to stop or diminish the scale of outbreak.

Patterns and Extent of Infectious Disease

Epidemiology is a science studying the origin, cause, and spread of disease. Endemic disease is normally found in a particular area. An epidemic is a relatively sudden increase in the number of cases of a particular disease in a particular place. A pandemic is an epidemic that encompasses the entire world, like the bird flu. Due to rapid international travel, pandemics could increase in the future.

Infectious diseases still account for 30%–50% of deaths in developing countries because of poor sanitation. By comparison, mortality from infectious diseases is 10 times less frequent in developed countries. Poorer sanitation procedures are responsible for the relatively high levels of infectious-disease mortality in these countries. The decreased mortality due to waterborne infectious diseases in the world is a consequence of a higher standard of living, better sanitation procedures, and better quality of environmental and civil engineering.

Cooperation of Epidemiologists and Civil Engineers

Environmental engineers are concerned with the prevention of airborne, waterborne, soilborne, and vectorborne infectious diseases through engineering solutions in ventilation and air treatment, water supply, waste treatment, excavations, sanitation, construction, and monitoring.

Environmental engineers often work together with epidemiologists as members of one team to identify the reason of disease, find its reservoir, define the major means of transmission and develop a way of stopping or limiting the spread of the infection. Purification of water, air, soil, sewage treatment, and disposal performed by environmental and civil engineers are the main conditions that prevent outbreaks of infectious diseases.

Infections via Vectors

When an infectious agent is spread by insects such as mosquitoes, fleas, lice, biting flies, or ticks, the insects are referred to as vectors. Examples of infectious agents spread in this way include bubonic plague spread by fleas, Lyme disease spread by ticks, and dengue fever and malaria spread by mosquitoes. Eradication of the vectors by tools of civil and environmental engineers, for example, through the absence of stagnant water pools in the living zone, is a major way for prevention of infections via vectors.

Factors of Microbial Pathogenicity

The ability to cause infectious disease depends on the following abilities of microorganisms:

1. Production of exotoxins, which are extracellular proteins. In this case, host damage can occur at sites far removed from a localized focus of infection. For example, the anaerobic bacteria *Clostridium tetani* can be introduced from soil into the body by deep puncture wounds. If the wound becomes anaerobic, the organism can grow and release tetanus toxin, which causes spastic paralysis.
2. Production of enterotoxins, which are exotoxins that act in the small intestine. These cause diarrhea, the secretion of fluid into the intestinal passage.
3. Production of endotoxins, which are lipopolysaccharides of the outer membrane of Gram-negative bacteria. Endotoxins are less toxic than exotoxins.
4. Formation of microstructures (fimbriae, flagellum) and macromolecules for specific adherence of microbial cells or viruses to a host cell.
5. Formation of cell structures (capsule) and macromolecules (O-antigen) protecting microbial cells from the reaction of a host macroorganism.

Exotoxins

Pathogens can produce highly toxic exotoxins, which are extracellular proteins. One example is tetanus toxin produced by *C. tetani* in human wounds and causes paralysis. Another example is botulinum toxin produced by *C. botulinum* in improperly processed food, which is so toxic that it is considered as biological warfare.

Enterotoxins

Enterotoxin is an exotoxin that acts in the small intestine. Enteroxins usually cause diarrhea, which is the secretion of fluid into the intestinal passage. The pathogens enter the body and reproduce in intestine after the consumption of contaminated food or water. For example, *Vibrio cholerae* causes a life-threatening diarrhea as a result of the consumption of infected water, growth of bacteria in intestine, and action of bacterial exotoxin.

Problem of Opportunistic Bacteria in Environmental Biotechnology

Some bacteria, for example, from the genera *Pseudomonas*, *Acinetobacter*, and *Mycobacterium*, are active biodegraders of organic substances during wastewater treatment and soil bioremediation, but they are also opportunistic pathogens. This needs to be taken into account by environmental and civil engineers but very often it is not.

Waterborne Pathogens

These pathogens include some viruses, bacteria, and protozoa. They primarily cause intestinal diseases, and exit the host in fecal material, contaminating the water supply, and then entering the recipient by ingestion. Their survival period in water varies widely and is influenced by many factors such as salinity and temperature. Sources of waterborne pathogens other than sewage outlets are wildlife watersheds, farm lots, garbage dumps, and septic-tank systems.

Detection of Waterborne Pathogenic Viruses

Enteric viruses survive less than 3 months in the environment but have been reported surviving up to 5 months in sewage. Tests for the presence of viruses in water supplies are difficult. Viruses may be detected using animal or plant tissue cultures by their cytopathic effect. Modern rapid methods for virus detection are using immunoelectron microscopy, immunofluorescence, enzyme-linked immunosorbent assay (ELISA), radioimmunoassay, PCR, and nucleic acid probes.

Waterborne Pathogenic Bacteria

A minimum infectious dose of several hundred to several thousand organisms is necessary to cause bacteriological infection. Pathogenic bacteria are usually eliminated from natural waters by competition and predation. Low temperatures, adsorption on sediments, and anoxic conditions occasionally prolong their survival.

Some pathogenic strains of *Escherichia coli* produce gastroenteritis and urinary infections. *Leptospira* spp. is a cause of leptospirosis, which begins as a wound infection, is an occupational disease among workers in close contact with water polluted by rodents, domestic pets, or wildlife. *V. cholerae* causes cholera. It is a serious, highly contagious disease causing dramatic and fatal loss of water and electrolytes. *Shigella* spp., *Salmonella* spp., and *Campylobacter* spp. cause gastrointestinal diseases. Most gastroenteric diseases, which are caused by these bacteria in developed countries, are foodborne ones but transmission by drinking water is the major route of infection in undeveloped countries.

Fate of Released Pathogens in the Environment

The natural death of pathogens in environment is the most important factor in the termination of infectious-disease outbreaks. Active biodegraders, which are used in environmental engineering, are often opportunistic pathogens or genetically modified strains. Therefore, the study of environmental fate, death rate, and survivability of microorganisms that are used in bioremediation of polluted soil or spills in marine environment and released into the environment is essential for determining process feasibility. A rule of thumb is that applied microorganisms must have some reasonable limits of lifetime in the treated and surrounding areas. This short lifetime can prevent accumulation or spread of unwanted microorganisms in the environment in case of accidental or intentional release of environmental engineering microorganisms (Figure 13.1).

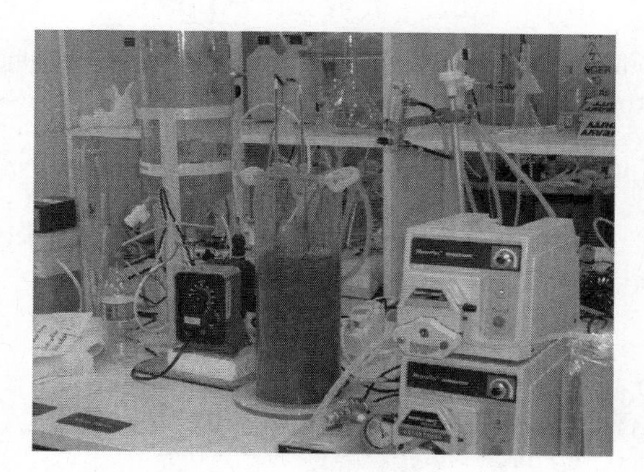

Figure 13.1 (See color insert following page 292.) This figure depicts a potentially dangerous environmental engineering experiment. Air is passed through an aeration tank containing bacterial biomass. The resulting contaminated air contains aerosols of pathogenic and opportunistic microorganisms and must be filtered before being released into the atmosphere.

Waterborne Pathogenic Protozoa

Protozoans entering the host body by ingestion are usually in cyst form. The protozoans of major concern as waterborne pathogens are *Giardia intestinalis* and *Entamoeba histolytica*. Cysts of *E. histolytica*, which cause amebic dysentery, survive for long periods of time at low temperatures and damp conditions in clean water.

Indicator Microorganisms

The variety of pathogenic organisms in water supplies is large but their concentrations are low. Testing for each organism is a highly expensive way for water quality monitoring. Therefore, indicator microorganisms are detected in water to indicate the possible presence of disease-causing constituents. The indicator microorganism is a microorganism whose presence is evidence that pollution associated with fecal contamination from man or other warm-blooded animals has occurred.

An indicator microorganism is a conventionally selected microorganism or group of microorganisms, which is used to determine the risk of waterborne infection associated with fecal contamination of water from humans or animals. An indicator microorganism must be of the same origin and have similar physiological properties as pathogens, and should be easily detected or enumerated in a water sample.

Coliforms as Indicators of Fecal Pollution

The common indicators of fecal pollution of water with enterobacteria, major agents of waterborne diseases, are the numbers of cells of *E. coli* (fecal coliforms) and *Streptococcus* spp. (fecal streptococci). The concentration of coliforms is usually less than 1 cell/mL in treated drinking water and more than a few million (10^6) cells/mL in sewage. The number of lactose-fermenting coliforms is considered a reliable indicator of the health risk related with enterobacterial pathogens. It is often determined by the total coliform MPN test.

Other Indicators of Fecal Pollution

Other indicating parameters, for example, total number of heterotrophic bacteria (heterotrophic plate count, HPC), are also used in the detection of water quality. The concentration of anaerobic bacteria from genera *Clostridium*, *Bifidobacterium*, or *Bacteroides* may be also considered as a good indicator of fecal pollution of water because their content in feces is some orders larger than the content of coliforms. However, their cultivation is a technically difficult procedure, so molecular-biological methods are more suitable for the detection or enumeration of these microorganisms in water. The standards for microbiological quality of drinking water used in the practice are related to the presence of specific indicator organisms.

Bacteriological Quality of Water

The concentration of coliforms is usually as follows:

- Treated drinking water ≤1 cell/mL
- River water ≤1000 cells/mL
- Domestic sewage ≤10,000,000 cells/mL

Protozoan and Viral Pathogens

There are no indicator organisms for protozoan cysts and viruses, because of the distinct release and survival pattern of each strain of pathogen. Therefore, these pathogens must be determined directly. Protozoans and viruses usually survive longer than *E. coli* and may also survive during disinfection. Therefore, most protozoans and viruses must be determined directly for the analysis of water quality and quality of disinfection.

Detection of Pathogens in the Environment

The direct detection and enumeration of numerous pathogens requires their preliminary concentration by centrifugation, filtration or adsorption, and finally the usage of modern molecular-biological methods. These methods are based on

- Genomics, i.e., detection of specific genes of microorganisms and viruses
- Proteomics, i.e., detection of specific proteins of microorganisms and viruses
- Transcriptomics, i.e., detection of specific RNAs of microorganisms

Specific genes of the pathogens are detected mainly by quantitative PCR, or using a genomics microarray, which is usually a slide with several thousands of microcells containing the fluorescent probes for specific genes. A new technique, suitable for environmental monitoring, is real-time loop-mediated isothermal amplification (LAMP) where the target sequence is amplified using several sets of primers.

Specific proteins of the pathogens are detected in the environment using immunochemical methods based on specific reactions between antibody and antigen, or using a proteomics microarray that is usually a slide with several thousands of microcells containing probes for specific proteins.

Specific RNAs of the pathogens are detected using a transcriptomics microarray, which is usually a slide with several thousands of microcells containing the fluorescent probes for specific RNAs.

Removal and Killing of Pathogens from the Environment

Pathogens can be removed or killed in environmental engineering systems by the following methods:

1. Bulk or membrane filtration and UV treatment for air and aerosol disinfection
2. Coagulation, aggregation, sedimentation, and slow filtration of water and wastewater to diminish the concentration of pathogens due to adsorption, sedimentation of microbial aggregates, or predation of protozoa
3. Chemical treatment by chlorine, chlorine dioxide, ozone, or UV light for water and wastewater-effluent disinfection
4. Chemical treatment by oxidants, organic solvents, surfactants, salts of heavy metals, or UV light for the disinfection of solid surfaces and microbial biofilms
5. Aeration, biotreatment, thermal treatment, acidification, hydrogen peroxide addition, and electromagnetic radiation to disinfect or to diminish the content of pathogens in solid waste or soil

Control of Microbial Death in Water

The control of unwanted microbial growth can be carried out by physical or chemical inhibition of growth, killing of the microorganisms, or their removal from the environment. Antimicrobial agents kill bacterial and fungal cells or inactivate viral particles; thus the terms bactericidal, fungicidal, and viricidal agents are used. The most sensitive targets of the microbial cell are as follows:

1. Integrity of cytoplasmic membrane
2. Active centers of enzymes
3. Secondary, tertiary, and quaternary structures of enzymes
4. Primary and secondary structures of nucleic acids

Thermal Treatment of Water

The rate of cell death under heat treatment is a function of the first order:

$$\frac{\mathrm{d}X_\mathrm{d}}{\mathrm{d}t} = k(X_\mathrm{o} - X_\mathrm{d}) \quad \text{or} \quad \ln\left(1 - \frac{X_\mathrm{d}}{X_\mathrm{o}}\right) = -kt, \tag{13.1}$$

where

X_d and X_o are the number of dead cells and initial number of viable cells, respectively
t is the time of exposure
k is a constant of decay

Another parameter, the decimal reduction time (d), which is the time required for a 10-fold reduction of the population, is used in practice:

$$d = \frac{\ln 0.1}{-k} = \frac{2.3}{k} \quad (13.2)$$

Pasteurization and Sterilization

Pasteurization is a weak thermal treatment, which kills bacterial, fungal, and protozoal vegetative cells. Vegetative cells of bacteria have a decimal reduction time in the range of 0.1–0.5 min at a temperature of 65°C. During bulk pasteurization, the liquid is exposed to 65°C for 30 min. During flash pasteurization, the liquid is heated to 71°C for 15 s and then rapidly cooled. Pasteurization reduces the level of microorganisms in the treated material but does not kill all of them.

Thermal sterilization kills all microorganisms in the treated water. It is often performed in an autoclave. Bacterial endospores cannot be killed at the temperature of boiling water. Therefore, the autoclave uses steam at a pressure of 1.1 atm, which corresponds to a temperature of 121°C. The time of exposure of the sterilized material in autoclave at 121°C must be from 10 to 20 min.

Sterilization by Electromagnetic Radiation

Electromagnetic irradiation such as microwaves, ultraviolet (UV) radiation, x-rays, gamma rays, and electrons are also used to sterilize materials. UV irradiation, which does not penetrate solid, opaque, or light-absorbing materials, is useful for disinfection of surfaces, air, and water. Gamma radiation and x-rays, which are more penetrating, are used in the sterilization of heat-sensitive biomedical materials.

Sterilization by Filtration

Sterilization by filtration of cells and viruses is performed by filtration of water through polymer membranes with a defined diameter of pores smaller than the cell size. Thus, microorganisms are trapped on the surface of the membrane. Sterilization of liquids by membrane filtration preserves sensitive biological substances such as DNA, RNA, proteins, antibiotics, and vitamins that are easily inactivated by heat.

Bulk filtration is adsorption on the fibers and is used mainly for the removal of fungi spores, bacterial cells, and virus particles from air.

Conservation

Conservation is the prevention of microbial spoilage of water, water sample, food, and other materials and samples due to

1. Low storage temperature
2. Low pH
3. Drying
4. Addition of salt or organic substances to decrease water activity
5. Addition of organic solvent (ethanol)
6. Formation of anaerobic conditions

Disinfectants and Antiseptics

Disinfectants are chemical antimicrobial agents that are used on inanimate objects. Typical disinfectants are chlorine gas, chloramine, ozone, and quaternary ammonium compounds.

Antiseptics are chemical antimicrobial agents that are used on living tissue. Typical antiseptics are iodine, 3% solution of hydrogen peroxide, and 70% solution of ethanol.

Antibiotics

Antibiotics are microbially or chemically synthesized substances that are used at a low concentration to treat infectious diseases, killing pathogenic microorganisms. The mode of action is specific for each antibiotic. The action of antibiotic is usually based on the inhibition of activity of a specific enzyme. There are thousands of known antibiotics but only a hundred are applicable in medicine because of the toxic effects of antibiotics on humans. Antibiotics are used in environmental engineering to select specific strains of microorganisms and in the construction of recombinant strains.

Disinfection

The ultimate technology to ensure bacteriological quality of water is disinfection. Disinfection is the chemical or physical treatment of water or wastewater-treatment plant effluent by strong oxidants such as chlorine, chloramines, chlorine dioxide, ozone, ferrate, or by UV, with the aim of diminishing the concentration of pathogenic microorganisms and viruses to some level, defined in the standards of water quality.

The rate of cell death during disinfection ideally should follow first-order kinetics:

$$\frac{dX}{dt} = -kt \quad \text{or} \quad \ln\left(\frac{X}{X_o}\right) = -kt, \tag{13.3}$$

where

X and X_o are the final and initial numbers of living cells
t is the time of exposure
k is the constant of decay that depends on the conditions of the disinfection

However, the actual kinetics of disinfection, the order of the equation, and the constant of decay change during the disinfection process due to the presence of microorganisms with varying degrees of resistance to the disinfectant.

Resistance of Different Microbial Groups to Disinfection

Generally, the resistance of microbes to disinfection follows this order: vegetative bacteria (most sensitive group) → viruses → spore-forming bacteria → protozoan cysts (most resistant).

Test (surrogate) microorganisms can be used in some cases to study water disinfection kinetics and to compare different disinfectants and regimes of disinfection instead of the pathogens. Indicator microorganisms are used often as test microorganisms. For example, cells of *E. coli* can be used as test organisms in the disinfection of media containing pathogenic enterobacteria. However, the correct approach to evaluate efficiency of disinfection is to use the most resistant surrogates, for example, bacterial endospores or protozoan cysts.

Comparison of Chemical Disinfectants

Disinfectants can be compared under the same conditions and microorganisms by k, a constant of decay, or other technical or economical parameters. A benefit of chlorination is that chlorine residue remains in the water during distribution, which protects against recontamination. An undesirable side effect is that chlorination forms trihalomethanes, some of which are suspected carcinogens, and bad-smelling chlorophenols.

The efficiency of disinfection by chlorine gas decreases with pH because non-dissociated hypochlorous acid (HOCl) is more active than hypochlorite ion (OCl^-).

Chloramines (NH_2Cl and $NHCl_2$) are weaker disinfectants than chlorine but are effective in the control of microbial biofilms with exopolysaccharide matrix.

Ozone is more effective against viruses and protozoa than chlorine, but it is more expensive because it is generated by electrical discharge in a dry air stream at the site of application. Another disadvantage is that there is no residual antimicrobial activity after ozonation. Therefore, ozonation and chlorination are commonly used in sequence.

UV Disinfection of Water

UV radiation at a wavelength of 260 nm damages microbial DNA. Microbial inactivation is proportional to the UV dose. Humic substances, phenolic compounds, and suspended solids interfere with UV transmission. Microbial cells can be reactivated by repairing DNA damages due to exposure of cells to visible light. Therefore, UV-treated water should not be exposed to light during storage.

Chemical Interference with Disinfection of Water

Ferrous and manganese ions, nitrites, sulfides, and organic substances reduce the concentration of oxidizing disinfectants, thus reducing the inactivation of microorganisms during disinfection.

Organic material in raw water occurring as naturally derived humic and fulvic acids, also contributes to water coloration. Humic and fulvic acids, reacting with chlorine, produce chloroform ($CHCl_3$) during disinfection. Therefore, these substances must be removed or pre-oxidized before disinfection.

Physical Interference with Disinfection of Water

Adsorption of cells to the particles of clay, silt, iron hydroxides, as well as aggregation of cells, their encapsulation in slime, in macroorganisms, or covering of microorganisms by the sheath, significantly reduce inactivation of microorganisms and viruses during disinfection because of a steep gradient of oxidant in aggregates and its low concentration in cells. Therefore, particles and aggregates must be removed from water by coagulation and filtration or aggregation prior to disinfection.

Organisms may aggregate together. The organisms that are on the interior of the aggregate are shielded from the disinfectant and are not inactivated. Organisms may also be physically embedded within larger organisms such as nematodes. So, to disinfect water adequately, the water is treated by coagulation and filtration to reduce the concentration of solid materials to an acceptably low level.

14

Biotechnological Processes

Biotechnology

Biotechnology deals with the use of microorganisms or their products in large-scale industrial processes. Microbial biotechnology is used in agriculture, food industry, production of materials, in mining, in pharmaceutics and medicine, in energy production, and in the maintenance of clean and sustainable environments (Figure 14.1).

Environmental Biotechnology

Environmental biotechnology is the application of biotechnology for monitoring, protection, and remediation of environment. It includes

- Water biotreatment
- Wastewater biotreatment
- Biotreatment of solid wastes
- Soil bioremediation
- Biotreatment of polluted air and exhaust gases
- Prevention of infectious diseases
- Environmental monitoring using biosensors and bioindicators

Comparison of Biotechnological Treatment and Other Methods

Pollution of water, soil, solid wastes, and air can be prevented and the pollutants can be removed by physical, chemical, physicochemical, or biological (biotechnological) methods. The advantages of biotechnological treatment of wastes are as follows:

- Biodegradation or detoxication of a wide spectrum of hazardous substances by natural microorganisms

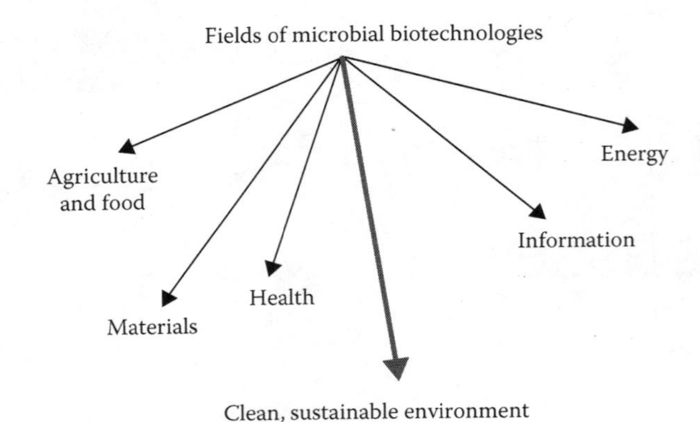

Figure 14.1 Applications of microbial biotechnology.

- Availability of a wide range of biotechnological methods for complete destruction of hazardous wastes
- Diverse set of conditions, which are suitable for biotechnological methods
- Microorganisms are "self-replicating reagents"

The disadvantages of biotechnological treatment of wastes are as follows:

- Nutrients and electron acceptors must be added to intensify the biotreatment.
- Optimal conditions must be maintained in the treatment system.
- There may be unexpected or negative effects of applied microorganisms, such as emission of cells, odors or toxic gases during the biotreatment, presence or release of pathogenic, toxigenic, or opportunistic microorganisms in environment.
- There may be unexpected problems in the management of biotechnological system because of the complexity and high sensitivity of biological processes.

Applicability of Environmental Biotechnology

Biotechnology can be applied for biotreatment of polluted environment or for waste treatment in the following cases:

- Technically and economically reasonable rate of biodegradability or detoxication of substances during biotechnological treatment
- Big volume of treated wastes
- Low concentration of pollutant in water or waste
- Ability of natural microorganisms to degrade substances
- For better public acceptance of environmental biotreatment

Combination of Biotechnology with Other Methods

The efficiency of application of an actual biotechnological method depends on its design, process optimization, and cost minimization. Many failures have been reported on the way from bench laboratory scale to full-scale biotechnological treatment because of instability and diversity of both microbial properties and conditions in the treatment system. In many cases, a combination of biotechnological and chemical or physical treatments may be more efficient than a single type of treatment.

Bioprocesses Used in Environmental Biotechnology

The following bioprocesses for transformation of substrate S into product P are used in environmental biotechnology:

$$\text{Growth} \quad S \rightarrow P\ (\text{biomass})$$

$$\text{Oxidation} \quad S - e^- \rightarrow P$$

$$\text{Reduction} \quad S + e^- \rightarrow P$$

$$\text{Degradation} \quad S\ (\text{organics}) \rightarrow CO_2$$

$$\text{Hydrolysis} \quad (S)_n + nH_2O \rightarrow P\ (\text{monomer})$$

$$\text{Biosynthesis} \quad nS \rightarrow P\ (\text{polymer})$$

$$\text{Biorecognition} \quad S + E\ (\text{enzyme}) \rightarrow [SE]\ (\text{substrate-enzyme complex})$$

Stages of Biotechnological Process

Any biotechnological process includes

- Preliminary step (upstream processes)
- Biotreatment step (core process)
- Posttreatment step (downstream processes)
- Monitoring and control

Upstream Processes in Environmental Biotechnology

Typical upstream processes include

- Pretreatment of waste
- Preparation of medium or waste for treatment
- Preparation of equipment
- Preparation of inoculum

Pretreatment in Biotechnology

Efficient pretreatment schemes, used prior to biotechnological treatment of wastes, include

- Crushing, grinding, and homogenization of the particles of solid wastes
- Homogenization (mechanically or by ultrasound) of suspended hydrophobic liquid (hydrocarbons)
- Chemical oxidation of hydrocarbons by hydrogen peroxide or ozone
- Chemical treatment with alkali or acids
- Preliminary washing of wastes by surfactants
- Thermal pretreatment
- Freezing pretreatment

Medium Preparation

A medium (pl. *media*) is an artificial environment for the cultivation of microorganisms in the form of solution, suspension, or solid matter. In environmental engineering, a medium can be a waste, which is biologically treated.

Preparation of the medium includes

- Mixing
- Concentrated nutrient additions
- Microelement additions
- pH and oxidation–reduction potential (ORP) adjustment

Additionally, there may be

- Addition of indicators of medium quality (pH, ORP, sterilization indicators)
- Sonication of medium
- Pre-oxidation
- Thermal pretreatment
- Disinfection
- Sterilization
- Conservation

To store the medium it can be cooled, frozen, dried, pasteurized, or sterilized.

Components of Medium

Major elements in the microbial cell are C, H, O, and N. The average content of elements in biomass may be shown by the empirical formula $CH_{1.8}O_{0.5}N_{0.2}$. There are macronutrients (C, H, O, N, P, S, K, Na, Mg, Ca, and Fe) and micronutrients (Cr, Co, Cu, Mn, Mo, Ni, Se, W, V, and Zn).

Growth factors are organic compounds such as vitamins, amino acids, and nucleosides that are required by some strains called auxotrophs. Prototrophic strains do not require growth factors.

A defined media is a mixture of pure mineral salts and organic substances. A medium with undefined content of substances is called a complex medium. The organic and inorganic substances are added as digests or extracts of natural ingredients.

Preparation of Equipment

Equipment preparation includes cleaning, washing, filling, sterilization (if needed), adjustment of temperature, aeration rate, stirring rate, and checking of the equipment and conditions.

Preparation of Inoculum

The organisms used to start up the core process are called inoculum by microbiologists or "seeds" by engineers. The inoculum could be a suspended, frozen, dried, or cooled biomass. The cultivation of inoculum is performed in batch or continuous processes. The volume of inoculum must range from 5% to 20% of the volume of the bioreactor in core process to the start-up period of batch culture.

Core Process in Environmental Biotechnology

Core process is the cultivation of microorganisms coupled with the biotreatment of liquid, solid, or gaseous wastes. It is performed in bioreactors. The cultivation is performed usually together with the selection of a population or enrichment culture.

The cultivation of microorganisms is performed at suitable conditions, usually at optimal temperature, pH, osmotic pressure, and concentration of gases (oxygen, carbon dioxide, hydrogen), on a semisolid or liquid medium containing all necessary substances for the growth of the strain.

The elemental composition of biomass can be shown approximately by the formula $CH_{1.8}O_{0.5}N_{0.2}$ but half of the known elements are used in the synthesis of microbial biomass and must be present in the medium. Suitable conditions and essential substances for the growth of some strains are not known yet and the cultivation of these microorganisms has not been successful to date. Additionally, some microorganisms live in strong symbiotic or parasitic relationships with other micro- or macro-organisms that cannot be cultivated apart from these organisms. Therefore, not all microorganisms can be isolated and cultivated.

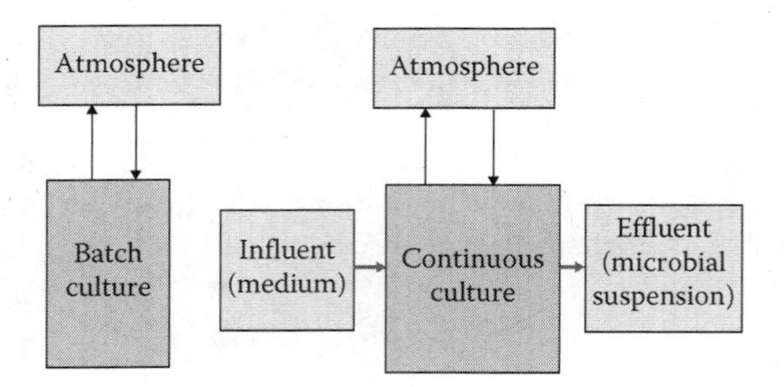

Figure 14.2 Semi-closed and open systems of cultivation.

Semi-Closed and Open Systems of Cultivation

The microorganisms are cultivated in semi-closed (exchanges with the environment only by the gases) and open systems (exchanges with environment by liquids and gases). Semi-closed system of cultivation is called batch culture and open system is called continuous culture (Figure 14.2).

Batch Culture

There is usually a supply of air, release of gaseous products, additions of acid or alkali to maintain pH, and addition of antifoam substance during batch cultivation. Due to exhaustion of nutrients and accumulation of biomass and metabolites, the following sequence of the phases is typical in batch culture:

1. Lag-phase or phase of cells adaptation and self-control of environment; its duration depends on the concentration of inoculated biomass and magnitude of the difference between previous and current conditions of cultivation.
2. Short log-phase or phase of exponential growth.
3. Transitional period between log-phase and stationary phase.
4. Stationary phase is characterized by slow growth or its absence due to exhaustion of nutrients or accumulation of metabolites.
5. Death phase is characterized by an increasing number of dead cells and their lysis.

Continuous Cultivation

Microbial continuous culture is open for exchange by gases and liquids. There is a large diversity of aerobic and anaerobic bioreactors for continuous cultivation, for example:

1. Bioreactors of complete mixing; the most common type is a chemostat where the dilution rate, which is a ratio between flow rate and working volume of the reactor, is maintained constant.

2. Plug-flow systems.
3. Consecutively connected bioreactors of complete mixing, which form a plug-flow system.
4. Fixed biofilm reactor or retained biomass reactor with the flow of medium through it; biomass is retained in the reactor due to adhesion, sedimentation, or membrane filtration.
5. Complete mixing or plug-flow continuous cultivation with recycling of microbial biomass.
6. Semi-continuous and sequencing batch cultivation, which is a continuous cultivation with the periodical addition of nutrients and removal of suspension.
7. Cultivation in membrane bioreactor, where biomass is retained in reactor by porous membrane.

Chemostat

Most common type of continuous culture is a chemostat. To perform continuous cultivation in chemostat the dilution rate (D), which is the ratio between flow rate (F) and working volume of the reactor (v), is maintained constant. If $d \leq \mu_{max}$, the stability of the system is maintained due to feed-back interactions between specific growth rate (μ), concentration of substrate (S), and biomass concentration (X) (Figure 14.3):

$$\frac{dX}{dt} = +f_1(m)$$

$$\frac{dS}{dt} = -f_2(X)$$

$$\frac{dm}{dt} = +f_3(S)$$

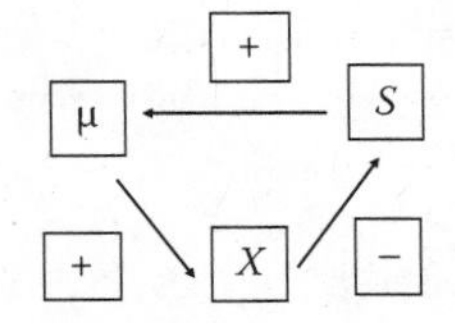

Figure 14.3 Interactions between specific growth rate (μ), concentration of substrate (S), and biomass concentration (X). Signs mean positive or negative effects of one parameter on another.

Plug-Flow System

Plug-flow system of continuous cultivation is also used in environmental engineering for deep biological purification of wastes. The parameters of this system are constant in time but not along the length of the reactor (Figure 14.4a). The series of the complete mixing reactors may be considered as plug-flow system of cultivation (Figure 14.4b).

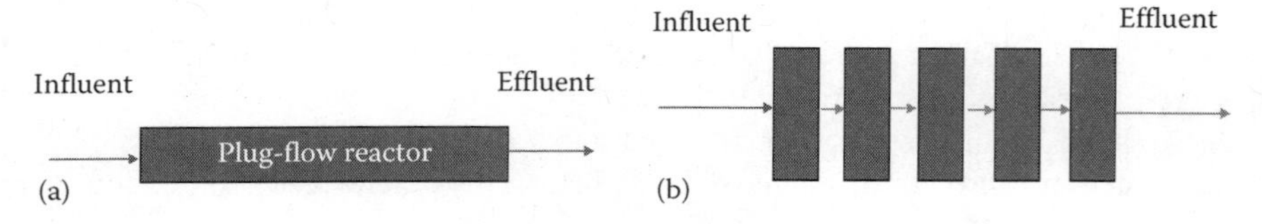

Figure 14.4 Plug-flow systems of continuous cultivation. (a) one plug-flow reactor; (b) the series of complete mixing reactors, which can be considered as plug-flow reactor.

Figure 14.5 (See color insert following page 292.) The industrial-scale reactor used in environmental biotechnology.

Suspended Biomass Bioreactors

Chemostat and plug-flow cultivation is usually performed in the bioreactors with suspended biomass. Volume of these bioreactors varies from 100 mL to several thousand m^3. The most popular reactors in industrial biotechnology are aerobic and anaerobic fermentors of complete mixing. The reactor may contain various devices to monitor and control the conditions. The reactors used in environmental biotechnology are constructed usually as simple aeration tank where mixing is performed due to aeration (Figure 14.5). The cultivation is performed without sterilization of wastewater and with less number of controlled parameters than in the industrial biotechnology.

Continuous Systems with Internal Recycle or Retention of Biomass

The retention or recycle of biomass in the reactor (Figure 14.6) in continuous process ensures following properties:

- The flow rate can be higher than specific growth rate of microorganisms, so slow growing microorganisms can be maintained in the reactor.
- The concentration of biomass and biodegradation rate in the reactor can be higher than that in the chemostat.

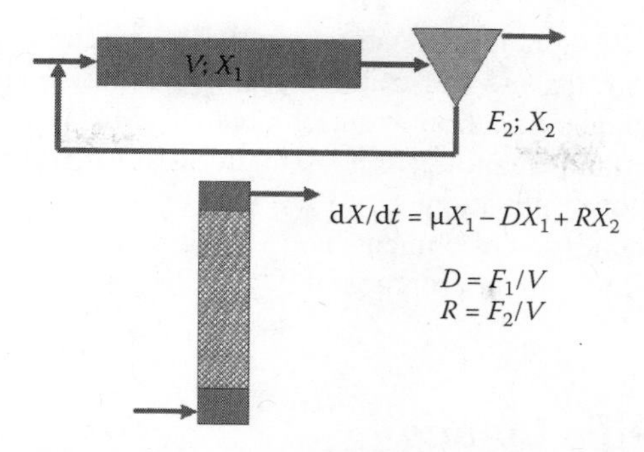

Figure 14.6 (See color insert following page 292.) Continuous systems with internal recycling or retention of biomass.

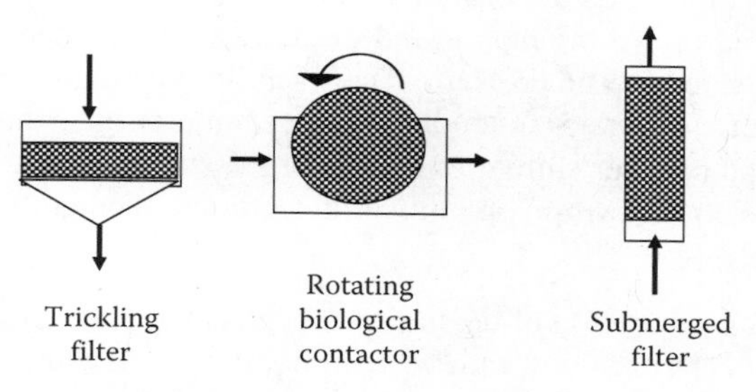

Figure 14.7 Types of biofilm reactors.

Fixed biofilm reactors can retain almost all biomass. In a fixed-film biological process, the microorganisms are attached to a solid surface. The support material can include gravels, stones, plastic, sand, or activated carbon particles.

Biofilm reactors include trickling filters, rotating biological contactors (RBC), and submerged filters (Figure 14.7). These reactors are used for oxidation of organic matter, nitrification, denitrification, and anaerobic treatment of wastewater.

A membrane bioreactor (MBR) retains biomass due to filtration of medium through a porous membrane inside or outside the reactor.

A sequencing batch reactor (SBR) retains biomass due to periodical settling of biomass followed with removal of liquid. It is a semi-continuous cultivation with the periodical settling of biomass, removal of liquid (effluent), and addition of medium (wastewater).

Microbiological Methods Used in Environmental Engineering

Specific microbiological methods are used to study microorganisms in environmental engineering systems:

1. Isolation, cultivation, identification, and quantification of pure cultures
2. Selection of strains and construction of recombinant microbial strains
3. Selection and quantification of enrichment cultures
4. Identification and quantification of microorganisms in environmental samples without cultivation
5. Extraction, cloning, enrichment, and identification of microbial genes and their products in environmental engineering systems

Isolation of Pure Culture

Isolation of pure culture (of a microbial strain) is usually performed by spreading a diluted microbial suspension on a Petri dish with a semi-solid medium to produce 10–50 colonies on the dish after several days of cultivation. Cells of one colony are picked up for the next round of cultivation on a semi-solid or liquid medium. Usually, there is no problem of cultivability for environmental engineering strains because the major function of selected microorganisms is biodegradation of different natural or similar to natural organic compounds.

Instead of isolation of strain on semi-solid medium other methods can be used, for example:

1. Mechanical separation of big microbial cells by micromanipulator
2. Sorting of cells or microbeads with immobilized cell using sorter of flow cytometer
3. Magnetic or immunomagnetic separation
4. Cell dielectrophoresis
5. Cell chromatography

Selection of Microorganisms

Microbiological methods that are used for selection of microorganisms needed for the performance of environmental engineering processes are as follows:

1. Selection and isolation of pure culture
2. Selection of enrichment cultures
3. Selection of artificial ecosystem

Selection is the screening of microbial variants with specific desirable characteristics within the population of one strain. These variants may include

1. Faster or more efficient growth (positive selection)
2. Faster or more efficient biochemical function (positive selection)
3. Slower or less efficient biochemical function (negative selection)
4. Better survival under harmful conditions (positive selection)
5. Weaker resistance to some factors of environment (negative selection)

The differences between variants are caused by natural spontaneous mutations, i.e., changes in the DNA sequences of genes. Mutagenic chemicals, ultraviolet rays, and ionizing radiation are used to increase the rate of mutagenesis and to increase the probability of desirable variant formation. The screening of the desirable variant can be replaced by the creation of selection pressure, i.e., conditions favorable for growth, survival, or development of desirable variant. Therefore, this variant will be accumulated in a microbial population and can be detected during the screening.

Theoretical Selection

The theoretical selection of a microbial group that is most suitable for the specific environmental engineering process is the first step of selection. This selection is used for the specification of conditions that are most suitable for the experimental selection of microorganisms for the defined environmental engineering process. However, there are often possible multiple choices of suitable microbial groups and, respectively, multiple choices of conditions for the experimental selection of microorganisms.

Selection of Enrichment Culture

Selection of enrichment culture or autoselection refers to the selection of the population with one dominant strain or one dominant microbial community, which is accumulated in the system of cultivation due to the preferred conditions (selection pressure) for this strain or community. Enrichment cultivation is often used in environmental engineering to select microorganism(s) capable of particular metabolic transformations.

Selective Pressure

Selective conditions (selection pressure) for the production of enrichment culture are as follows:

1. Source of energy
2. Source of carbon
3. Sources of nitrogen and phosphorus
4. Temperature
5. pH
6. Concentration of heavy metals
7. Presence of specific antibiotic in a medium
8. Concentration of dissolved oxygen
9. Osmotic pressure of a medium
10. Spectrum and intensity of light
11. Settling rate, etc.

Mechanisms of Autoselection

The mechanisms of autoselection of enrichment culture are as follows:

1. Faster or more efficient growth of one or several strains (positive growth-related autoselection)
2. Faster or more efficient some biochemical functions of one or several strains (positive metabolic autoselection)
3. Slower or less efficient growth of one or several strains (negative growth-related autoselection)
4. Slower or less efficient biochemical function (negative metabolic autoselection)
5. Better survival under harmful conditions (positive survival-related autoselection)
6. Weaker resistance to some factors of environment (negative survival-related autoselection)
7. Stronger or more specific adherence of cells to surface (positive or negative cell adherence-related autoselection)

Instability of Autoselected Features

Autoselected features of the enrichment culture can be genetically unstable and could disappear after several generations of cells when the selection pressure will be absent. This is known for cell surface hydrophobicity, which can be enhanced several times by the retention of cells on hydrophobic carrier during several cell generations, or can disappear during several cell generations if there will be selection pressure in medium. This feature is important in cases of microbial remediation of oil spills, where cells must float at the water surface, or in the case of formation of microbial granules, which can replace conventional flocs of activated sludge in municipal wastewater treatment.

Selection of Ecosystem

Selection of artificial microbial ecosystem is similar to the selection of enrichment culture, but there may be several alternative or changed selective factors (selection pressures) ensuring dominance of several microbial communities with different, even alternative, physiological functions. For example, in selected artificial microbial ecosystems, there may be simultaneously aerobic and anaerobic microbial communities, which were selected due to the presence of both aerobic and anaerobic conditions in the environmental engineering system. The set of the mechanisms of selection and the selective pressures is a specific factor for the selection of an artificial microbial ecosystem. Additional mechanisms of selection of an artificial ecosystem can be positive or negative interactions between the microbial communities of the ecosystem: commensalistic, mutualistic, amensalistic, antagonistic, or parasitic relationships.

Construction of Genetically Engineered Microorganisms

Microorganisms, suitable for the biotreatment of wastes, can be isolated from the natural environment. However, their ability for biodegradation can be modified and amplified by artificial alterations of the genetic (inherited) properties of these microorganisms.

Natural recombination of the genes (units of genetic information) occurs during DNA replication and cell reproduction, and includes the breakage and rejoining of chromosomal DNA molecules (separately replicated sets of genes) and plasmids (self-replicating mini-chromosomes containing several genes).

Steps of Artificial Recombination of DNA

Recombinant DNA techniques or genetic engineering can create new, artificial combinations of genes, and increase the number of desired genes in the cell. Genetic engineering of recombinant microbial strains for biotreatment involves the following steps:

1. DNA is extracted from cell and is cut into small sequences by specific enzymes.
2. Small sequences of DNA can be introduced into DNA vectors.
3. A vector, which is a virus or plasmid, is transferred into the cell and self-replicated to produce multiple copies of the introduced genes.
4. Cells with newly acquired genes are selected based on activity (e.g., production of defined enzymes, biodegradation capability) and stability of acquired genes.

Applications of Genetic Engineering in Environmental Engineering

Genetic engineering of microbial strains can create the ability to biodegrade xenobiotics or amplify this ability through the amplification of related genes.

Another approach is the construction of hybrid metabolic pathways to increase the range of biodegraded xenobiotics and the rate of biodegradation. The desired genes for biodegradation of different xenobiotics can be isolated and then cloned into plasmids. Some plasmids have been constructed containing multiple genes for the biodegradation of several xenobiotics simultaneously. The strains containing such plasmids can be used for the bioremediation of sites heavily polluted by a variety of xenobiotics.

The main problem in environmental engineering applications is maintaining the stability of the plasmids in these strains. Other technological and public concerns include the risk of application and release of genetically modified microorganisms in the environment.

Physical Boundary of Artificial Ecosystem

An important element of selected ecosystem is the boundary between an ecosystem and its surrounding environment, which could be a steep gradient of physical or chemical properties. The physical boundary is formed by an interface between solid and liquid phases, solid and gas phases, and liquid and gas phases. For example, the microbial ecosystem of an aerobic tank for wastewater treatment is separated from the environment by the reactor walls and air–water interfaces. The boundaries of this ecosystem are as follows:

1. Side walls of the equipment with a fixed microbial biofilm
2. Bottom of the equipment with the sediment of microbial aggregates
3. Gas–liquid interface with accumulated hydrophobic substances (lipids, hydrocarbons, aromatic amino acids) and cells or aggregates with high hydrophobicity of their surface or cells and aggregates containing gas vesicles

Chemical Boundary of Artificial Ecosystem

The steep gradient of chemical substances, for example, oxygen, ferrous ions, hydrogen sulfide, etc., forms a chemical barrier. Such barriers separate, for example, aerobic and anaerobic ecosystems in a lake. The steep gradient of conditions can be also created by cell aggregation in flocs, granules, or biofilms. The main function of the boundary is to maintain integrity of an ecosystem by controlled isolation from the environment and to protect an ecosystem from the destructive effects of the environment. A defined ecosystem cannot exist, cannot be selected, and maintained without a defined boundary.

Macro- and Microenvironments

A microenvironment is characterized by gradients of parameters that change at the micrometer scale and that are determined by the microbial activity. The typical scale of such microbial microenvironments as aggregate, biofilm, or microbial mat is between 0.1 and 100 mm. An artificial microenvironment is created due to adhesion of cells on the carrier or cell incorporation into the carrier. Cells can be concentrated not only on liquid–solid interface but also on liquid–gas interface because the nutrient concentrations are higher there than in the bulk of liquid. Macroenvironment is characterized by the gradients of parameters that are changing in meters-scale distance, for example, in bioreactor or in natural ecosystem.

Effects of Nutrients on Growth Rate

Kinetic limitation means that specific growth rate depends on the concentration of limiting nutrient. Usually, it is one limiting nutrient but there may be simultaneous limitation by some nutrients. If one nutrient limits the specific growth rate, this dependency is often expressed by Monod's equation:

$$\mu = \mu_{max}\left[\frac{S}{S + K_S}\right]$$

where
μ_{max} is maximum of specific growth rate
S is concentration of the nutrient (substrate) limiting growth rate
K_S is a constant

However, there are many other known models describing μ as $f(S_i)$. For example, a double limitation of μ by donor of electrons and oxygen is typical for the cases when the initial step of catabolism is catalyzed by oxidase or oxygenase incorporating atom(s) of oxygen into a carbon molecule or energy source.

Downstream Processes

Typical downstream processes include

- Separation of biomass from liquid
- Concentration of biomass
- Value-added by-production
- Drying/dewatering of biomass
- Packing/disposal of secondary waste

Microbial Aggregates

A multicellular aggregate is formed and separated from its surrounding environment due to

- Aggregation by hydrophobic force, electrostatic interactions, or salt bridges
- Loose polysaccharide or inorganic matrix (iron hydroxide as example) combining the cells altogether by mechanical embedding, chemical bonds, hydrogen bonds, electrostatic forces, or hydrophobic interactions

- Formation of mycelia, which is a net of branched cell filaments
- Polysaccharide matrix with a filamentous frame
- Coverage by a common sheath of organic (polysaccharides, proteins) or inorganic origin (iron hydroxide, silica, calcium carbonate); the common sheath can be made also from dead cells of aggregate ("skin" of microbial aggregate)

Structure of Microbial Aggregate

Usually, the matrix of aggregate is structured with the layers or sub-aggregates. Therefore, a microbial aggregate can be considered as a multicellular organism because its parts have some extent of coordination and synchronization of physiological functions, i.e., synchronous growth, motility, sexual interactions, assimilation of atmospheric nitrogen, production of extracellular polysaccharides, transport and distribution of nutrients, oxidation of electron donors, and reduction of electron acceptors.

Sedimentation of Cellular Aggregates

The sedimentation velocity (V) of cells and cell aggregates with diameter D and density d_p can be described by the Stokes' law:

$$V = D^2 \frac{(d_p - d_l)}{18 \mu g} \qquad (14.1);$$

where

d_l is liquid density
μ is the viscosity of liquid
g is gravitational acceleration

Because of the small size of bacterial cells, their sedimentation rate is too small to be used in practice. Therefore, microbial aggregates are selected during cultivation to ensure separation of biomass from water using sedimentation. Three major types of suspended microbial aggregates with sizes from 50 μm to several mm are used in environmental engineering: (1) floc, which is a suspended aggregate of irregular shape that is not dense, and (2) granule, which is a dense suspended aggregate of regular shape.

So, the selection of microorganisms able to form aggregates can be performed by settling rate or filtration. A simple selection of a microbial strain that is able to form cellular aggregates can be performed on Petri dish based on the size of the colonies of pure culture. The colonies of the biggest size were most probably originated not from one cell but from the aggregate of several cells.

Selection of Microbial Aggregates by Settling Time

Settling of aggregate for 20–30 min and retaining or recycling of this settling aggregate in a bioreactor is used for the selection of flocs. Intensive aeration for the mechanical compaction of aggregate by air bubbles, settling of aggregate for 2 min, and retaining of this settling aggregate in bioreactor is used for the selection of granules.

15

Aquatic Systems and Water Biotreatment

Microorganisms of Hydrosphere, Lithosphere, and Atmosphere

Microorganisms live in the hydrosphere (seawater, freshwater, groundwater), lithosphere (soil, rocks), and survive in atmosphere. Bacterial mass on the earth amounts to $10{,}000 \times 10^9$ metric ton. For comparison, the mass of *Homo sapiens* (human beings) on earth is 0.3×10^9 metric ton.

Microorganisms live in the hydrosphere at the water surface, to a depth of 11,000 m. The major biotope of microorganisms in lithosphere is soil, the surface layer of lithosphere to a depth of 0.3 m. Microorganisms do not live in the atmosphere, but the spores of bacteria and fungi can be found up to a height of 70,000 m from the earth's surface.

Functions of Microorganisms in Hydrosphere

Microorganisms in hydrosphere perform the following functions:

- Biogeochemical cycling of carbon and other elements, i.e., cyclic chemical changes of the compounds such as oxidation/reduction or synthesis/biodegradation
- Mediate physical changes of the compounds such as volatilization, dissolution, and precipitation
- Purification of water in environment by removal of organics, nutrients, and heavy metals
- Assimilation of photo- and chemical energy, i.e., transformation of chemical or light energy into biomass
- Specific accumulation of metabolites, for example, sulfur, iron, and oil deposits
- Conversion of atmosphere from anaerobic to aerobic
- Positive and negative interactions of microorganisms with aquatic plants and animals

All these functions are also used in environmental engineering, in the field, or in industrial engineering systems. All processes that are developed and used in environmental engineering are natural biogeochemical processes.

Biogeochemical Carbon Cycle

The biogeochemical cycle of carbon includes assimilation of CO_2 from atmosphere, for example, by oxygenic photosynthesis

$$CO_2 + H_2O + \text{energy of light} \rightarrow CH_2O \text{ (organic matter)} + O_2$$

and mineralization of organic matter, for example, by aerobic oxidation

$$CH_2O + O_2 \rightarrow CO_2 + H_2O$$

Photosynthetic organisms are called as primary producers. Organic matter produced through photosynthesis is partially transformed into the biomass of plants and then animals. The dead biomass is mineralized after the death of plants and animals by heterotrophic bacteria and fungi.

Biogeochemical Nitrogen Cycle

Nitrogen can occur in a variety of oxidation states such as ammonium (NH_4^+), organic nitrogen (RNH_2), nitrogen gas (N_2), nitrogen oxides (NOx), nitrite (NO_2^-), and nitrate (NO_3^-). Microbiological processes of the nitrogen cycle $N_2 \rightarrow RNH_2—NH_4^+ \rightarrow NO_2^- \rightarrow NO_3^- \rightarrow N_2$ are shown below.

Nitrogen fixation is performed under anaerobic conditions by symbiotic and free-living nitrogen fixing bacteria:

$$N_2 \text{ (from atmosphere)} + [2H = \text{electron donor}] + x\text{ATP} \rightarrow RNH_2 \text{ (organic nitrogen)}$$

Ammonification is a step in the decay of organic compounds:

$$R(COO)NH_2 + \text{chemical energy} + [2H = \text{bioreductant}] \rightarrow RCOOH + NH_3$$

Nitrification is performed by nitrifying bacteria under aerobic conditions in the presence of pH buffer ($CaCO_3$) because of acidification of environment during nitrification:

$$NH_4^+ + 2O_2 \rightarrow NO_3^- + 2H^+ + H_2O$$

$$CaCO_3 + 2H^+ \rightarrow Ca^{2+} + CO_2\uparrow + H_2O$$

Denitrification is performed by denitrifying bacteria under anaerobic conditions:

$$2NO_3^- + [10H = \text{bioreductant, organic, or inorganic substance}]$$
$$\rightarrow N_2\uparrow + 2OH^- + 4H_2O$$

Oxidation of ammonia to nitrite under aerobic conditions can be combined with oxidation of remaining ammonia with produced nitrite under anaerobic conditions. It is the so-called anammox (anaerobic ammonia oxidation) process performed by strictly anaerobic bacteria and used in industry:

$$NH_4^+ + 1.5O_2 \rightarrow NO_2^- + 2H^+ + H_2O$$

$$NH_4^+ + NO_2^- \rightarrow N_2 + 2H_2O$$

Biogeochemical Phosphorus Cycle

The main processes of the phosphorus cycle are as follows:

- Biomineralization of organic phosphorus by hydrolysis with the production of orthophosphate.
- Bioassimilation of phosphate as a component of nucleic acids and phospholipids
- Precipitation of phosphates controlled by pH and by the presence of Ca^{2+}, Mg^{2+}, Fe^{3+}, and Al^{3+} with the formation of insoluble compounds are $Ca_{10}(PO_4)_6(OH)_2$, $Fe_3(PO_4)_2 \cdot 8H_2O$, $AlPO_4 \cdot 2H_2O$.
- Solubilization of phosphates from the complexes with Ca, Mg, Fe, and Al due to the bioproduction of nitric, sulfuric, and organic acids.
- Biological accumulation of intracellular granules of polyphosphate under aerobic conditions and biological release of phosphate from the cells under anaerobic conditions. These processes are used in the removal of phosphate from wastewater. Phosphate is stored in intracellular granules of polyphosphate called volutin.

Biogeochemical Sulfur Cycle

Sulfur can occur in a variety of oxidation states such as sulfide/sulfhydryl (S^{2-}), organic sulfur (RSH), elemental sulfur (S), sulfite (SO_3^{2-}), and sulfate (SO_4^{2-}). Microbiological processes of the sulfur cycle $S^{2-} \rightarrow S \rightarrow SO_3^{2-} \rightarrow SO_4^{2-} \rightarrow S^{2-}$ are shown below.

Aerobic oxidation by acidophilic sulfur-oxidizing bacteria and archaea is accompanied by acidification of medium due to formation of sulfuric acid:

$$H_2S + 2O_2 \rightarrow H_2SO_4$$

$$S + 1.5O_2 + H_2O \rightarrow H_2SO_4$$

This acidification is a reason for biocorrosion, production of acid mine drainage, and leaching of heavy metals from sulfides.

Microaerophilic oxidation by filamentous sulfur-oxidizing bacteria is accompanied by accumulation of liquid sulfur inside cells:

$$H_2S + 0.5O_2 \rightarrow S + H_2O$$

Anaerobic phototrophic oxidation is performed by anoxygenic phototrophic bacteria and accompanied by accumulation of liquid sulfur inside cells:

$$H_2S + CO_2 + \text{light} \rightarrow S + H_2O + CH_2O \text{ (carbohydrates)}$$

Anaerobic mineralization of organic sulfur (RSH, where R is the organic radical) is a process of the decay of organic matter:

$$RSH + [2H = \text{electron donor}] \rightarrow RH + H_2S$$

Assimilative sulfate reduction is performed by all organisms for inclusion of sulfur in organic compounds:

$$SO_4^{2-} + 3[2H] + RH + x\text{ATP} \rightarrow RSH + 2H_2O + 2OH^-$$

Dissimilative sulfate reduction is performed by sulfate-reducing bacteria for the production of biological forms of energy:

$$SO_4^{2-} + 4[2H] \rightarrow H_2S + 2H_2O + 2OH^-$$

Biogeochemical Iron Cycle

The cycle of iron includes oxidation of Fe^{2+} and reduction of Fe^{3+}.

The oxidation of Fe^{2+} under neutral pH is performed by neutrophilic iron-oxidizing bacteria (Figure 15.1):

$$4Fe^{2+} + O_2 + 10H_2O \rightarrow 4Fe(OH)_3 + 8H^+$$

Actually, these microaerophilic bacteria do not oxidize ferrous ions directly but they produce oxygen radicals destroying organic acids, which protect ferrous ions from chemical oxidation by oxygen.

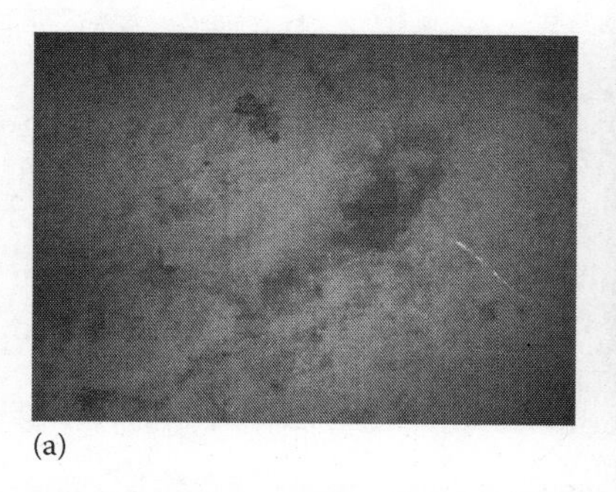
(a)

(b)

Figure 15.1 (See color insert following page 292.) Iron hydroxide precipitates occur naturally in a stream (b) due to oxidation of ferrous chelates by neutrophilic iron-oxidizing bacteria; (a) shows a closeup of the precipitates.

Oxidation of Fe^{2+} under low pH is performed by acidophilic iron-oxidizing bacteria:

$$4Fe^{2+} + O_2 + 4H^+ \rightarrow 4Fe^{3+} + 2H_2O$$

Reduction of Fe^{3+} under anaerobic conditions is performed by iron-reducing bacteria

$$Fe^{3+} + [H = \text{bioreductant}] + OH^- \rightarrow Fe^{2+} + H_2O$$

Stratification in Aquatic Ecosystem

The aquatic environment often develops vertical density gradients due to thermal and geochemical stratification. Geochemically, there are anaerobic, anoxic, and aerobic (photosynthetic) zones in aquatic systems.

The bottom sediments may become anaerobic because organic matter, which is produced through photosynthesis in the surface layer, sinks to the bottom, where heterotrophic bacteria degrade it and consume oxygen. There are productions of CH_4, H_2S, NH_4^+, and Fe^{2+} in the anaerobic zone due to the activity of methanogenic, sulfate-reducing, iron-reducing, and other anaerobic bacteria.

The reduced substances migrate to the upper layer, where they are used as electron donors by anoxic bacteria. These bacteria oxidize products of fermentation such as alcohols, organic acids, and hydrogen using such electron acceptors as nitrate, ferric, and sulfate ions, which are produced in the upper aerobic photosynthetic zone. In the aerobic photosynthetic zone, CO_2 is reduced to organic matter using light energy.

Beneath this zone, there is a microaerophilic zone where methanotrophic, sulfide-oxidizing, nitrifying, and iron-oxidizing bacteria oxidize the remaining products of anaerobic bacterial activity such as methane, as well as sulfide, ammonium, and ferrous ions.

Sources of Water Pollution

The sources of water pollution are as follows:

- *Agriculture*: fertilizers, pesticides, solid wastes, and chemical and biological pollutants from soil
- *Entertainment*: golf courses surrounding reservoirs and lakes—fertilizers and pesticides used to grow grass, which are washed out into reservoirs
- *Farms*: wastewater, manure, and insects
- *Wild life*: biowaste and dead biomass
- *Domestic wastewater*: organic and inorganic pollutants, viruses, and microorganisms
- *Industrial effluent*: organic and inorganic pollutants
- *Solid wastes*: organic and inorganic pollutants if not properly managed and disposed of
- *Air*: sulfur oxide, nitrogen oxide, and dust

Eutrophication of Water in Reservoirs

Eutrophication is nitrogen and phosphorous nutrient enrichment of a water body, derived from human sources, following rapid growth of phytoplankton and benthic plants causes lowered levels of oxygen with the consequent effect of suffocation of higher organisms.

Lakes, rivers, and coastal waters can be polluted with runoff enriched with N and P from fertilizers, sewage, or intensive aquaculture (Figure 15.2). Bloom of phototrophs is a reason of foul smelling odors, bad taste of water, and accumulation of allergens and toxins in water. It also creates problems in water filtration by clogging of the slow sand filters.

Signs of Eutrophication

Eutrophication deteriorates the quality of water. The signs of this deterioration are as follows:

- Many species of cyanobacteria and algae produce toxins that affect animals and humans
- Increased turbidity of water
- Increased concentration of pathogenic microorganisms in water
- Increased concentration of toxic microorganisms, for example, dinoflagellates causing red tide and death of fish in coastal waters
- Formation of surface scum that creates a physical and biological barrier for oxygen transfer from air to water
- Increased concentration of anaerobic microorganisms and their products

Figure 15.2 (See color insert following page 292.) Eutrophication of urban reservoirs.

- Increased concentration of toxic, allergenic, or bad smelling microbial products
- Decreased dissolved oxygen concentration
- Decreased biodiversity in aquatic system

Treatment of Stormwater in Artificial (Constructed) Wetlands

Natural wetlands are major clarifiers of freshwater in the environment due to precipitation, by microbial oxidation/reduction of organic and inorganic substances in bottom sediments in the bulk of water and by biomass attached to plants, as well as consumption of nutrients by wetland plants. Because of the shortage of the arable land, natural wetlands are becoming extinct.

Therefore, artificial models of wetlands with higher efficiency and load than natural ones are constructed to remove pollutants from surface water, agricultural drainage water, runoff, and urban drain water. The major application of constructed wetland is removal of nutrients from stormwater. Usually, constructed wetlands consist of a shallow pond of 0.5–1.0 m depth, with plants typical for natural wetlands, with the input of polluted water routed properly through the wetland, where microorganisms and plants uptake nutrients resulting in water purification.

We proposed to enhance consumption of nutrients in constructed wetland with the use of biofilters (Figure 15.3) that oxidize ammonium and produce ferrous ions to precipitate phosphate.

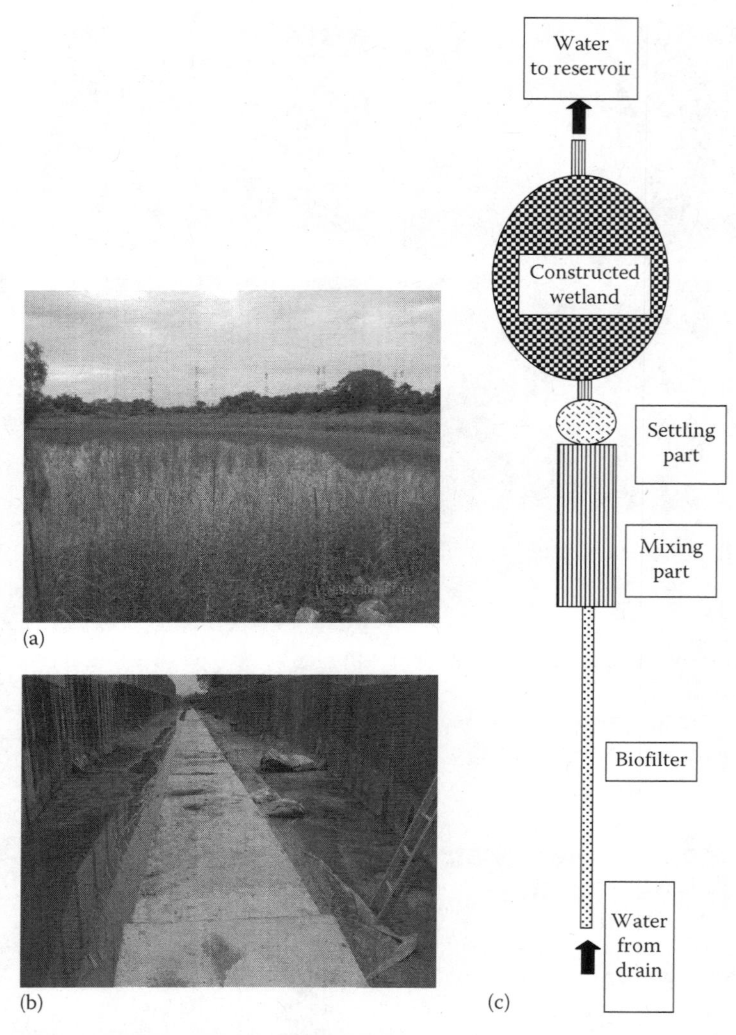

Figure 15.3 (See color insert following page 292.) Combination of (a) constructed wetland with (b) biofilter inside drain. Schematic of this combination is shown in (c).

One of these biofilters treated 520 m/day of dry weather flow and removed more than 50% of P and N from drain water with concentrations of 0.3 mg P/L and 1.1 mg N/L. With the triggering level of P for eutrophication equaling 0.05 mg/L, this biofilter prevents 660,000 m^3 of freshwater from eutrophication annually. The area of constructed wetland enhanced with the biofilter was 500 m^2. To perform the same treatment by conventional constructed wetland, its area must be 5,000 m^2, and the area of natural wetland for the same biotreatment must be 20,000 m^2.

Stages of Raw Drinking-Water Treatment

To ensure good quality of drinking water, the source of this water must be protected from industrial, agricultural, aquacultural, recreational, air-born pollution, and eutrophication. Then, raw water from clean sources of water is conveyed by pipelines to the water treatment plant where it is chemically and biologically treated, filtered, and disinfected. Typically, the major bacteriological characteristic of drinking water is that the total coliform bacteria counts per 100 mL of water determined after incubation in special medium at 35°C for 24 h must be lower than 1.

Pollutants of Water That Can Be Removed by Biotreatment

The following water quality parameters can be improved using biotechnological methods:

- Concentration of nitrate, NO_3^-
- Concentration of iron (Fe^{2+}, Fe^{3+}, organic iron)
- Concentration of manganese (as Mn^{2+} or Mn^{4+})
- Concentrations of other heavy metals (As, Cd, Cu, Pb, Zn)
- Biological instability
- Concentration of organic micropollutants

Biological Instability of Water

Biological treatment is most widely used to remove biodegradable electron donors (also called biological instability) from water. These electron donors include biodegradable organic matter (BOM), ammonium, ferrous iron, manganese (II), and sulfides. The presence of small quantities of these biologically instable components can foster the growth of bacteria in water distribution systems.

Additional biological instability of water may be generated during water distribution because decay of monochloramine (NH_2Cl) produces NH_4^+, or corrosion of Fe^0 in the pipes produces Fe^{2+}. The growth of nitrifying (NH_4^+-oxidizing) or iron-oxidizing bacteria is accompanied by the release of bacterial organics in water, which promotes the growth of heterotrophic bacteria.

Chemical Removal of Biological Instability

The common approach for removal of biological instability is to maintain substantial chlorine residuals in the distribution systems to prevent growth of bacteria using energy of reduced substances. However, the presence of chlorine residuals can generate significant concentrations of disinfection by-products, e.g., chloroform, chloroacetic acid, trihalomethanes, which have known or suspected health risks. If bacterial growth in the pipelines cannot be suppressed with high chlorine residuals, or if relatively high concentration of disinfection by-products must be prevented, the distributed water must be made biologically stable by oxidation of electron donors before supply into the distribution network.

Biological Removal of Biological Instability

Bacteria can oxidize all components of biological instability:

- BOM $CH_2O + O_2 \rightarrow CO_2 + H_2O$
- Ammonium $NH_4^+ + 2O_2 \rightarrow NO_3^- + 2H^+ + H_2O$
- Ferrous iron $4Fe^{2+} + O_2 + 4H^+ \rightarrow 4Fe^{3+} + 2H_2O$
- Manganese (II) $2Mn^{2+} + O_2 + 4H^+ \rightarrow 2Mn^{4+} + 2H_2O$
- Sulfides $H_2S + 2O_2 \rightarrow H_2SO_4$

Disadvantages of Bioremoval of Biological Instability

Bacteria that grow due to the oxidation of water instability components can produce the following negative effects:

- Increase of heterotrophic plate count (HPC)
- Increase of coliforms concentration
- Increase of water turbidity
- Formation of bad taste and odor in water
- Increase of nitrite concentration
- Corrosion in water distribution system

Biodegradable Organic Matter in Water

Most BOM in water supplies are products of decay of natural biopolymers from dead biomass of Archaea, bacteria, fungi, plants, and animals. This BOM is humic-like organic matter with molecular mass from 200 to several thousand daltons. Ozonation of water oxidizes BOM and produces rapidly biodegradable by-products.

Chemical Oxygen Demand, Biological Oxygen Demand, and Total Organic Carbon

Chemical oxygen demand (COD) is a parameter that measures the organic pollution of wastewater. COD is defined as the amount of oxygen required for chemical oxidation of organic matter in water. The biological oxygen demand (BOD) is defined as the amount of oxygen required for biological oxidation of organic matter in water for 5 (BOD_5) or 20 days (BOD_{20}). The dimension for both the parameters is mg O_2/L, but they provide a measure of the concentration of organic matter.

COD values show not just the quantity of organic matter, but also the energetic value. For example, the chemical oxidation of hydrocarbons

$$C_3H_8 + 5O_2 \rightarrow 3CO_2 + 4H_2O$$

requires 3.64 g O_2/g of MOM, while the chemical oxidation of carbohydrates

$$CH_2O + O_2 \rightarrow CO_2 + H_2O$$

requires 1.07 g O_2/g of MOM.

BOD_5 in water and wastewater are as follows:

- Raw reservoir water, 1–15 mg/L
- Raw sewage, 20–400 mg/L
- Effluent after municipal wastewater treatment plant, 10–100 mg/L
- Industrial wastewater, 1–10,000 mg/L

BOD indicates the quantity of biodegradable organic matter. The ratio of BOD/COD in water or wastewater samples is a measure of biodegradability of water pollutants. It is 1.0 for 100% biodegradable substances and below 0.1 for slow degradable substances.

Total organic carbon (TOC) represents the total organic carbon in a water sample. It is independent of the oxidation state of the organic matter. There are automatic TOC analyzers based on the oxidation of the organic carbon to carbon dioxide and measurement of produced CO_2 by infra-red analyzer.

Measurement of BOM

BOM in water varies from 0.1 (0.26 mg BOD/L) to 20 mg C/L (52 mg BOD/L). BOM measurement involves BOD measurement or batch incubation of water sample with a bacterial inoculum followed by the measurement of bacterial growth by plate count or removed dissolved organic carbon.

BOM can be also measured as BOD, but the lowest limit for BOD measurement is 0.5 mg/L.

Plate count (Figure 15.4) can be converted to a concentration of assimilable organic carbon (AOC) using calibration with water polluted with known organic substance. Water is considered as biologically stable and can be distributed without chlorine residuals if AOC is lower than 10 μg C/L (=26 μg BOD/L).

The treatment goal for AOC of drinking water depends on disinfection practices for the distributed water. In Europe, where many countries use small or zero chlorine residuals, the treatment goal can be below 10 μg/L as AOC. In the

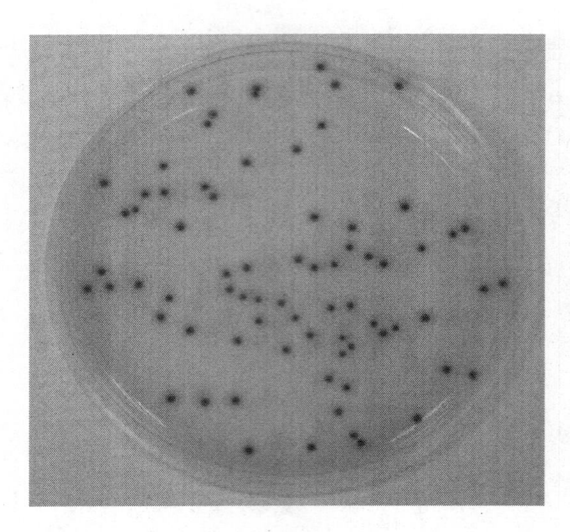

Figure 15.4 Plate count of bacteria in water sample.

United States, where larger chlorine residuals are common, acceptable AOC may be higher.

Bioremoval of Biological Instability

Bioremoval of biological instability is performed by oligotrophic bacteria because concentrations of electron donors are very low. Usually, it is performed in a biofilter because of the short hydraulic retention time (HRT) in this bioreactor due to the high concentration of biomass attached to the surface of the carrier in the biofilter. Bioremoval is performed aerobically to avoid adverse quality impacts of the anaerobic process like bad taste and odor, black color, and solubilization of metals.

Bioremoval of organic matter is performed by oligotrophic aerobic bacteria. Ammonium is oxidized by nitrifying bacteria. Oxidations of organic chelates of Fe^{2+} to Fe^{3+} and Mn^{2+} to Mn^{4+} at neutral pH is catalyzed by specific iron-oxidizing and manganese-oxidizing bacteria that precipitate these metals as hydroxides. Sulfide in water is oxidized by microaerophilic and aerobic sulfur-oxidizing bacteria.

Fixed-Bed Biofilter

Biomass of bacteria is attached to the carrier in this biofilter, for example, expanded clay or expanded polystyrene beads with size from 1 to 4 mm with specific surface from 200 to 1000 m^2/m^3 forming an aerated layer of 2–4 m depth. Such a biofilter can sustain at BOM loads up to 0.15 g BOM/m^2 day and ammonium loads up to 0.40 g N/m^2 day.

These biofilters are used in water biotreatment and designed as rapid or slow filters with HRT of several minutes to several hours, respectively.

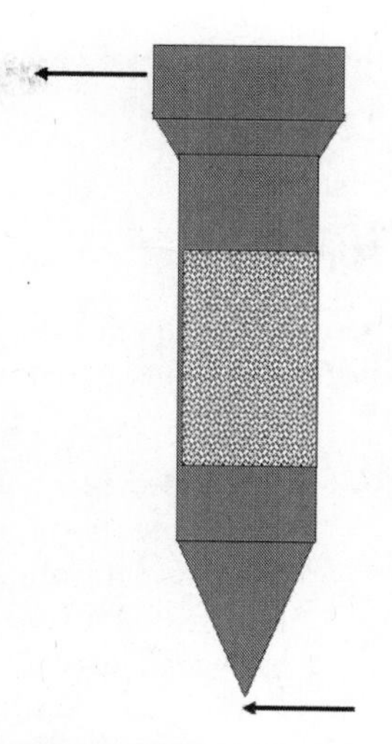

Figure 15.5 A fluidized-bed reactor.

Fluidized-Bed Biofilm Reactor

A fluidized-bed reactor (Figure 15.5) includes a suspended carrier for microbial biofilm, like suspended sand, polymer beads, and floating plastic foam. The advantages of fluidized-bed reactor are intensive mass transfer and absence of clogging of the pores. Such a biofilter can sustain at BOM loads up to 10 g BOM/m^2 day and ammonium loads up to 30 g N/m^2 day.

The biofilm on the carrier in a biofilter gets partially detached and releases microorganisms to the effluent, decreasing water quality, turbidity, and biological stability.

Biofiltration of Water through Soil

Soil contains a lot of microorganisms. So, slow filtration of river water or groundwater through the natural layer of soil or sand for several days could improve water quality due to oxidation of organic and inorganic substances in soil, thus enhancing water stability and diminishing the concentration of pollutants before main treatment and disinfection of water.

Organic Micropollutants of Water

BOM can contain not only natural organic matter but also toxic chemicals or bad-tasting pollutants:

- Aliphatic, monocyclic, and polycyclic hydrocarbons
- Oxygenated derivatives of hydrocarbons
- Halogenated derivatives of hydrocarbons
- Surfactants, for example, alkylbenzenesulfonate and alkylphenol ethoxylates
- Pesticides
- Endocrine disruptors
- Nitrogen compounds
- Biotoxins
- Heavy metals

In the majority of cases, all these micropollutants can be removed from water using biofiltration.

Bioremoval of Endocrine Disruptors from Water

An exposure to some chemicals may result in the disruption of endocrine (hormonal) systems in humans and in wildlife populations. These endocrine disruptors are

- Natural and synthetic estrogens (the hormones that stimulate the development of secondary sexual characteristics; the sources of water pollution are hormone replacement therapy residues, birth control pill residues, and human and animal wastes)
- Phyto- and mycoestrogens (substances in plants and fungi that have an estrogenic effect; the source of water pollution is soil)
- Chemicals (diethylstilbestrol, dioxins, polychlorinated biphenyls; the sources of water pollution are pesticide residues and components of plastics)

Almost all natural and synthetic endocrine disruptors can be degraded by aerobic or anaerobic microorganisms or by a combination of these.

Bioremoval of Petroleum Hydrocarbons and Their Derivatives from Water

Petroleum hydrocarbons are widely present as micropollutants in groundwater and surface-water supplies. The most prevalent are the soluble aromatic compounds of gasoline: benzene, toluene, ethylbenzene, and xylenes (BTEX).

Methyl tertiary butyl ether (MBTE) is a fuel oxygenate added to gasoline to enhance its combustion (Figure 15.6a). MBTE is more soluble than BTEX but is not an easily degradable substance and may persist in the environment. It is possibly a human carcinogen and a common pollutant of groundwater due to leakage from underground storage tanks. It can be removed from water using biofiltration.

Polynuclear aromatic hydrocarbons (PAHs), such as naphthalene and phenanthrene are pollutants of water and can be removed after oxygenation, incorporation of oxygen into molecule, and cleavage of aromatic ring. Toxic and cancerogenic metabolites can be accumulated in water if the biodegradation of PAHs is not complete.

Bioremoval of Halogenated Compounds from Water

The halogenated methanes and ethanes are widely used as organic solvents and pollute surface water and groundwater. These substances, chloroform ($CHCl_3$),

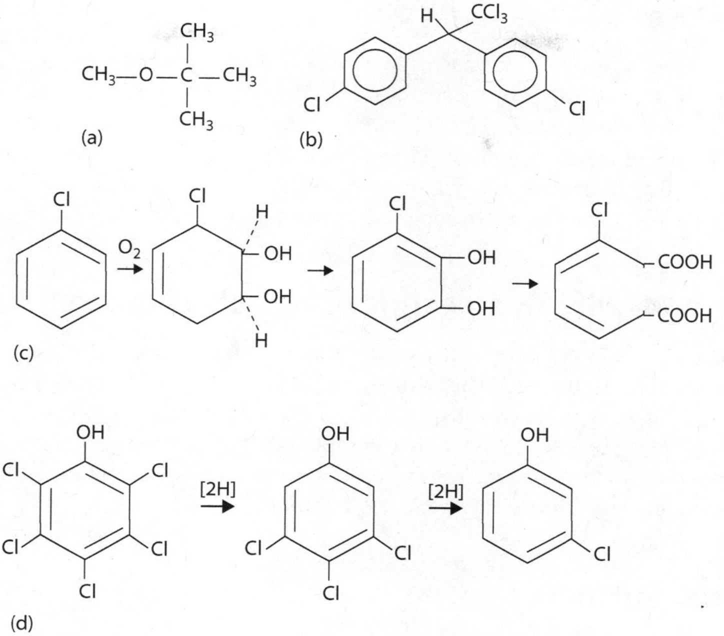

Figure 15.6 (a) Methyl-*tert*-butyl ether (MBTE), a common water pollutant. (b) Example of chlorinated pollutant of water, DDT, dichlorodiphenyltrichloroethane. (c) Oxygenation and ring cleavage of chlorobenzene. (d) Reductive dechlorination of pentachlorophenol.

trichloroethane, tetrachloroethane, and carbon tetrachloride, are resistant to biodegradation in aerobic systems because of electronegative atoms of chlorine in molecule. However, they can be dechlorinated and biodegraded using co-oxidation with methane or ammonium.

Halogenated aromatic compounds that are used as solvents, lubricants, and pesticides are also common contaminants of soil and groundwater. DDT (dichlorodiphenyltrichloroethane), polychlorinated biphenyls (PCBs), and pentachlorophenol (Figure 15.6b) were widely used in industry and agriculture. Chlorophenols have a strong odor often felt in water. Many microorganisms can even increase this odor by methylation of chlorophenols and form substances with a threshold odor concentration of less than 1 ng/L.

Aromatic chlorinated compounds can be biologically degraded using biofiltration with microorganisms that are able to perform the following reactions:

1. Dehalogenation of the ring structure by oxygenolytic, hydroxylic, or reductive dechlorination (Figure 15.6d)
2. Oxygenation of ring
3. Ring cleavage (Figure 15.6c)

Sucralose

Sucralose, a chlorinated disaccharide (4-chloro-4-deoxy-α,D-galactopyranosyl-1,6-dichloro-1,6-dideoxy-β, D-fructofuranoside), $C_{12}H_{19}Cl_3O_8$, is a zero-calorie sugar substitute artificial sweetener widely used in food products. Sucralose is a hardly biodegradable compound. Therefore, it is present in municipal wastewater (3000–8000 ng sucralose/L) and surface water (4–4000 ng sucralose/L) and is considered as a tracer for anthropogenic contamination of water.

Taste-and-Odor Compounds of Microbial Origin

Cyanobacteria and actinomycetes produce substances such as geosmin, methylisoborneol, and trichloroanisole, with earthy and musty odors in drinking water. Anaerobic bacteria produce substances with fishy and rotten odors (amines, fatty acids), swamp odors (reduced sulfur compounds), and antiseptic odors (chlorinated phenols and benzenes). All these substances can be biodegraded under the aerobic conditions of drinking-water biotreatment.

Arsenic in Water

Drinking water rich in arsenic leads to arsenic poisoning or arsenicosis: color changes on the skin; hard patches on the palms; skin cancer; and cancers of the bladder, kidney, and lung. Natural arsenic contamination is a cause for concern in Argentina, Bangladesh, Chile, China, India, Mexico, Thailand, and the United States. WHO's guideline value for arsenic in drinking water is 0.01 mg/L. In Bangladesh, 27% of shallow tube wells have been shown to have high levels of arsenic (above 0.05 mg/L). It has been estimated that 35–77 millions of the total population of 125 millions of Bangladesh are at a risk of drinking arsenic-contaminated water. Arsenic is released to groundwater together with ferrous iron produced by iron-reducing bacteria. One of the most effective ways for bioremoval of arsenic is its adsorption on ferric hydroxide produced during oxidation of organic chelates of Fe^{2+} in water by iron-oxidizing bacteria.

Nitrate and Nitrite in Water

Reservoirs and aquifers are polluted with nitrate and nitrite due to infiltration or runoff from areas exposed to agricultural fertilizers or manure. Nitrate in urban reservoirs is supplied from the soil of the parks and golf courses. This pollution occurs also in new residential areas that are constructed on displaced farms and orchards. The permitted maximum NO_3^- concentration in drinking water is 10 mg NO_3^--N/L in most parts of the world, while Europe also maintains a 0.1 mg/L standard for NO_2^--N. These standards are set to prevent methemoglobinemia in infants, in whose guts NO_3^- is reduced to NO_2^-, which binds to hemoglobin and causes suffocation. A risk for adults is related to the formation of carcinogenic nitrosamines produced from nitrite at low pH in stomach or from nitric oxide at neutral pH in intestine. NO_2^- can be formed as a partial reduction intermediate when the electron donor is depleted.

Denitrification of Drinking Water

Nitrite and nitrate can be removed from water by denitrification in a fixed-bed biofilter, fluidized-bed bioreactor, and membrane bioreactor. The electron donor for reduction of nitrate could be an organic substance (methanol, ethanol, acetate, or glucose) or inorganic substance (S, H_2, or Fe^{2+}).

Denitrification is an anaerobic (anoxic) process and can be performed by many facultative anaerobic bacteria. Denitrifiers are common among the Gram-negative *Proteobacteria*, such as *Pseudomonas*, *Alcaligenes*, *Paracoccus*, and *Thiobacillus*.

Methanol, ethanol, and acetate are often used in the practice of denitrification. Ethanol and acetate are alternative organic electron donors; ethanol is substantially less expensive than acetate, but may have political or regulatory liabilities as an additive in drinking-water treatment because it is an alcohol. Bacteria *Paracoccus denitrificans* are usually dominant in enrichment cultures containing nitrate and ethanol:

$$5CH_3CH_2OH + 8NO_3^- + 8H^+ \rightarrow 4N_2 + 5CO_2 + 19H_2O$$

Hyphomicrobium spp. is auto-selected in mediums containing nitrate and methanol:

$$5CH_3OH + 6NO_3^- + 6H^+ \rightarrow 3N_2 + 5CO_2 + 13H_2O$$

However, methanol is acutely toxic to humans and is a regulated substance.

Some Gram-positive bacteria, including representatives of the genus *Bacillus*, can perform denitrification. There are a few halophilic Archaea, such as *Halobacterium*, which are able to perform denitrification.

The electron donor for denitrification can also be organic substances such as H_2, S, H_2S, or Fe^{2+}. This is the so-called autotrophic denitrification (CO_2 is used as the source of carbon). The oxidation of hydrogen is the most environmentally friendly process and residual electron donor is absent in drinking water:

$$5H_2 + 2NO_3^- + 2H^+ \rightarrow N_2 + 6H_2O$$

Sulfur is the cheapest electron donor for denitrification, but this process can be used only for denitrification of wastewater discharged in coastal seawater.

$$5S + 6NO_3^- + 2H_2O \rightarrow 3N_2 + 5SO_4^{2-} + 4H^+$$

Raw water contains mainly reduced nitrogen (NH_4^+) but groundwater contains mainly oxidized nitrogen (NO_3^-). This is because of the strong adsorption of positively charged ammonium to negatively charged clay particles of soil and migration of negatively charged nitrate ions through soil to groundwater. To remove nitrogen from surface water, denitrification (an anaerobic process) is often coupled to nitrification (an aerobic process), which is needed to create the oxidized nitrogen. This combination can be performed if there are anaerobic and aerobic zones or layers in the microbial biofilm in a biofilter.

Conditions for Denitrification

Denitrification is inhibited by dissolved oxygen. Very low concentrations of electron donors or very high concentrations of dissolved oxygen can lead to the accumulation of the denitrification intermediates: NO_2^-, NO_2^-, and N_2O. The latter two are greenhouse gases whose release should be avoided.

pH values outside the optimal pH range of 7–8 can also lead to the accumulation of intermediates. In low-alkalinity waters, pH control must be performed because denitrification produces a strong base.

For the denitrification of drinking water, the residual electron donor should be minimal in order to keep the water biologically stable. Therefore, the amount of electron donor added must be less than the stoichiometric requirement for complete denitrification.

Bioreduction of Perchlorate in Drinking Water

Perchlorate (ClO_4^-) is used in jet fuel as an oxidant, and it contaminates water around some manufacturing plants and rocket-test facilities, causing pollution of ground and surface waters in some cases. Many denitrifying bacteria are able to use perchlorate as an alternate electron acceptor, reducing perchlorate to chloride ion in the presence of electron donors such as acetate or H_2.

Iron and Manganese in Water

Ferrous (Fe^{2+}) and manganous (Mn^{2+}) ions are chemically similar and cause similar problems being present in water: the intoxication effects of heavy metals, staining of laundry, deterioration of flavor and color of food and beverages, and deposition in pipelines and water heaters. When oxidized chemically or biologically, reduced forms of iron and manganese form insoluble ferric iron (Fe^{3+}) and manganic manganese (Mn^{4+}). The rate of oxidation depends on pH, alkalinity, and presence of oxygen and oxidizing bacteria in water.

Iron- and manganese-oxidizing neutrophilic bacteria are microaerophilic bacteria producing oxygen radicals that oxidize iron and manganese or their chelates with organic acids stable for chemical oxidation by oxygen in air. These bacteria can clog pipes and wells due to formation of iron- and manganese hydroxide precipitates. Biological oxidation can be used in slow filters. Its advantage is the simultaneous removal of arsenic and many other metals including radionuclides from water due to adsorption and coagulation by forming hydroxides.

The chemical precipitation of iron and manganese with phosphate or oxidation with oxygen, ozone, or permanganate are faster methods for purification of water but can be ineffective for chelates of iron and manganese with organic acids. The chemical oxidation of manganese is facilitated by increasing the pH to about 8.5.

16

Anaerobic and Anoxic Treatment of Wastewater

Oxygen and Energy Generation

The evolution of the Earth's atmosphere from an anaerobic state of the earth to an aerobic one resulted in the creation of:

- Anaerobic (living without oxygen) microorganisms
- Facultative anaerobic (living under either anaerobic or aerobic conditions) microorganisms
- Microaerophilic (living under low concentrations of dissolved oxygen) microorganisms
- Obligate aerobic (living only in the presence of oxygen) microorganisms

Anaerobes produce energy from

- Fermentation (destruction of organic substances without external acceptor of electrons)
- Anaerobic respiration using electron acceptors such as NO_3^-, NO_2^-, Fe^{3+}, SO_4^{2-}, and CO_2
- Anoxygenic ($H_2S \rightarrow S$) photosynthesis

Microaerophiles and aerobes produce energy from

- Aerobic oxidation of organic matter
- Oxygenic photosynthesis

The sequence of increasing production of biological energy per mole of transferred electrons is as follows: fermentation → CO_2 respiration (hydrogenotrophic methanogenesis) → dissimilative sulfate-reduction → dissimilative iron reduction (iron respiration) → nitrate respiration (denitrification) → aerobic respiration.

Anaerobic Digestion of Organic Matter

There are many applications of anaerobic processes in the treatment of polluted soil, solid waste, and wastewater. One of the major applications is anaerobic digestion of activated sludge produced in municipal wastewater treatment plants. Anaerobic digestion consists of a series of microbiological processes that convert organic compounds of activated sludge (biomass of aerobic microorganisms with empirical formula $CH_{1.8}O_{0.5}N_{0.12}$) to methane and carbon dioxide:

$$CH_{1.8}O_{0.5}N_{0.12} \rightarrow X_1CH_4 + X_2CO_2 + 0.12NH_3$$

The anaerobic digestion process is used to treat and stabilize waste sludges from conventional sewage treatment prior to disposal.

Similar processes of anaerobic degradation of organic matter are performed in the bottom sediments of aquatic systems, in landfills, and in the rumen of the ruminant animals such as cows, deers, and camels.

Microbiology of Anaerobic Digestion of Biopolymers

Anaerobic digestion of biomass is performed by Bacteria and Archaea. Five physiological groups of prokaryotes are involved in the anaerobic digestion process (Figure 16.1):

1. *Hydrolytic bacteria.* These bacteria break down biopolymers such as cellulose and other polysaccharides, proteins, nucleic acids, as well as lipids

Figure 16.1 (See color insert following page 292.) Anaerobic digesters on municipal wastewater treatment plant.

into water soluble monomers such as amino acids, nucleosides, glucose and other sugars, long-chain fatty acids, and glycerol. Extracellular enzymes such as cellulases, amylases, proteases, DNAses and RNAses, and lipases catalyze the hydrolysis of the complex molecules. The hydrolytic phase can be relatively slow and can be the limiting step in anaerobic digestion. The monomers produced by hydrolytic bacteria are then directly available for fermenting bacteria. Commonly, hydrolytic bacteria are also active fermenting species.

2. *Acidogenic bacteria.* These acidogenic (producing acid) bacteria, commonly from the genus *Clostridium*, ferment monomers, i.e., sugars, amino acids, nucleosides, into volatile fatty acids (VFA) such as formic, acetic, propionic, lactic, and butyric acids as well as producing ethanol, H_2, and CO_2.
3. *Acetogenic bacteria.* Acetogenic (producing acetate) bacteria, commonly from the genera *Syntrobacter* and *Syntrophomonas*, convert products of fermentation mainly ethanol, propionic, and butyric acids into acetate, hydrogen, and carbon dioxide that are used immediately by the acetotrophic and hydrogenotrophic methanogens:

$$CH_3CH_2OH + H_2O \rightarrow CH_3COOH + 2H_2$$

$$CH_3CH_2COOH + H_2O \rightarrow CH_3COOH + CO_2 + 3H_2$$

$$CH_3CH_2CH_2COOH + H_2O \rightarrow 2CH_3COOH + 2H_2$$

The physiological group of acetogens requires low hydrogen tensions for acetogenesis; therefore, there is a symbiotic relationship between the acetogens and methanogens. Methanogens consume H_2 and help achieve the low hydrogen tension required by the acetogens.

4. Acetotrophic methanogens convert acetate to methane:

$$CH_3COOH \rightarrow CH_4 + CO_2$$

It is the so-called acetoclastic reaction performed by Archaea from genera *Methanosarcina*, *Methanosaeta*, and *Methanothrix*.

5. Hydrogenotrophic methanogens produce methane from CO_2 and H_2:

$$CO_2 + 4H_2 \rightarrow CH_4 + 2H_2O$$

A variety of morphological types of methanogens have been isolated and studied. There are single cocci and aggregates of cocci, short and long rods, as well as filamentous cells. All methanogens are obligate anaerobes with a low maximum specific growth rate at $0.05\,h^{-1}$.

Biogas Collection and Use

Biogas produced during anaerobic digestion of organic wastes is valuable fuel. Typical content of biogas is as follows (vol.%): CH_4, 55–70; CO_2, 30–45; H_2, 5–10;

N_2, 0–2; H_2O, 0.3; CO, 0.2; H_2S, 0.01–0.2; and NH_3, 0.01. Biogas is collected in the gas holder and supplied for incineration under pressure. Energy released in incineration is used for electricity generation or water heating. In all cases, this energy is used directly on the municipal wastewater treatment plant for air compression or heating of the bioreactors. The production and use of biogas methane is cost-saving but cannot be considered as a way of energy generation for industry or households because the cost of energy of biogas is significantly higher than the cost of supplied electrical energy or conventional fuel.

Optimal Conditions for Anaerobic Digestion of Organic Wastes

Mesophilic anaerobic digestion in municipal wastewater treatment plants is carried out at temperatures from 25°C to 40°C, with an optimum at approximately 35°C. Thermophilic anaerobic digestion is performed at temperatures from 50°C to 65°C.

Because of the low specific growth rate of methanogens, the hydraulic retention time (HRT) of liquids in anaerobic digesters is 10–40 days.

The optimum pH is at a pH range of 6.7–7.4. If there is intensive production of fatty acids at the stage of acidogenesis and pH drops below 6, the anaerobic digestion process may fail. The required pH can be restored by increasing the concentration of proteins, releasing alkaline NH_3 during digestion, or by an addition of $Ca(OH)_2$ (lime), NaOH, or $NaHCO_3$.

The optimal C:N:P ratio for anaerobic bacteria is 700:5:1. Trace elements such as iron, cobalt, molybdenum, and nickel are also necessary.

Methanogenesis is more sensitive than fermentation to oxygen, ammonia, chlorinated hydrocarbons, and surfactants.

Interaction between Methanogenesis and Sulfate Reduction

Methanogens and sulfate-reducing bacteria compete for the same electron donors, acetate, and hydrogen. Sulfate-reducing bacteria have a higher affinity for acetate than methanogens. This means that sulfate-reducing bacteria can outcompete methanogens under low acetate conditions. This competitive inhibition results in the shunting of electrons from methane generation to sulfate reduction. Sulfate reducers and methanogens are most competitive at COD/SO_4^{2-} ratios of 1.7–2.7. An increase of this ratio is favorable to methanogens, while a decrease of this ratio is favorable to sulfate reducers. Sulfate reduction increases the content of bad-smelling, toxic, and corrosive dihydrogen sulfide in biogas. Additionally, high content of H_2S in biogas will increase the content of harmful-for-environment sulfur oxides in incineration exhaust gas.

Comparison of Anaerobic and Aerobic Digestions of Organic Waste

1. Anaerobic digestion requires no oxygen, the supply of which adds substantially to wastewater treatment costs.
2. Anaerobic digestion produces, in average, five times lower amounts of sludge (biomass).
3. Anaerobic digestion produces a fuel, methane. This fuel can be burned in wastewater treatment plants to provide heat for digesters or to generate electricity.
4. Anaerobic digestion is suitable for high-strength industrial wastes.
5. The activity of anaerobic microorganisms can be preserved for a long time. In practice, it is important to have this store of anaerobic biomass to start up the reactor after accidental failure.

However, anaerobic digestion is a slower process than aerobic treatment and it is more sensitive to toxicants than the aerobic process.

Bioreactors Used in Anaerobic Wastewater Treatment

Different bioreactors are used in anaerobic treatment of organic waste and wastewater:

- Stirred tank reactor
- Anaerobic biofilter, which is a biofilter without aeration
- Upflow anaerobic sludge blanket (UASB) reactor
- Fluidized bed reactor
- Septic tank
- Lagoon covered with an impermeable plastic membrane
- Covered landfill

Stirred Tank Reactor for Anaerobic Digestion of Organics

Stirred tank reactor can be used as a single-stage anaerobic digester, which is a tank with mechanical mixing, heating, gas collection, sludge addition and withdrawal, and supernatant outlets. In reactors with volumes of several thousand m^3, stirring is absent and mixing inside the reactors is performed by the movement of biogas. The sludge forms several functional layers in this tank.

Upflow Anaerobic Sludge Blanket Reactor

The upflow anaerobic sludge blanket (UASB) reactor uses the upward flow of wastewater for auto-selection of fast settled biomass to retain sludge. The UASB reactor consists of a bottom layer of sludge, a sludge blanket, and an upper liquid layer. Wastewater effluent passes out around the edge of the reverted funnel, while gases pass out through the hole in the middle of the funnel. The capacity of UASB reactors is high due to the high concentration of retained biomass, which forms microbial granules that range from 0.5 to 5 mm in diameter. The granules have diverse microbial communities. The core of the granule consists of acetotrophic methanogens. The middle layer consists of acetogens and hydrogenotrophic methanogens. The outer layer of the granule consists of fermenting and hydrolytic bacteria.

Septic Tank

The septic tank system consists of a tank, where anaerobic digestion of organic matter is performed for 1–5 days, and an absorption field. The accumulated septage from the septic tank is disposed of regularly. The effluent from the septic tank flows to an absorption field through a system of perforated pipes for percolation through soil to groundwater.

Anaerobic Processes in Landfills

Landfilled organic and inorganic wastes are slowly transformed by indigenous microorganisms in the wastes. Organic matter is hydrolyzed by bacteria and fungi. Amino acids are degraded via ammonification with the formation of toxic organic amines and ammonia. Amino acids, nucleotides, and carbohydrates are fermented or anaerobically oxidized with the formation of organic acids, CO_2, and CH_4. Biogas can be collected in the landfill through a pipeline system and used as fuel. Xenobiotics in the landfill matter may be reduced and degraded by fermentation of anoxic respiration. Heavy metals in the landfill matter may be reduced and subsequently dissolved. These bioprocesses may result in the formation of toxic landfill leachate, which can be detoxicated by aerobic biotechnological treatment to oxidize organic hazards and to immobilize dissolved heavy metals.

Anaerobic Degradation of Xenobiotics by Fermenting Bacteria

Anaerobic fermenting bacteria, for example, from the genus *Clostridium*, perform three important functions in environmental engineering processes:

- Hydrolyzing biopolymers
- Fermenting monomers
- Performing reductive dechlorination of xenobiotics due to formation of hydrogen in fermentation

Many toxic substances, for example, chlorinated solvents, phthalates, phenols, ethyleneglycol, and polyethylene glycols, can be degraded by anaerobic fermenting or anoxic microorganisms.

Anoxic Bioprocesses

Anaerobic respiration is more effective than fermentation, in terms of output of energy per mole of transferred electrons. Anaerobic respiration can be performed by different groups of prokaryotes with such electron acceptors as NO_3^-, NO_2^-, Fe^{3+}, SO_4^{2-}, and CO_2. Therefore, if the concentration of one such acceptor in the hazardous waste is sufficient for the anaerobic respiration and oxidation of the pollutants, the activity of the related bacterial group can be used for the treatment.

Anoxic Biotechnological Methods

The following biotechnological methods are related to anoxic microorganisms:

- Biotechnological methods of nitrate bioreduction
- Biotechnological methods of sulfate bioreduction
- Biotechnological methods of ferric bioreduction
- Biotechnological methods of carbon dioxide bioreduction
- Biotechnological methods of manganese, selenium, and arsenic bioreduction
- Biotechnological methods of dechlorination
- Anoxic oxidation of ammonia
- Anoxic bioreactors
- Anoxic biofilms
- Anoxic biofouling, biocorrosion, and biodeterioration
- Anoxic soil bioremediation
- Anoxic soil biocementation
- Combination of anoxic and aerobic processes
- Combination of anoxic and anaerobic processes
- Extremophilic anoxic microorganisms and related processes

Nitrate Reduction in Water and Wastewater Treatment

Denitrifying bacteria, which are usually facultative anaerobic prokaryotes, are able to reduce nitrate as the terminal electron acceptor in anaerobic respiration:

$$4NO_3^- + 5CH_2O \rightarrow 2N_2 + 5CO_2 + 3H_2O + 4OH^-$$

Denitrification of water and wastewater can be performed in biofilters with *Pseudomonas denitrificans* or *Paracoccus denitrificans* and the addition of such donor of electrons as methanol, ethanol, and acetate, or H_2, S, H_2S, and Fe^{2+} (see details in Chapter 15).

Nitrate Reduction in Microbially Enhanced Oil Recovery

Formation of toxic, bad-smelling, and corrosive dihydrogen sulfide by sulfate-reducing bacteria is very common in oil fields, where seawater is used to maintain pressure in the oil deposit:

$$SO_4^{2-} + CH_4 + \text{sulfate-reducing bacteria} \rightarrow H_2S + CO_2 + 2OH^-$$

This negative effect of sulfate-reducing bacteria can be eliminated by nitrate-reducing bacteria because they over complete sulfate-reducing bacteria in oxidation of hydrocarbons:

$$1.6NO_3^- + CH_4 + \text{nitrate-reducing bacteria} \rightarrow 0.8N_2 + CO_2 + 1.2H_2O + 1.6OH^-$$

The addition of nitrate to seawater, which is pumping into oil deposit results in

- production of gas and creation of higher pressure enhancing oil recovery
- absence of corrosive and toxic H_2S in oil and water

Nitrate Reduction in Soil Bioremediation

Soil polluted with xenobiotics (conventionally, CH_2O) can be purified with the addition of nitrate as electron acceptor:

$$0.8NO_3^- + CH_2O \rightarrow 0.4N_2 + CO_2 + 0.6H_2O + 0.8OH^-$$

The advantages of this process are as follows:

- Fast bioremediation.
- Can be performed by indigenous microorganisms.
- Bioremediation can be performed under anaerobic conditions in situ.

The disadvantages are as follows:

- High cost of treatment
- Increase of pH

Nitrate Reduction in Geotechnical Improvement of Soil

Saturation of sandy soil with water is a reason for sand liquefaction, eventually destroying buildings. Water from sand can be removed by gas production during denitrification:

$$0.8NO_3^- + CH_2O + \text{nitrate-reducing bacteria} \rightarrow 0.4N_2 + CO_2 + 0.6H_2O + 0.8OH^-$$

Increase of pH during denitrification can be used for soil biocementation with calcium salts:

$$Ca^{2+} + 2OH^- + CO_2 \rightarrow CaCO_3 \downarrow + H_2O$$

Iron Reduction in Water, Wastewater, and Groundwater Treatment

Fe^{3+} is an environmentally friendly electron acceptor. It is naturally abundant in clay minerals, magnetite, limonite, goethite, and iron ores, but its compounds are usually insoluble and it diminishes the rate of oxidation in comparison with dissolved electron acceptors. Iron-reducing bacteria can produce dissolved Fe^{2+} ions from insoluble Fe^{3+} minerals:

$$4Fe^{3+} + CH_2O + H_2O \rightarrow 4Fe^{2+} + CO_2 + 4H^+$$

$$Fe^{2+} + 6H_2O + 2CO_2 \rightarrow 2Fe(OH)_2 + 2FeCO_3 + 8H^+$$

$$4Fe^{3+} + CH_2O + 7H_2O \rightarrow 2Fe(OH)_2 + 2FeCO_3 + 12H^+$$

Fe^{3+} bioreduction can be also used in the following processes:

1. Water and wastewater dephosphorylation

$$Fe^{2+} + HPO_4^{2-} \rightarrow FeHPO_4$$

2. Adsorption of arsenic from groundwater on $Fe(OH)_2$ and $Fe(OH)_3$
3. Coagulation of water and wastewater using Fe^{2+}
4. Removal of sulfide from groundwater and wastewater

$$Fe^{2+} + S^{2-} \rightarrow FeS \downarrow$$

Iron-reducing bacteria are able to oxidize many xenobiotics using Fe^{3+} as the terminal electron acceptor:

- Aliphatic hydrocarbons
- Aromatic hydrocarbons
- Phenols

Anaerobic biodegradation and detoxication of wastewater can be significantly enhanced by precipitation of toxic organics, acids, phenols, or cyanide with Fe(II) produced by iron-reducing bacteria.

Iron-reducing bacteria are also able to reduce Mn(VI), U(VI), Tc(VII), Co(III), Cr(IV), Au(III), Hg(II), As(V), and Se(VII) using organic acids and alcohols as electron donors.

The final stage of treatment with iron-reducing bacteria must be aerobic oxidation of remaining Fe^{2+} to avoid discharge of iron into the environment. It can be done chemically or using oxidation by neutrophilic iron-oxidizing bacteria.

Sulfate Reduction in Water and Wastewater Treatment

Sulfate-reducing bacteria are obligate anaerobes that use organic compounds as energy sources. Sulfate or elemental sulfur serves as the terminal electron acceptor in anaerobic respiration, and hydrogen sulfide is the toxic product of this process:

$$SO_4^{2-} + CH_2O + 6H^+ \rightarrow H_2S + CO_2 + 3H_2O$$

These organisms are widespread in sulfate-rich environments and can use almost all organic compounds as electron donors. Acceptable donors for environmental engineering applications are cellulose and organic acids producing by fermenting bacteria in cellulose fermentation.

Sulfate reducers can be used for the removal of heavy metals from wastewater:

$$CH_3COOH + SO_4^{2-} \rightarrow 2CO^2 + H_2S + 2OH^-$$

$$Cu^{2+} + H_2S \rightarrow CuS + 2H^+$$

However, toxic and corrosive H_2S creates a lot of problems in environmental engineering.

Dehalogenating Bacteria

Halogenated organic compounds are widespread environmental pollutants because of their use as solvents, pharmaceutical products, biocides, plasticizers, hydraulic and heat transfer fluids, and intermediates for chemical synthesis. First, the most essential step in their biodegradation, is the removal of chlorine atoms from the molecules. Dehalogenating bacteria are able to oxidize such electron donors as formate, acetate, pyruvate, lactate, and H_2 due to anaerobic reductive dechlorination and can be used for degradation of chlorinated organic compounds in soil or wastewater. The ability to reduce Fe(III), Mn(VI), Se(VI), and As(V) for anaerobic reductive dechlorination is a common property of several anoxic bacteria.

Combined Anaerobic/Aerobic Biotreatment of Wastes

A combined anaerobic/aerobic biotreatment can be more effective than an aerobic or an anaerobic treatment alone. The simplest approach for this type of treatment is the use of aerated stabilization ponds, aerated and non-aerated lagoons, and natural and artificial wetland systems, whereby aerobic treatment occurs in the upper part of these systems and anaerobic treatment occurs at the bottom end. A typical organic loading is 0.01 kg BOD/m^3·day and the retention time varies from a few days to 100 days. A more intensive form of biodegradation can be achieved by combining aerobic and anaerobic reactors with controlled conditions or by integrating anaerobic and aerobic zones within a single bioreactor.

Combinations or even alterations of anaerobic and aerobic treatments are useful in the following situations:

1. Biodegradation of chlorinated aromatic hydrocarbons including anaerobic dechlorination and aerobic ring cleavage
2. Sequential nitrogen removal including aerobic nitrification and anaerobic denitrification
3. Anaerobic reduction of Fe(III) and microaerophilic oxidation of Fe(II) with production of fine particles of iron hydroxide for adsorption of organic acids, phenols, ammonium, cyanide, radionuclides, and heavy metals

Biotechnological Treatment of Heavy Metals–Containing Waste and Radionuclides-Containing Waste

Microbial metabolism generates products such as hydrogen, oxygen, H_2O_2, and reduced or oxidized iron that can be used for oxidation/reduction of metals. Reduction or oxidation of metals is usually accompanied by metal solubilization or precipitation.

Solubilization or precipitation of metals may be mediated by microbial metabolites. Microbial production of organic acids in fermentation or the production of inorganic acids (nitric and sulfuric acids) in aerobic oxidation will promote formation of dissolved chelates of metals.

Microbial production of phosphate, H_2S, and CO_2 stimulates precipitation of non-dissolved phosphates, carbonates, and sulfides of heavy metals, for example, arsenic, cadmium, chromium, copper, lead, mercury, and nickel; the production of H_2S by sulfate-reducing bacteria is especially useful in removing heavy metals and radionuclides from sulfate-containing mining drainage waters, liquid waste of nuclear facilities, drainage from tailing pond of hydrometallurgical plants, wood straw, or saw dust. Organic acids, produced during the anaerobic fermentation of cellulose, may be preferred as a source of reduced carbon for sulfate reduction and further precipitation of metals.

The surface of microbial cells is covered by negatively charged carboxylic and phosphate groups and positively charged amino groups. Therefore, depending on pH, there may be a significant adsorption of heavy metals onto the microbial surface. Biosorption, for example, by fungal fermentation residues, is used to accumulate uranium and other radionuclides from waste streams.

Metal-containing minerals, for example, sulfides, can be oxidized and metals can be solubilized. This approach is used for the bioleaching of heavy metals from sewage sludge before landfilling or biotransformation.

Some metals, arsenic and mercury, for example, may be volatilized by methylation due to activity of anaerobic microorganisms. Arsenic can be methylated by methanogenic Archaea and fungi to volatile toxic dimethylarsine and trimethylarsine or can be converted to less toxic, nonvolatile methanearsonic and dimethylarsinic acids by algae.

In some cases, the methods may be combined. Examples would include the biotechnological precipitation of chromium from Cr(VI)-containing wastes from electroplating factories by sulfate reduction to precipitate chromium sulfide. Sulfate reduction can use fatty acids as organic substrates with no accumulation of sulfide. In the absence of fatty acids but with straw as organic substrate, the direct reduction of chromium has been observed without sulfate reduction.

Hydrophobic organotins are toxic to organisms because of their solubility in cell membranes. However, many microorganisms are resistant to organotins and can detoxicate them by degrading the organic part of organotins.

Dissolved acceptors of electrons such as NO_3^-, NO_2^-, Fe^{3+}, SO_4^{2-}, and HCO_3^- can be used in the treatment system if oxygen transfer rates are low. The choice of the acceptor is determined by economical and environmental reasons. Nitrate is often proposed for bioremediation because it can be used by many microorganisms as an electron acceptor. However, it is relatively expensive and its supply to the treatment system must be thoroughly controlled because it can also pollute the environment.

Sulfate and carbonate can be applied as electron acceptors in strictly anaerobic environments only. Another disadvantage of these acceptors is that these anoxic oxidations generate toxic and foul smelling H_2S or the "greenhouse" gas CH_4.

17

Aerobic Treatment of Wastewater

Domestic Wastewater

The major application of aerobic treatment is the biotreatment of domestic sewage, which contains human feces and urine and the so-called gray water from washing, bathing, and cooking. Domestic sewage wastewater typically contains 800 mg of total solids/L, including 440 mg of organic matter (total volatile solids)/L, 240 mg of suspended solids/L, 200 mg BOD/L, 35 mg N/L, and 7 mg P/L. The typical domestic wastewater flow per capita is 450 L/day.

The BOD/N/P weight ratio required for biological treatment is 100/5/1 and raw domestic wastewater has a ratio of 100/17/3. So, there is no need to add nitrogen, phosphorus, and other nutrients for microbial growth in sewage. Growth yield of biomass microorganisms on raw sewage is about 0.4 g BOD of biomass/BOD of sewage.

Industrial Wastewater

The content of industrial wastewater is unique to each industry. For exemption of food processing wastewater, aerobic biotreatment of industrial wastewater usually requires addition of inorganic nutrients and adjustment of pH. If the BOD_5:COD ratio is less than 0.01, it means that the components of industrial wastewater are relatively non-biodegradable. A BOD:COD ratio greater than 0.1 means that the components of industrial wastewater are relatively degradable.

Aeration

Because of the low solubility of oxygen in water, about 7 mg/L at 35°C, it will be consumed for several minutes at an oxygen consumption rate of 50–500 mg O_2/g of dry biomass·h and biomass concentration of 3–10 g of dry biomass/L, which are typical for aerobic wastewater treatment. So, compressed air or even pure oxygen must be supplied permanently in the reactor at the rate of 0.1–1.0 L/L of liquid/min during aerobic treatment. The cost of aeration is approximately 20%–40% of aerobic treatment of wastewater.

Objectives of Aerobic Biotreatment of Wastewater

The objectives of aerobic wastewater treatment are as follows:

1. Removal of organics (BOD, COD, TOC) from wastewater
2. Removal of suspended particles
3. Removal of toxic organic compounds
4. Removal of heavy metals
5. Removal of inorganic nutrients
6. Removal/inactivation of pathogens

Aerobic Treatment of Wastewater

Aerobic treatment of wastewater is a part of wastewater treatment, which usually consists of three steps:

1. Primary treatment, including screening, to separate debris and coarse materials that may clog treatment equipment followed by sedimentation to remove suspended solids.
2. Secondary treatment, including the stage of aerobic biotreatment, in aerated tank with suspended biomass called "activated sludge" or aerated biofilters with attached biomass such as submerged biofilter, trickling filter, rotating biological contactor (RBC), and stage of biomass and treated wastewater separation using sedimentation, flotation, centrifugation, or coagulation.
3. Tertiary or advanced treatment, including chemical or biological treatment, to remove nutrients, toxic chemicals, and pathogens.

Conventional Biotreatment in Aerobic Tank

Wastewater is supplied in aeration tank ensuring aeration and retention of microbial biomass, historically called "activated sludge," along with liquid, within a defined hydraulic retention time (HRT), typically from 1 to 8 h. It is the reciprocal to dilution rate (D).

Due to the separation of biomass from treated wastewater and recycling of the portion of settled biomass from the settling tank (clarifier) into the aerated tank

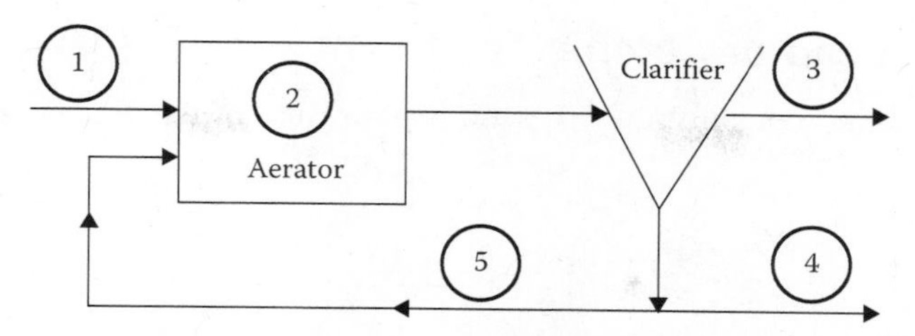

Figure 17.1 Aerobic biological treatment of wastewater: (1) influent, (2) aeration tank, (3) effluent, (4) waste sludge, and (5) recycled sludge.

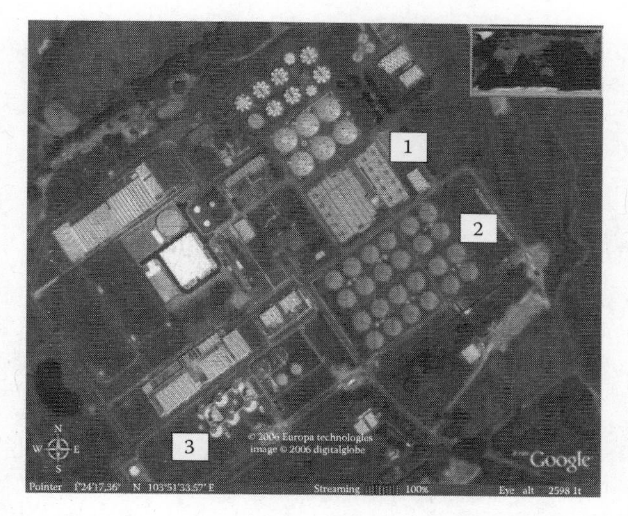

Figure 17.2 Typical municipal wastewater treatment plant: (1) aeration tanks, (2) settling tanks, and (3) anaerobic digesters.

(see Figures 17.1 and 17.2), the retention time for biomass called "volatile solids retention time" (VSRT) is not the same value as HRT and is in the range of 1–20 days. This time is also called "sludge age" = $(MLVSS \times V)/(SS3 \times Q_3 + SS_4 \times Q_4)$, where SS_3 is suspended solids in wastewater effluent, Q_3 is the flow rate of effluent, SS_4 is the suspended solids in wasted sludge after the settling tank, and Q_4 is the flow rate of wasted sludge after the settling tank.

The process consists of an aerobic oxidation of organic matter to CO_2, H_2O, NH_3, and the formation of new biomass of the so-called activated sludge. The growth yield of biomass is about 40% of BOD and the growth rate is 0.1–1 h^{-1}.

The liquid suspension of microorganisms in an aeration basin is generally referred to as "mixed liquor," and the concentration of biomass is called in environmental engineering as "mixed liquor volatile suspended solids" (MLVSS).

Organic Load of Aerobic Treatment

The food-to-microorganism (F/M) ratio is a measure of the organic load in the activated sludge:

$$\frac{F}{M} = \frac{Q_i \times BOD_i}{MLVSS \times V} \frac{kg\ O_2}{kg\ VSS} \cdot day$$

where

Q_i is the influent flow rate
V is the working volume of aeration tank

For conventional aeration tanks, F/M is 0.2–0.5 day^{-1}. Microorganisms are starved at low F/M ratio, diminishing the efficiency of wastewater treatment and creating preferable conditions for filamentous microorganisms in aeration tanks.

Activated-Sludge Settleability

The sludge volume index (SVI) is a measure of sludge settleability. It is the volume occupied by 1 g of sludge after 30 min of sedimentation. The normal range of SVI for sludge in aeration tank is 50–150 mL/g of biomass (MLVSS). When the concentration of biomass in an aeration tank is 3 g/L and the SVI is 70 mL/g of biomass, the volume of settled activated sludge in the settling tank is 500 mL/L = 50% of the volume of the settling tank; the concentration of biomass in the bottom part of the settling tank is twice bigger than that in the aeration tank. When the concentration of biomass is 5 g/L and the SVI is 200 mL/g of biomass, the volume of settled activated sludge in the settling tank is 1 L/L = 100% of the volume of settling tank, which means that there is no separation of biomass and treated wastewater.

Activated-Sludge Recycling

A portion of activated sludge, usually 10%–30%, is constantly returned back into the aeration tanks. This recycling of biomass ensures

1. The maintenance of a higher concentration of biomass in the aeration tank
2. The retention and finally the selection of fast settling organisms in the aeration tank
3. The retention of slow-growing organisms, for example, nitrifying bacteria, in the aeration tank

Biomass balance at steady state is as follows:

$$\mu X_1 - DX_1 + rX_2 = 0$$

$$r = \frac{Q_2}{Q_1}$$

$$D = \mu + \frac{rX_2}{X_1}$$

where
- μ is the specific growth rate of biomass
- D is the dilution rate = influent flow rate (Q_1)/volume of aeration tank
- X_1 and X_2 are concentrations of biomass in aeration tank and in the recycled liquid
- r is a portion of wastewater, which is recycled from the settling tank to the aeration tank
- Q_2 is the flow rate from the settling tank to the aeration tank

If the portion of recycled wastewater is 0.05 (5%) and the concentration of biomass in the settling tank is five times higher than the concentration of biomass in the aeration tank, the dilution rate in the aeration tank at steady state can be 25% higher than the specific growth rate of biomass in the aeration tank.

Microorganisms, which are capable of aggregation and rapid settling of the flocs of aggregated cells in the clarifier, are recirculated back to the aeration tank for further growth. Microorganisms, which do not aggregate and do not form settling flocs, are discharged with the effluent from the settling tank. The settling and recycling of flocs to the aeration tank ensures the selection of floc-forming strains of bacteria. This improves the separation of biomass from the wastewater effluent and the quality of treated wastewater.

Microorganisms of Activated Sludge

The number of viable bacterial cells in activated sludge is from 10^9 to 10^{11} cells/g of suspended solids. Theoretically, 1 g of dry biomass can contain 10^{12} cells. So, a floc of activated sludge contains 1%–10% of live bacterial biomass and 90%–99% of extracellular material.

The Gram-negative bacteria are major biocomponents of activated-sludge flocs. Culture-based techniques revealed the presence of the rod-shaped and coccus-shaped cells from the bacterial genera *Pseudomonas*, *Flavobacterium*, *Alcaligenes*, *Achromobacter*, *Acinetobacter*, *Zoogloea*, as well as cells from the genera of filamentous bacteria *Sphaerotilus*, *Nocardia*, and *Beggiatoa*.

Aerobic, facultative anaerobic, and occasionally even anaerobic bacteria are present in the flocs of activated sludge, because of the gradient of dissolved oxygen inside flocs. There are also aggregates of nitrifying bacteria, oxidizing ammonium to nitrate, inside activated-sludge flocs.

Fungi, for example, from the genera *Geotrichum* and *Penicillium*, may be present in the aeration tank under pH below 6, because the range of optimum pH for the growth of fungi is wider than this for bacteria. The same can be found in the case when the BOD:N ratio of wastewater is above 100:4 and the BOD:P ratio is above 100:0.5, because fungi have lower nitrogen and phosphorous requirements than bacteria.

The biomass of protozoa is approximately 5% (w/w) of activated sludge. Protozoa ingest and digest suspended bacterial cells. Therefore, amoebas, flagellates, and free-swimming ciliates are present when BOD of treated wastewater and number of non-aggregated bacterial cells are high and efficiency of wastewater treatment is low. Stalked ciliates, attached to the surface of flocs, are present in activated sludge at low concentrations of non-aggregated bacterial cells and low BOD of treated wastewater. Protozoa reduce the concentration of non-aggregated bacterial cells and viruses in treated wastewater.

Activated-Sludge Flocs

The biomass of activated sludge is selected as suspended flocs of aggregated bacterial cells. The size of the flocs is 50–1000 mm (=1 mm). The aggregation of the microbial cells in the flocs permits sedimentation of flocs by gravity in 10–30 min.

Flocs are cell aggregates formed by hydrophobic interactions, electrostatic and salt bridges between the surfaces of cells, as well as cell embedding in microbial slime. A wide range of macromolecular products, secreted from cells as an extracellular slime, may be implicated in floc formation. However, slime-producing bacteria form only small-sized and mechanically unstable flocs.

Filamentous Bacteria in Activated-Sludge Flocs

The filamentous bacteria from genera *Nocardia*, *Thiothrix*, *Sphaerotilus*, *Beggiatoa*, and *Microthrix* appear to be responsible for the enlargement and physical stability of the flocs. However, a high content of filamentous bacteria in aeration tanks causes "filamentous bulking" of activated sludge and its low settling velocity, so bacterial biomass cannot be separated from wastewater in the settling tank. The dominance of filamentous bacteria is promoted by

1. Low concentrations of dissolved oxygen in aeration tanks due to a low aeration rate or a high organic load (bacteria from genus *Sphaerotilus*).
2. Low concentrations of nutrients due to low F/M ratio or presence of lipids (bacteria from genus *Nocardia* and *Microthrix*).

3. Presence of sulfides in aeration tanks due to their production in anaerobic digesters (microaerophilic bacteria from genera *Beggiatoa* and *Thiothrix* with liquid sulfur granules accumulated in cells).

Control of Filamentous Bulking of Activated Sludge

1. To overcome low dissolved-oxygen bulking, a bulk dissolved oxygen (DO) concentration must be increased above 2 mg/L by applying a higher aeration rate or lower organic loading.
2. To overcome low F/M bulking, the selector tank must be placed before the aeration tank. A selector is a mixing tank where raw sewage and recycled activated sludge from the settling tank are mixed for 15–30 min prior to the aeration tank. Floc-forming aggregating bacterial cells accumulate the energy-storage polymer polyhydroxybutyrate (PHB), which is used in aeration tanks and gives preference to floc-forming bacteria under low F/M ratio. Filamentous bacteria do not accumulate PHB in selector and consequently starve in the aeration tank.
3. To overcome bulking caused by the presence of sulfides, the source of sulfides must be eliminated, or sulfides must be oxidized by hydrogen peroxide before the activated-sludge process. Addition of chlorine into the aeration tank can also selectively kill filamentous bacteria because of the high resistance of bacterial cells, embedded into slide, to chlorine.

Foaming

Foaming diminishes the efficiency of wastewater treatment and deteriorates work space hygiene. The formation of foam in aeration tanks could be due to

1. The presence of non-degraded surfactants in wastewater
2. The presence of lipids in wastewater, forming surfactants after biohydrolysis
3. The formation of biosurfactants by *Actinomycetes*, mainly from the genera *Nocardia* and *Rhodococcus*
4. The long solids retention time (SRT) and HRT in the aeration tank and accumulation of foam-producing substances

Suspended Microbial Aggregates (Granules) Used in Aerobic Treatment of Wastewater

Microbial granules with diameters from 1 to 10 mm are formed in 10–14 days of cultivation, when aerobically grown in a sequencing batch reactor (SBR)

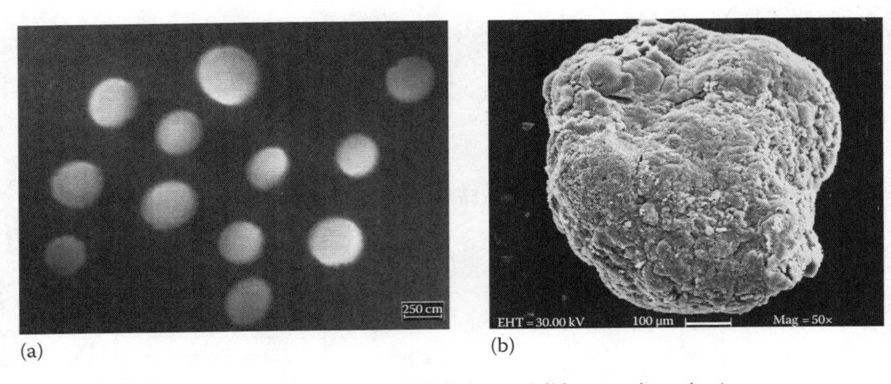

Figure 17.3 Microbial granules under (a) light and (b) scanning electron microscopes.

(Figure 17.3). These microbial aggregates have short settling times and the ability to treat high-strength or toxic wastewater and can be used instead of conventional microbial flocs.

Granules are formed due to

- Intensive aeration, which hammers flocs, forming more dense aggregates
- Selection of aggregating cells due to the retention of aggregates able to settle down in short period of time, usually from 2 to 10 minutes.
- Washing of non-aggregated cells and flocs from the reactor (Figure 17.4)

Due to the dense aggregation of cells, the rate of mass transfer of nutrients and metabolites between bulk medium and granular matrix may not be sufficient to ensure normal cell metabolism in the granule interior. The layers of aerobic, facultative anaerobic (Figure 17.5), and obligate anaerobic bacteria in the matrix of aerobically grown microbial granules perform different functions and communicate with the environment through channels and pores (Figure 17.6).

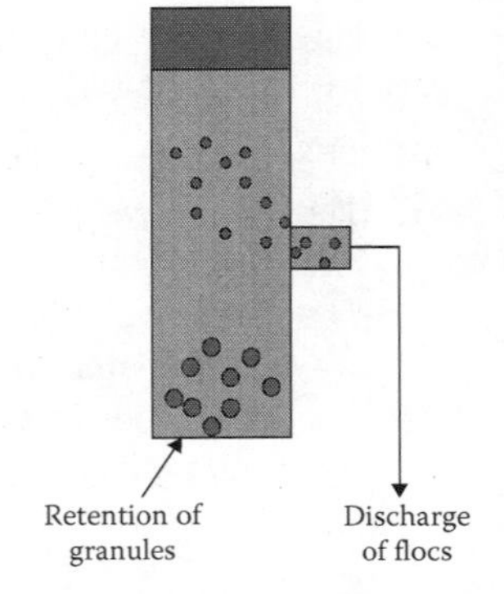

Figure 17.4 Retention of granules and discharge of flocs during discharge of effluent in column sequencing batch reactor.

Removal of Nutrients

Nitrogen and phosphorus, major environmental pollutants triggering eutrophication of aquatic systems at concentrations of 1 mg N/L and 0.05 mg P/L, can be removed biologically in the activated-sludge process.

Nitrogen is removed biologically by aerobic nitrification of ammonium followed by anaerobic denitrification of nitrate. Ammonia removal from wastewater is also important because it is toxic to aquatic animals at concentrations of about several mg N/L. The first

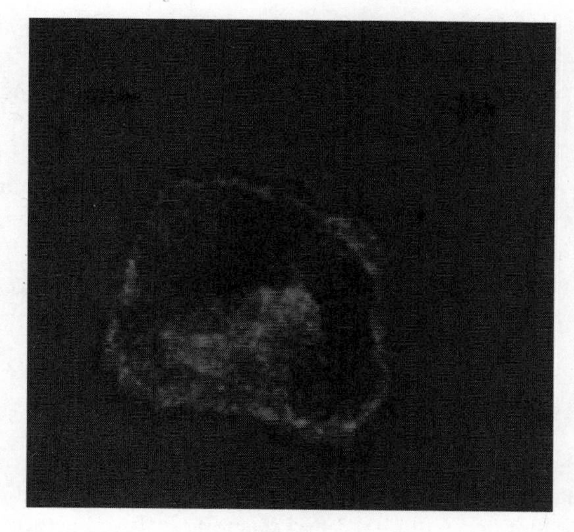

Figure 17.5 Distribution of facultative anaerobic enterobacteria (red fluorescence) in microbial granule.

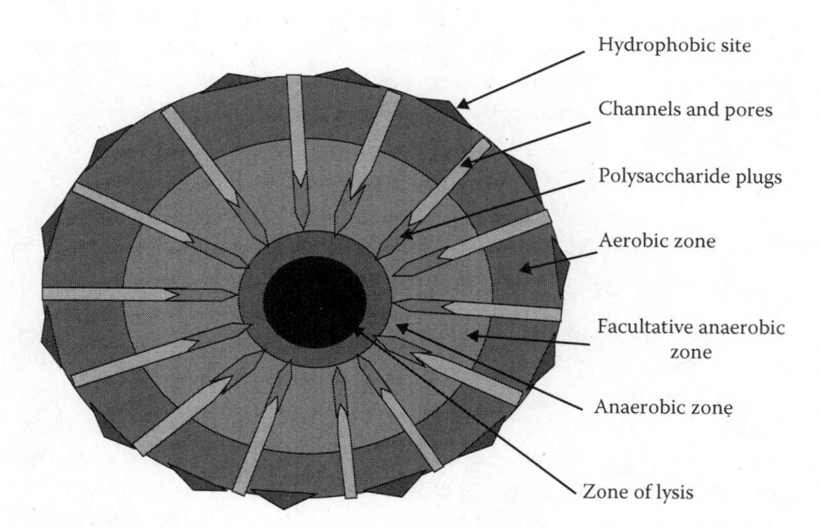

Figure 17.6 (See color insert following page 292.) Schematic of microbial granule structure.

stage of nitrification is the oxidation of ammonium to nitrite by ammonium-oxidizing bacteria and the second stage is the oxidation of nitrite to nitrate performed by nitrite-oxidizing bacteria. To maintain these slow-growing bacteria in aeration tank "sludge age" = "solids retention time" must be several days.

If nitrification in aerobic tank is active enough, nitrate is reduced in the settling tank by facultative anaerobic bacteria and produces N_2 gas bubbles that can attach to sludge particles, which can float to the top of the settling tank. This rising sludge can decrease the quality of treated wastewater.

The most effective technologies for the removal of nutrients include the alternation of the aerobic and anaerobic stages with the recycling of wastewater and biomass between them. For nitrogen removal, it is the combination of nitrification with denitrification and partial nitrification to nitrite with anammox process. For phosphorus removal, it is the combination of anaerobic stage where energy for enhanced consumption of phosphate is accumulated in cells followed by the aerobic stage, where phosphate is accumulated in biomass as polyphosphate and removed from wastewater in the settling tank altogether with activated sludge.

Aerobic Reactors

Wastewater can be treated aerobically in suspended biomass stirred-tank bioreactors, plug-flow bioreactors, rotating-disk contactors, packed-bed fixed biofilm reactors (or biofilters), fluidized-bed reactors, diffused aeration tanks, airlift bioreactors, jet bioreactors, membrane bioreactors, and upflow-bed reactors.

The exhaust air may contain volatile hazardous substances or intermediate biodegradation products. Therefore, the air must be treated as secondary hazardous wastes by physical, chemical, physicochemical, or biological methods. Secondary waste is also considered as biomass of microorganisms, which can accumulate hazardous substances. This hazardous liquid or semisolid waste must be properly treated, incinerated, or disposed.

Aerobic Treatment of Wastewater with a Low Concentration of a Hazardous Substance

Industrial wastewater with low concentrations of hazardous substances may reasonably be treated using biotechnological methods such as granular activated carbon (GAC) fluidized-bed reactors or cometabolism. GAC or other adsorbents ensure sorption of hydrophobic hazardous substances on their particle surfaces. Microbial biofilms can also be concentrated on the surface of these particles and can biodegrade hazardous substances with higher rates compared to situations when both substrate and microbial biomass are suspended in the wastewater.

Cometabolism refers to the simultaneous biodegradation of hazardous organic substances (which are not used as a source of energy) and stereochemically similar substrates, which serve as a source of carbon and energy for microbial cells. Biooxidation of the hazardous substance is performed by the microbial enzymes due to stereochemical similarity between the hazardous substance and the substrate. The best known applications of cometabolism are in the biodegradation/detoxication of chloromethanes, chloroethanes, chloromethylene, and chloroethylenes by bacterial enzyme systems using the oxidization of methane or ammonia as the main source of energy. In practice, the bioremediation is achieved by adding methane or ammonia, oxygen (air), and the biomass of methanotrophic or nitrifying bacteria to groundwater polluted by toxic chlorinated substances.

Aerobic Digestion of Activated Sludge

The excess of activated sludge produced on municipal wastewater treatment plants can be removed not only by anaerobic digestion, but also by using aerobic digestion of biomass, which have a high biodegradation rate and produce an odorless stabilized sludge. However, the process consumes large amounts of energy needed to supply air for the oxidation of activated sludge (biomass):

$$CH_{0.8}O_{0.5}N_{0.12} + 1.02O_2 \rightarrow CO_2 + 0.54H_2O + 0.12NH_3$$

Combinations of Aerobic Treatment of Wastewater with Other Treatments

To intensify the aerobic treatment of industrial wastewater the following pretreatments can be used:

1. Mechanical disintegration or suspension of hydrophobic substances to improve the reacting surface in the suspension and increase the rate of biodegradation.
2. Removal from wastewater or diminishing the concentration of hazardous substances by sedimentation, centrifugation, filtration, flotation, adsorption, extraction, ion exchange, evaporation, distillation, freezing separation, as well as preliminary oxidation by H_2O_2, ozone, and Fenton's reagent to produce active oxygen radicals; preliminary photooxidation by UV; and electrochemical oxidation of hazardous substances.

Application of Microaerophilic Microorganisms in Biotechnological Treatment

Some aerobic microorganisms prefer low concentrations of dissolved oxygen in the medium for growth, for example, concentrations below 1 mg/L. These microorganisms include filamentous sulfide-oxidizing bacteria from the genus *Beggiatoa*; pathogenic bacteria from the genera *Campylobacter, Streptococcus*, and *Vibrio*; microaerophilic spirilla from the genus *Magnetospirillum*; and neutrophilic iron-oxidizing bacteria. Iron-oxidizing bacteria can produce sheaths or stalks that act as organic matrices upon which the deposition of ferric hydroxides can occur. Some microaerophiles are active biodegraders of organic pollutants in post-accident sites, while other microaerophiles form H_2O_2 to oxidize xenobiotics. Ferric hydroxide–containing sheaths of neutrophilic iron-oxidizing bacteria adsorb heavy metals and radionuclides from water.

Aerobic Biofilm Reactors

Aerobic biofilm reactors are used for the oxidation of organic matter, nitrification, denitrification, and anaerobic digestion of wastewater.

The trickling filter has been in use for a century. It consists of a tank with a carrier (medium) with attached microbial biofilm, distributor of wastewater for uniform hydraulic load to surface of trickling filter, drain collector of treated wastewater, and settling tank for separation of particles from treated wastewater. The present-day carrier (medium) is made from PBC or polypropylene, has a low bulk density, and a surface area up to $200\,m^2/m^3$. Permanent recirculation of treated wastewater increases the efficiency of the biotreatment.

RBCs are disks submerged up to 30%–40% of their height in wastewater and slowly rotated for aeration and increased mass transfer between biofilm attached to the disks and wastewater.

The submerged packed-bed biofilter is filled with different plastic media such as structured packing, dump packing, or fiber mash pad, where the microbial biofilm is attached. It is aerated with air during the upflow of wastewater through the biofilter. The most important parameters of the plastic medium are specific surface area (m^2/m^3), void fraction (%), plugging potential (days of biofilter exploitation), and specific weight (kg/m^3). Every medium requires the selection of a specific microbial community adapted to the medium.

Development of Biofilm in Aerobic Biofilm Reactors

The surface of the carrier is colonized with natural microorganisms, introduced microbial communities, and selected enrichment or pure cultures. Colonization is due to hydrophobic forces, electrostatic interactions, salt bridges, or binding with microbial polysaccharides. The biofilm grows during wastewater treatment and later the inner layer of this biofilm becomes anaerobic and can die because of the production of organic acids and the low concentration of nutrients supplied from wastewater into biofilm. Dead cells are subsequently sloughed off the surface. Due to strong gradients of oxygen and nutrients in the biofilm, its microbial community is very diverse and can include aerobic, facultative anaerobic, anaerobic bacteria, nitrifying and denitrifying bacteria, and protozoa attached to biofilm and feeding with suspended bacterial cells.

18

Value-Added By-Products of Environmental Engineering

Organic Wastes as a Raw Material for Biotechnological Transformation

Many organic wastes may be used in further biotechnological transformation for value-added products such as enzymes, single-cell proteins, fuels, chemicals, biodegradable plastics, and pharmaceutical preparations. For instance, the inedible parts of plants, agricultural residues from the harvesting, and food-processing waste can be used as raw materials for biotransformation, amounting to more than 13×10^9 ton/year. These organic wastes contain valuable components, such as starch in potato and cassava pulp, pectin in apple pomace, sucrose in molasses, cellulose and hemicellulose in wood, garden waste, and rice hull. The European Union alone produces 10^9 ton of agricultural waste, 0.5×10^9 ton of garden and forestry waste, and 0.25×10^9 ton of organic waste from food-processing industry. A significant part of these wastes can be transformed into value-added products.

Biotechnological Products from Organic Wastes

Vegetable and fruit processing wastes contain mainly starch, cellulose, and organic acids. These substances can be used for biotechnology as raw material. Table 18.1 lists various biotechnological products that can be produced from different food-processing wastes.

Production of Enzymes from Wastes

An enzyme is a biological catalyst that increases the rate of specific biochemical reactions severalfold at room temperature. The classification of enzymes

TABLE 18.1 Value-Added By-Products from Food-Processing Wastes

Waste	Value-Added Product	Microorganism
Fruit and berries pomace	Enzyme polygalacturonase	*Lentinus edodes*
Wheat bran, sunflower flour, coffee husk, soybean meal, rice bran, corn bran, rice hull, aspen wood, sweet potato residue, waste hair	Enzymes	*Aspergillus* sp., *Penicillium* sp., *Rhizopus* sp., *Bacillus* sp., *Trichoderma* sp.
	Proteases (acidic, neutral, and alkaline)	
The mixture of sugarcane bagasse with wheat bran or orange bagasse	Enzyme pectin lyase and polygalacturonase	*Thermoascus aurantiacus*
Lemon pulp	Enzyme pectinase	*Trichoderma viride*
Wheat bran, rice bran, apple pomace	Enzyme pectinase	*Bacillus* sp.
Wheat bran, rice bran, coconut oil cake and corn flour	Enzyme inulinase	*Staphylococcus* sp. *Kluyveromyces marxianus*
Lignocellulosic wastes, sugarcane bagasse	Enzymes cellulases β-glucosidase	*Aspergillus ellipticus* *Aspergillus fumigatus*
Banana waste	Enzymes α-amylase and cellulases	*Bacillus subtilis*
Tea waste	Enzyme glucoamylase	*Aspergillus niger*
Wheat bran, rice straw, and minerals	Enzymes cellulases	*Aspergillus ustus, Botrytis* sp., *Trichoderma* sp. *Sporotrichum pulverulentum*
Copra paste and spent coffee	Enzyme β-mannanase	*Aspergillus oryzae* *Aspergillus niger*
Carrot-processing waste	Lactic acid	*Rhizopus oryzae*
Hemicellulosic hydrolysate from *Pinus taeda* chips	Lactic acid	*Rhizopus oryzae*
Kumara	Citric acid	*Aspergillus niger*
Pineapple waste	Citric acid	*Aspergillus foetidus*
Sugarcane bagasse	L-Glutamic acid	*Brevibacterium* sp.
Sweet potato residue	Antibiotic tetracycline	*S. viridifaciens*
Soybean curd residue, okara	Antibiotic iturin A	*Bacillus subtilis*
Wheat bran	Plant growth hormone Gibberellic acid	*Gibberella fujikuroi*
Cassava flour, sugar cane bagasse	Plant growth hormone Gibberellic acid	*Gibberella fujikuroi*
Hydrolyzed tomato pomace	Vitamin B_{12}	*Propionibacterium shermanii*
Prawn-shell waste	Single-cell protein	*Candida* spp.
Olive pomace	Poultry feed enriched by protein	*Candida utilis*

TABLE 18.1 (continued) Value-Added By-Products from Food-Processing Wastes

Waste	Value-Added Product	Microorganism
Apple pomace	Animal feed enriched by protein	*Candida utilis*
Hydrolyzed potato starch waste	Exopolysaccharide pullulan	*Aureobasidium pullulans*
Grape skin pulp extract, starch waste, olive oil waste effluents, and molasses	Exopolysaccharide pullulan	*Aureobasidium pullulans*
Sent malt grains, apple pomace, grape pomace, and citrus peels	Exopolysaccharide xanthan	*Xanthomonas campestris*
Olive mill wastewaters	Exopolysaccharide xanthan	*Xanthomonas campestris*
Coconut waste	Bioinsecticide	*Bacillus thuringiensis*
Sugar beet pulp	Flavor vanillin	*Pycnoporus cinnabarinus*
Cassava bagasse	Fruity aroma compounds	*Kluyveromyces marxianus*
Waste material of the pineapple juice production	Ethanol	*Zymomonas mobilis*
Pineapple cannery waste	Ethanol	*Saccharomyces cerevisiae*
Corn fiber (after acidification)	Butyrate	*Clostridium tyrobutyricum*
Sugar refinery wastewater	Hydrogen	*Rhodobacter sphaeroides*
Starch-manufacturing wastes (sweet potato starch residue)	Hydrogen	*Clostridium butyricum*
Waste water of a starch factory	Hydrogen	Anaerobic microflora
Organic waste (wheat grains)	Hydrogen	*Bacillus licheniformis*
Organic waste (whey)	Hydrogen	*Rhodopseudomonas*
Olive pomace	Biogas	Anaerobic microflora
Molasses	Biosurfactant	*Bacillus subtilis*
Potato processing wastes	Biosurfactant	*Bacillus subtilis*
Potato processing waste	Poly-β-hydroxybutyrate	*Alcaligenes eutrophus*
Organic waste	Compost	Microbial consortium

by the International Union of Biochemists includes six functional classes: (a) oxidoreductases act on different chemical groups to add or remove hydrogen atoms; (b) transferases transfer functional groups between donor and acceptor molecules; (c) hydrolases add water across a bond hydrolyzing it; (d) lyases add water, ammonia, or carbon dioxide across double bonds or remove these molecules to produce double bonds; (e) isomerases carry out all kinds of isomerization; (f) ligases catalyze reactions in which two chemical groups are joined with the use of energy from ATP.

Enzymes are widely used in food industry, textile industry, as components of detergents, in medicine, and cosmetics. The industrial production of many enzymes includes cultivation of microorganisms on a medium. Microorganisms can produce intracellular enzymes (inside cell) and extracellular enzymes

(enzymes that are excreted from cells into the medium). Intracellular and extracellular enzymes must be extracted from the biomass or medium after cultivation and then purified.

Food-processing waste can be used as a substrate for the low-cost microbial production of amylolytic enzymes like α-amylase, which is widely used in the food, textile, and paper industries.

Pectinolytic enzymes are widely used to increase yields, improve liquefaction, clarification, and filterability of fruit juices for better maceration and extractability of plant tissues in the fruit processing industry, and for clarification of wine. Food-processing waste such as orange peels, orange finished pulp, wheat bran, sugar beet pulping waste, apple pomace, cranberry pomace, and strawberry pomace can be used for the production of microbial polygalacturonase.

Cellulases are a group of enzymes that catalyze the bioconversion of cellulose to the soluble sugar, glucose. Cellulases can be produced from wheat brain, rice straw, and banana residues.

Production of Organic Acids

Organic acids are used in the food and pharmaceutical industries as preservatives or chemical intermediates. There is also a growing demand for these acids in the production of biodegradable polymers such as polylactic acid and polyhydroxyalkanoates (PHAs). Such acids as formic, acetic, propionic, lactic, pyruvic, succinic, fumaric, maleic, malic, itaconic, tartaric, citric and isocitric acids, and valeric and isovaleric acids can be obtained by the batch or continuous aerobic or anaerobic cultivation in liquid medium using suspended microbial culture or immobilized biomass. After biomass separation, organic acids can be isolated and purified by liquid extraction, chromatography, evaporation, ultrafiltration, reverse osmosis, dialysis, electrodialysis with bipolar membrane, crystallization, precipitation, and drying.

L(+)-Lactic acid (2-2-hydropropionic acid), $CH_3CHOHCOOH$, is an important chemical used as acidulant, flavor, and preservative in the food, cosmetics, pharmaceuticals, leather, and textile production. The biggest application of lactic acid is the production of polylactic acid, which is a biodegradable plastic. The worldwide biotechnological production of L-lactic acid is 80,000 ton.

Lactic acid bacteria are used for the biological production of lactic acids. Originally, lactic acid bacteria are inhabitants of biosurfaces and leaves of plants. They do not have the ability to synthesize some vitamins and amino acids, and therefore demand those factors for their growth. The most important producers of lactic acid are the strains of *Lactobacillus helveticus*, *Lactobacterium plantarum*, *L. lactis*, *L. delbrueckii*, and *L. bulgaricus*. The majority of lactic acid producers are mesophilic bacteria with an optimal temperature for growth at 30°C ± 2°C, but there are some thermophilic strains such as *L. bulgaricus*, *L. delbrueckii*, and *Streptococcus thermophilus*, which grow at temperatures higher than 40°C. Some lactic acid bacteria produce only lactic acid (homofermentation), and others can convert sugars into lactic acid, carbon dioxide, and ethanol (heterofermentation). Different wastes have been proposed to be used for lactic acid production. For example, whey permeate, an inexpensive and abundant by-product of the dairy industry, especially from ultra filtration–based cheese manufacturing, can be used for the large-scale production of lactic acid.

An inexpensive substrate for lactic acid production may be agricultural waste, for example, wheat straw, containing cellulose and hemicellulose, which can be converted into soluble sugars by chemical or enzymatic hydrolysis for further use for microbial synthesis of L(+)-lactic acid. Europe's wheat production is 184 million tons and the average yield of straw is 1.3–1.4 kg/kg of grain. Wheat straw consists of cellulose (35%–40%), hemicellulose (30%–35%), and the hard-to-biodegrade polymer, lignin. To use lignocellulosic material as a substrate for biotechnological processes, it is necessary to separate cellulose and hemicellulose from lignin, and then to produce free sugars from cellulose and hemicellulose through their depolymerization. Glucose is produced from cellulose, and a mixture of monosaccharides including hexoses (mannose, galactose, glucose) and pentoses (arabinose and xylose) can be obtained from hemicellulose. The presence of xylose, the predominant monosaccharide released from hemicellulose, which is not used by the majority of microorganisms, complicates the usage of hemicellulose as a raw material for biotechnology.

Production of Flavors

The world market of aroma chemicals, fragrances, and flavors is about $10 billion. Most of the flavoring compounds are presently produced by chemical synthesis or extraction from natural materials. Microbiological bioconversion of food waste and agricultural residues is a cost-effective alternative method for natural aroma compounds production. The fungi *Ceratocystis fimbriata* produces a fruity aroma on sugarcane bagasse supplemented with a synthetic medium containing glucose, whereas the addition of leucine or valine amino acids results in a strong banana aroma. If these fungi grow on coffee husk supplemented with glucose, a strong pineapple aroma is produced. The yeast *Kluyveromyces marxianus* produces fruity aromas from cassava bagasse and palm bran. The fungi *Aspergillus niger* and *Pycnoporus cinnabarinus* produce vanillin from wheat bran.

Production of Polysaccharides

The polysaccharide pullulan is a water-soluble exocellular homopolysaccharide produced by fungi *Aureobasidium pullulans*. Pullulan forms solutions with a high viscosity at a relatively low concentration and is used for the production of oxygen-impermeable films and fibers, which are biodegradable, transparent, oil resistant, and impermeable to oxygen. Different organic wastes can be used for the production of pullulan, such as grape skin pulp extract, starch waste, olive oil waste effluents, molasses, and potato processing waste.

Xanthan gum is another widely used water-soluble heteropolysaccharide and is the most important microbial polysaccharide from a commercial point of view, with a worldwide annual production of 30,000 ton. It is used in food processing for emulsification, stabilization, temperature stability, improvement of rheological properties, and is an important ingredient in dietary, cosmetic, and pharmaceutical products. Xanthan is also used in the petroleum industry for preparation of drilling fluids and enhanced oil recovery. The production of xanthan by the bacteria *Xanthomonas campestris* from food-processing waste such as spent malt grains, apple pomace, grape pomace, citrus peels, olive mill wastewaters, and waste sugar beet pulp has been proposed.

Edible Mushroom Production

The worldwide commercial mushroom production constitutes approximately 5×10^6 ton of fresh weight and costs $1.2 billion annually. Different organic and lignocellulosic wastes can be used as substrates for cultivation of edible mushrooms. *Pleurotus ostreatus* (oyster mushrooms), *Agaricus campestris*, and *Agaricus bisporus* (field mushrooms) are cultivated all over the world, while *Lentinus edoides* (shiitaki), *Volvariella volvacea* (edible straw mushroom), and *Ganoderma lucidum* (red reishi) are cultivated mainly in Asia. In total, at least 20 mushrooms species are cultivated for commercial purposes such as food consumption as well as medicinal food.

The cultivation of mushrooms includes two major steps: preparation of the compost or solid medium and mycelium growth until fructification. For production of compost or medium for mushroom cultivation, different waste materials can be used such as wood chips, sawdust, hay, maize waste, paddy straw, cassava bagasse, waste paper, cottonseed hulls, water hyacinth, apple pomace, oil palm bunch, rice husk, banana leaves, cheese whey, horse manure, chicken manure, and others.

The wastes are mixed and composted for 8 days. The compost is then packed in boxes and pasteurized. After inoculation and mycelium growth, compost is covered with a mixture of soil, peat, and chalk. The optimal temperature for mycelium growth is 24°C, and the optimal temperature for fruiting body production is usually 14°C–18°C. The yield of mushrooms is 0.5–1 kg per 1 kg of compost dry matter.

Production of Biodegradable Plastics

About 100 million tons of petrochemical plastics are produced in the world annually. The use of biodegradable plastic of biological origin can diminish environmental pollution caused by disposed plastics, but the cost of biodegradable plastic is significantly higher than any petrochemical one. Therefore, bioplastics are used for manufacturing of high-value medical items.

Cheap and renewable raw materials and inexpensive biotechnological methods can make the price of microbial biodegradable plastics competitive compared to petrochemical plastics. For example, starch-containing food-processing wastes can be used for the production of the biopolymers poly-β-hydroxybutyrate (PHB) and polyhydroxyalkanoates (PHAs). The bacteria *Alcaligenes eutrophus* are used often, but the technologies based on pure cultures are significantly more expensive than the technologies based on enrichment microbial cultures. Lactic acid can be produced from food-processing wastes and used as a raw material for the synthesis of polylactic acid, which is a biodegradable plastic.

Production of Animal Feed

The amount of agricultural lignocellulose waste produced annually is 123×10^6 ton. The microbial bioconversion of lignocellulose wastes into protein-containing animal feed is performed using chemical or enzymatic hydrolysis and growth of different species of yeasts in the produced solution. Biomass of dry yeast contains 50% of protein with a high content of essential amino acids and vitamins.

Use of Organic Waste for Production of Fungi for Soil Bioremediation

Lignin-degrading white-rot fungi have the unique ability to degrade almost all organic substances, including toxic environmental pollutants such as munitions waste, pesticides, polychlorinated biphenyls, polycyclic aromatic hydrocarbons, bleach-plant effluents, synthetic dyes, synthetic polymers, and wood preservatives. White-rot fungi including several edible mushroom species produce extracellular peroxidases, forming oxygen radicals, which randomly degrade organic compounds including lignin. Straw and wood sawdust from wheat and other crops are used for cultivation of edible mushroom but the mycelium remaining after gathering of the fruit bodies of edible mushrooms is the most cost-effective and safe biological agent for soil bioremediation.

Solid-State Fermentation of Agricultural and Food-Processing Wastes

Solid-state fermentation (SSF) is a process in which microorganisms grow on or within solid substrates in the absence of free water. The basic component of SSF is the "solid substrate bed" in tray or rotating drum containing the solids with voids. Solid substrate bed contains water for growth of microorganisms. The solid material in this process acts both as a physical support and as a source of nutrients. To simplify product isolation from the medium, an inert material, for example, polyurethane foam, may be used instead of natural raw material such as wheat bran.

SSF has been conventionally more applicable for filamentous fungi, but yeast and even bacteria can also be used for biotechnological production by solid-state fermentation. SSF is of special interest to countries with an abundance of agro-industrial residues that can be used as inexpensive raw materials. SSF has many advantages in processing agro-industrial residues as compared with submerged fermentation: lower energy requirements, process simplicity, cheaper aeration, absence of rigorous control of fermentation parameters, and production of smaller quantity of wastewater. SSF was studied for the production of such value-added products as alkaloids, plant growth factors, enzymes, organic acids, biopesticides, biosurfactants, biofuel, aroma compounds, and the enhancement of crop residues using protein and vitamins.

The most important parameters for solid-state fermentation are water content, temperature, pH, and aeration. Water content for solid-state fermentation varies from 10% to 80%. The minimum water content for molds and yeast growth is 10%–20%. Bacteria usually can grow at higher water content than that needed by molds and yeasts. The optimal water content is specific for different microorganisms and for different physiological processes such as growth, sporulation, and production of primary or secondary metabolites.

Production of Fuel Ethanol

Production of ethanol as an alternative liquid fuel for transportation can reduce oil consumption for fuel production and protect the environment. All automobile manufacturers produce vehicles that can readily use 10% ethanol fuel blends for fuel. Starch, the most commonly used substrate, is the main raw material, cost-wise, in fuel ethanol production. The biotechnological production of ethanol is based on enzymatic hydrolysis of starch and fermentation of produced sugars into alcohol by the yeast *Saccharomyces cerevisiae*. Alternative substrates include different food-processing wastes and cellulose-containing agricultural residues. Cellulose-containing wastes are transformed into sugars using chemical hydrolysis with acids or enzymatic hydrolysis, followed by alcohol fermentation with yeasts.

Production of Hydrogen, Methane, and Biodiesel

Hydrogen is the most promising energy source for future use. Microbial production of hydrogen may be performed by phototrophic bacteria, for example,

Rhodobacter sphaeroides, using light energy or anaerobic bacteria, for example, *Clostridium thermocellum*, fermenting organic wastes.

Methane is produced by the anaerobic digestion of waste biomass of plants, animals, and microorganisms in anaerobic digesters on municipal wastewater treatment plants, sanitary landfills, covered lagoons, and wetlands (see chapter 16 on anaerobic digestion of organic wastes).

Algae cultivated in special ponds using light energy, CO_2 from incinerating or energy-generating plants, and nutrients from municipal wastewater treatment plant can accumulate oils inside cells up to 80% of biomass. This oil can later be transformed into biodiesel by chemical esterification of oil fatty acids with methanol and using direct thermochemical pyrolysis of biomass into biodiesel.

Production of Organic Fertilizer

Composting is the main environmental technology that has been used for several thousands of years for the bioconversion of organic wastes of human, farming, and agricultural origins into fertilizers. It is the aerobic, mesophilic, or thermophilic microbial decomposition of organic and inorganic constituents into a humus- and nutrient-rich stable organic fertilizer. The details of the composting are described in the next chapter.

Biorecovery of Metals from Mining and Industrial Wastes

Mine-drainage water and many industrial wastewaters contain metals that can be recovered. Methods such as bioleaching, bioprecipitation, and biosorption are used for biological recovery of metals from waste.

In bioleaching technology, metals are solubilized from poor ores or anaerobic sewage sludge due to microbial oxidation of sulfides or sulfur into solution of sulfuric acid, dissolving minerals.

In biosorption technology, metal ions can be adsorbed on the surface of microbial cells. Microbial biomass has also been successfully applied for metal recovery from industrial effluents. This biomass may be the waste after different biotechnological methods, for example, the biomass of the yeasts *S. cerevisiae*, waste from beer fermentation industry, biomass of *Penicillium chrysogenus*, waste of antibiotic production, biomass of *A. niger*, waste from citric acid production, or that obtained by microbial cultivation on food-processing wastes and agricultural residues. To enhance the adsorption capacity of biomass and metal uptake, modification by chemical treatment is carried out.

In bioprecipitation, metals such as Cu, Ni, Cr, Zn, and Fe are removed from the waste streams due to the formation of dihydrogen sulfide during sulfate bioreduction and precipitation of sulfides of metals. Sulfate-reducing bacteria couple oxidation of organic compounds with the reduction of sulfate. Preferable carbon sources for bacterial sulfate reduction (more correctly, dissimilative sulfate reduction) are organic acids such as lactic, pyruvic, propionic, formic, succinic, fumaric, malic, benzoic acids, and alcohols such as ethanol, propanol, and phenol. There is a large phylogenetic and physiological diversity of sulfate-reducing bacteria. They can be found in any anaerobic environment (redox potential must be lower than −100 mV), which is rich in organic matter, living at temperatures from 0°C to 70°C and pH of 5–9.5. H_2S that is produced by the biological reduction of sulfate reacts with heavy metal ions to form an insoluble metal sulfide:

$$SO_4^{2-} + CH_3COOH + 2H^+ + \text{sulfate-reducing bacteria} \rightarrow H_2S + 2CO_2 + 2H_2O$$

$$H_2S + Fe^{2+} \rightarrow FeS \downarrow + 2H^+$$

Recovery of chromium can be performed by the reduction of Cr(VI) to Cr(III) associated with its precipitation. Chromate, containing Cr(VI), is soluble and toxic, whereas Cr(III) is less toxic and tends to form insoluble hydroxide. Chemical Cr(VI) reduction by H_2S is mediated by sulfate-reducing bacteria. The biological reduction of Cr(VI) is mediated by chromium-reducing bacteria.

Recovery of Phosphate and Ammonia

Aluminum and iron salts, usually alum, $Al_2(SO_4)_3 \cdot 18H_2O$, sodium aluminate, $NaAlO_2$, polyaluminum chloride, ferric chloride, $FeCl_3$, ferrous sulfate, $FeSO_4$, and ferric sulfate, $Fe_2(SO_4)_3$, are used for the precipitation of phosphate from wastewater. Iron ore, an inexpensive source of iron, can be used for phosphate recovery from wastewater using iron-reducing bacteria producing dissolved ferrous ions for precipitation of phosphate as $FeHPO_4$. This biotechnological method is used for the recovery of phosphate from the effluent of anaerobic digesters of municipal wastewater treatment plants and from stormwater.

Chemical recovery of ammonium and phosphate from wastewater is performed by the formation of struvite, magnesium ammonium phosphate hexahydrate ($MgNH_4PO_4 \cdot 6H_2O$), but it requires the addition of Mg salt and must be performed at a pH higher than 9. A product of the precipitation process is the slow-releasing ammonium phosphate fertilizer.

19

Biotreatment of Industrial Hazardous Wastes

Donors and Acceptors of Electrons in Biotreatment

Biotreatment of industrial pollutants can be performed by hydrolysis, solubilization/precipitation/volatilization, or by oxidation-reduction of the pollutants. Donors of electrons in oxidation-reduction are organic or reduced inorganic substances such as H_2S, S, NH_4^+, and Fe^{2+}. Acceptors of electrons are external organic compounds, or are a part of oxidized organic molecules in fermentation, and such inorganic compounds as O_2, NO_3^-, NO_2^-, Fe^{3+}, SO_4^{2-}, and CO_2.

Energetic Efficiency of Biooxidation–Bioreduction

The energetic efficiency of biodegradation per mole of transferred electrons and growth yield is increased in the following sequences of the couples of donors/acceptors of electrons used for biological energy generation:

CO_2/H_2 (methanogenesis)
NH_4^+/NO_2^- (anammox process)
CH_2O/CH_2O (fermentation)
Fe^{2+}/NO_3^- (denitrifying iron reduction)
S/NO_3^- (autotrophic denitrification)
CH_2O/SO_4^{2-} (dissimilative sulfate reduction)
CH_2O/Fe^{3+} (iron reduction)
CH_2O/NO_3^- (denitrification)
H_2S/O_2 (sulfide oxidation)
Fe^{2+}/O_2 (ferrous oxidation)
NH_4^+/NO_2^- (nitrification)
CH_2O/CH_2O (aerobic respiration)

Aerobic Treatment of Xenobiotics

Such xenobiotics as aliphatic hydrocarbons and derivatives, chlorinated aliphatic compounds (methyl-, ethyl-, methylene-, and ethylene-chlorides), aromatic hydrocarbons and derivatives (benzene, toluene, phthalate, ethylbenzene, xylenes, and phenol), polycyclic aromatic hydrocarbons, halogenated aromatic compounds (chlorophenols, polychlorinated biphenyls, dioxins and relatives, and DDT and relatives), AZO dyes, compounds with nitro groups (explosive-contaminated waste and herbicides), and organophosphate wastes can be treated effectively by aerobic microorganisms.

Secondary Hazardous Wastes

The biological treatment of non-hazardous wastewater can release volatile hazardous substances in the exhaust air. Therefore, air must be treated as secondary hazardous wastes by physical, chemical, physicochemical, or biological methods.

Secondary hazardous wastes may include the biomass of microorganisms that accumulate hazardous substances or intermediate products of their biodegradation. This hazardous liquid or semi-solid waste must be properly treated, incinerated, or disposed.

Treatment of Wastewater with a Low Concentration of a Hazardous Substance

Wastewater with low concentrations of hazardous substances may be treated using a granular activated carbon (GAC) fluidized-bed reactor. GAC or other adsorbents ensure the sorption of hazardous substances on the surface of GAC or other adsorbent particles. Microbial biofilms can also be concentrated on the surface of these particles and can biodegrade hazardous substances with higher rates compared to situations when both substrate and microbial biomass are suspended in the wastewater.

Biodegradation Using Co-Oxidation/Co-Metabolism

Co-metabolism or co-oxidation refers to the simultaneous biodegradation of two organic or inorganic substances, one of which is a hazardous organic substance that is not used as a source of energy and the other is a stereochemically similar substance, which serves as a source of energy for microbial cells. Biooxidation of the hazardous substance is performed by the microbial enzyme due to stereochemical similarity between the co-oxidized hazardous substance and the main substance.

The best known applications of co-metabolism are the biodegradation/detoxication of chloromethanes, chloroethanes, chloromethylene, and chloroethylenes by enzymes oxidizing methane (methylotrophic bacteria) or ammonia (nitrifying bacteria). In practice, the bioremediation is achieved by adding methane or ammonia, oxygen (air), and biomass of methanotrophic or nitrifying bacteria to soil and groundwater.

Combinations of Aerobic Biotreatment with Other Treatments

To intensify the biotreatment of hazardous liquid waste, the following pretreatments can be used:

- The mechanical disintegration/suspension of hazardous hydrophobic substances to improve the reacting surface in the suspension and increase the rate of biodegradation.
- The removal from wastewater or concentration of hazardous substances by sedimentation, centrifugation, filtration, flotation, adsorption, extraction, ion exchange, evaporation, distillation, and freezing separation.
- Preliminary oxidation by H_2O_2, ozone, or Fenton's reagent to produce active oxygen radicals.
- Preliminary photooxidation by UV and electrochemical oxidation of hazardous substances.

Biotreatment of Hazardous Waste by Anaerobic Fermenting Bacteria

Anaerobic fermenting bacteria (e.g., from genus *Clostridium*) perform two important functions in the biodegradation of hazardous organics: they hydrolyze different natural polymers and ferment monomers with the production of alcohols, organic acids, and CO_2. Many hazardous substances, for example, chlorinated solvents, phthalates, phenols, ethyleneglycol, and polyethylene glycols can be degraded by anaerobic microorganisms.

Fermenting bacteria also perform anaerobic dechlorination thus enhancing further biodegradation of chlorinated organics. There are different biotechnological systems to perform anaerobic biotreatment of wastewater: biotreatment by suspended microorganisms, anaerobic biofiltration, and biotreatment in upflow anaerobic sludge blanket (UASB) reactors.

Landfilling of Hazardous Solid Wastes

Landfilled organic and inorganic wastes are slowly transformed by indigenous microorganisms in the wastes. Organic matter is hydrolyzed by bacteria and fungi. Amino acids are degraded via ammonification with formation of toxic organic amines and ammonia. Amino acids, nucleotides, and carbohydrates are fermented or anaerobically oxidized with the formation of organic acids, CO_2, and CH_4. Xenobiotics and heavy metals may be reduced and subsequently dissolved or immobilized. These bioprocesses result in the formation of a toxic landfill leachate, which can be detoxicated by aerobic biotechnological treatment to oxidize organic hazards and to immobilize dissolved heavy metals.

Combined Anaerobic/Aerobic Biotreatment of Hazardous Industrial Wastes

A combined anaerobic/aerobic biotreatment can be more effective than aerobic or anaerobic treatment alone. The simplest approach for this type of treatment is the use of aerated stabilization ponds, aerated and non-aerated lagoons, and natural and artificial wetland systems, whereby aerobic treatment occurs in the upper part of these systems and anaerobic treatment occurs at the bottom. A more intensive form of biodegradation can be achieved by combining aerobic and anaerobic reactors with controlled conditions or by integrating anaerobic and aerobic zones within a single bioreactor.

Combinations or even alterations of anaerobic and aerobic treatments are useful in the following situations:

- Biodegradation of chlorinated aromatic hydrocarbons, including anaerobic dechlorination and aerobic ring cleavage
- Sequential nitrogen removal, including aerobic nitrification and anaerobic denitrification
- Anaerobic reduction of Fe(III) and microaerophilic oxidation of Fe(II) with production of fine particles of iron hydroxide for adsorption of organic acids, phenols, ammonium, cyanide, radionuclides, and heavy metals

Optimization of Hazardous Wastes Biodegradation

Several key factors are critical for the successful application of biotechnology for the treatment of hazardous wastes:

- Environmental factors, such as pH, temperature, and dissolved oxygen concentration, must be optimized.
- Contaminants and nutrients must be available for action or assimilation by microorganisms.
- Content and activity of essential microorganisms in the treated waste must be sufficient for the treatment.

Optimum Temperature for Biodegradation of Hazardous Wastes

Psychrophilic microorganisms have optimal growth temperatures below 15°C. These organisms may be killed by exposure to temperatures above 30°C. Mesophilic microorganisms have optimal growth temperatures in the range between 20°C and 40°C. Thermophiles grow best above 50°C. Some bacteria can grow up to temperatures where water boils; those with optimal growth temperatures above 75°C are categorized as extreme thermophiles. Therefore, the biotreatment temperature must be maintained at optimal growth temperatures for effective biotreatment by certain physiological groups of microorganisms. The heating of the treated waste can come from microbial oxidation or fermentation activities if sufficient heat is generated and good thermal insulation of treated waste from the cooler environment is maintained. The bulking agent added to solid wastes may also be used as an internal thermal insulator.

Optimum pH for Biodegradation of Hazardous Wastes

The pH values of natural microbial biotopes vary from 1 to 11. For example, volcanic soil and mine drainage have pH values between 1 and 3; plant juices and acid soils have pH values between 3 and 5; freshwater and seawater have pH values between 7 and 8; alkaline soils and lakes, solutions of ammonia, and rotten organics have pH values between 9 and 11.

Most microbes grow most efficiently within the pH range 6–8. They are called neutrophiles. Species that have been adapted to grow at pH values lower than 4 are called acidophiles. Species which have been adapted to grow at pH values higher than 9 are called alkalophiles. Therefore, the pH of the treatment medium must be maintained at optimal values for effective biotreatment by certain physiological groups of microorganisms.

The optimum pH may be maintained physiologically by addition of pH buffer or pH regulator as described:

- Control of organic acid formation in fermentation
- Prevention of formation of inorganic acids in aerobic oxidation of ammonium, elemental sulfur, hydrogen sulfide, or metal sulfides
- Assimilation of ammonium, nitrate, or ammonium nitrate, leading to decreased pH, increased pH, or neutral pH, respectively
- pH buffers such as $CaCO_3$ or $Fe(OH)_3$ can be used in large-scale waste treatment
- Solutions of KOH, NaOH, NH_4OH, $Ca(OH)_2$, HCl, or H_2SO_4 can be added automatically to maintain the pH of liquid in stirred reactors

The maintenance of optimum pH in treated solid waste or bioremediated soil may be especially important if there is a high content of sulfides in waste or acidification/alkalization of soil in the bioremediation process.

Enhancement of Biodegradation by Nutrients

The major elements that are found in microbial cells are C, H, O, N, S, and P. An approximate elemental composition corresponds to the formula $CH_{1.8}O_{0.5}N_{0.2}$. Therefore, nutrient amendment may be required if the waste does not contain sufficient amounts of these macroelements. The waste can be enriched with carbon (depends on the nature of the pollutant that is treated), nitrogen (ammonium is the best source), phosphorus (phosphate is the best source), and/or sulfur (sulfate is the best source). Other macronutrients (K, Mg, Na, Ca, and Fe) and micronutrients (Cr, Co, Cu, Mn, Mo, Ni, Se, W, V, and Zn) are also essential for microbial growth and enzymatic activities and must be added into the treatment systems if present in low concentrations in the waste. The best sources of essential metals are their dissolved salts or chelates with organic acids. The source of metals for the bioremediation of oil spills may be lipophilic compounds of iron and other essential nutrients that can be accumulated at the water–air interface where hydrocarbons and hydrocarbon-degrading microorganisms are also concentrated.

Enhancement of Biodegradation by Growth Factors

In some biotreatment cases, growth factors must also be added into the treated waste. Growth factors are organic compounds such as vitamins, amino acids, and nucleosides that are required in very small amounts and only by some strains of microorganisms called auxotrophic strains. Usually, those microorganisms that are commensals or parasites of plants and animals require growth factors. However, sometimes these microorganisms may have the unique ability to degrade some xenobiotics.

Increase of Bioavailability of Contaminants

Hazardous substances may be protected from microbial attack by physical or chemical envelopes. These protective barriers must be destroyed mechanically or chemically to produce fine particles or waste suspensions to increase the surface area for microbial attachment and subsequent biodegradation.

Another way to increase the bioavailability of hydrophobic substances is by washing of waste or soil with water or a solution of surface-active substances. The disadvantage of this technology is the production of secondary hazardous waste because chemically produced surfactants are usually resistant to biodegradation. Therefore, only easily biodegradable or biotechnologically produced surfactants can be used for the pretreatment of hydrophobic hazardous substances.

Enhancement of Biodegradation by Enzymes

Extracellular enzymes produced by microorganisms are usually expensive for the large-scale biotreatment of organic wastes. However, enzyme applications may be cost-effective in certain situations. For example, toxic organophosphate waste can be treated using the enzyme parathion hydrolase produced and excreted by a recombinant strain of *Streptomyces lividans*. The cell-free culture fluid contains enzymes that can hydrolyze organophosphate compounds. Future applications may be related with cytochrome P450-dependent oxygenase enzymes that are capable of oxidizing different xenobiotics.

Enhancement of Biodegradation by Aeration and Oxygen Supply

The maximum concentration of dissolved oxygen is low (7–8 mg/L), and it can be rapidly depleted during waste biotreatment with oxygen consumption rates ranging from 10 to 2000 g O_2/L h. Therefore, oxygen must be supplied continuously in the system. Supply of air in liquid waste treatment systems is achieved by aeration and mechanical agitation. Different techniques are employed to supply sufficient quantities of oxygen in fixed-biofilm reactors, in viscous solid wastes, in underground layers of soil, or in aquifers polluted by hazardous substances. Very often, the supply of oxygen is the critical factor in the successful scaling-up of bioremediation technologies from laboratory experiments to full-scale applications. Air sparging in situ is a commonly used bioremediation technology, which volatilizes and enhances aerobic biodegradation of contamination in groundwater and saturated soils. Application of pure oxygen can increase the oxygen transfer rate by up to five times, and this can be used in situations with strong acute toxicity of hazardous wastes and low oxygen transfer rates, to ensure an adequate oxygen transfer rate in polluted waste.

Enhancement of Biodegradation by Oxygen Radicals

In some cases, hydrogen peroxide has been used as an oxygen source because of the limited concentrations of oxygen that can be transferred into the groundwater using above-ground aeration followed by reinjection of the oxygenated groundwater into the aquifer or subsurface air sparging of the aquifer. However, because of several potential interactions of H_2O_2 with various aquifer material constituents, its decomposition may be too rapid, making effective introduction of H_2O_2 into targeted treatment zones extremely difficult and costly.

Pretreatment of wastewater by ozone, H_2O_2, TiO_2-catalyzed UV-photooxidation, and electrochemical oxidation can significantly enhance biodegradation of halogenated organics, textile dyes, pulp-mill effluents, tannery wastewater, olive-oil mills, surfactant-polluted wastewater, and pharmaceutical wastes, and it can diminish the toxicity of municipal landfill leachates. In some cases, oxygen radicals generated by Fenton's reagent ($Fe^{2+} + H_2O_2$ at low pH) and iron peroxides (Fe(VI) and Fe(V)) can be used as oxidants in the treatment of hazardous wastes.

Many microorganisms can produce toxic metabolites of oxygen such as hydrogen peroxide (H_2O_2), superoxide radical (O_2^-), and hydroxyl radical ($OH^\bullet$) and release them to the environment. Lignin-oxidizing "white-rot" fungi can degrade lignin and all other chemical substances due to intensive generation of oxygen radicals that oxidize the organic matter by random incorporation of oxygen into molecule. Not much is known about the biodegradation ability of H_2O_2-generating microaerophilic bacteria.

Enhancement of Biodegradation by Electron Acceptors Other than Oxygen

Dissolved acceptors of electrons such as NO_3^-, NO_2^-, Fe^{3+}, SO_4^{2-}, and HCO_3^- can be used in the treatment system when oxygen transfer rates are low. The choice of the acceptor is determined by economical and environmental reasons. Nitrate is often proposed for bioremediation because it can be used by many microorganisms as an electron acceptor. However, it is relatively expensive and its supply to the treatment system must be thoroughly controlled because it can also pollute the environment. Fe^{3+} is an environmentally friendly electron acceptor. It is naturally abundant in clay minerals, magnetite, limonite, goethite, and iron ores, but its compounds are usually insoluble and it diminishes the rate of oxidation in comparison with dissolved electron acceptors. Sulfate and carbonate can be applied as electron acceptors in strictly anaerobic environments only. Another disadvantage of these acceptors is that these anoxic oxidations generate toxic and foul-smelling H_2S or the "greenhouse" gas, CH_4.

Bioenhancement of the Treatment of Hazardous Wastes

Addition of microorganisms (inoculum) to start-up or to accelerate the biotreatment process is a reasonable strategy under the following conditions:

1. If microorganisms that are necessary for hazardous waste treatment are absent or their concentration is low in the waste.
2. If the rate of bioremediation performed by indigenous microorganisms is not sufficient to achieve the treatment goal within the prescribed duration.
3. If the acclimation period is too long.
4. To direct the biodegradation/biotreatment to the best pathway from many possible pathways.
5. To prevent growth and dispersion of unwanted or indeterminate microbial strains that may be pathogenic or opportunistic in waste treatment systems.

Application of Acclimated Microorganisms

A simple way to produce a suitable microbial inoculum is the production of an enrichment culture, which is a microbial community containing one or more dominant strains naturally formed during cultivation in a growth medium modeling the hazardous waste under defined conditions. If the cultivation conditions are changed, the dominant strains in the enrichment culture may also be changed.

Another approach involves the use of a portion of the treated waste containing active microorganisms as inoculum to start-up the process. The application of acclimated microorganisms in an enrichment culture or in biologically treated waste may significantly decrease the start-up period for biotechnological treatment.

In cases involving treatment of toxic substances and high death rates of microorganisms during treatment, regular additions of active microbial cultures may be useful to maintain constant rates of biodegradation.

Selection and Use of Pure Culture

Notwithstanding the common environmental engineering practice of using a portion of the treated waste as inoculum, applications of defined, pure, starter cultures have the following theoretical advantages:

- Greater control over desirable processes
- Lower risk of release of pathogenic or opportunistic microorganisms during biotechnological treatment
- Lower risk of accumulation of harmful microorganisms in the final biotreated product

Pure cultures that are most active in biodegrading specific hazardous substances can be isolated by conventional microbiological methods, quickly identified by molecular-biological methods, and tested for pathogenicity and biodegradation properties. The biomass of pure culture can be produced in large scale in commercial fermentors, then concentrated and dried for storage before field application. Therefore, it is not only the biodegradation abilities of pure cultures, but also the suitability for industrial production of dry biomass that must be taken into account in the selection of pure culture for the biotechnological treatment of industrial hazardous waste. Generally, Gram-positive bacteria are more viable after drying and storage than Gram-negative bacteria. Spores of Gram-positive bacteria can be used as a superstable inocula.

Construction of Microbial Community

A pure culture is usually active in the biodegradation of one type of hazardous substance. Wastes containing a variety of hazardous substances must be treated by a microbial consortium comprising a collection of pure cultures most active in the degradation of the different types of substances. However, even in cases involving a single hazardous substance, degradation rates may be higher for a collection of pure cultures acting mutually (symbiotically) than for single pure cultures.

Mutualistic relationships between pure cultures in an artificially constructed or a naturally selected microbial community may be based on the sequential degradation of xenobiotic, mutual exchange of growth factors or nutrients between these cultures, mutual creation of optimal conditions (pH, redox potential), and gradients of concentrations. Mutualistic relationships between the microbial strains are more clearly demonstrated in dense microbial aggregates such as biofilms, flocs, and granules used for biotechnological treatment of hazardous wastes.

Construction of Genetically Engineered Microorganisms

Microorganisms, suitable for the biotreatment of hazardous substances, can be isolated from the natural environment. However, their ability for biodegradation can be modified and amplified by artificial alterations of the genetic (inherited) properties of these microorganisms. The description of the methods is given in many books on *environmental microbiology and biotechnology*.

Natural genetic recombination of the genes (units of genetic information) occurs during DNA replication and cell reproduction, and includes the breakage and rejoining of chromosomal DNA molecules (separately replicated sets of genes) and plasmids (self-replicating mini-chromosomes containing several genes).

The desired genes for biodegradation of different xenobiotics can be isolated and then cloned into plasmids. Some plasmids have been constructed that contain multiple genes for the simultaneous biodegradation of several xenobiotics. The strains containing such plasmids can be used for the bioremediation of sites heavily polluted by a variety of xenobiotics. The main problem in these applications is maintaining the stability of the plasmids in these strains. Other technological and public concerns include the risk of application and release of genetically modified microorganisms in the environment.

Application of Microbial Aggregates and Immobilized Microorganisms

Self-aggregated microbial cells of biofilms, flocs and granules, and artificially aggregated cells immobilized on solid particles are often used in the biotreatment of hazardous wastes. Advantages of microbial aggregates in hazardous waste treatment are as follows:

- Upper layers and matrix of aggregates protect cells from toxic pollutants due to adsorption or detoxication; therefore microbial aggregates or immobilized cells are more resistant to toxic xenobiotics than suspended microbial cells.
- Different or alternative physiological groups of microorganisms (aerobes/anaerobes, heterotrophs/nitrifiers, sulfate reducers/sulfur oxidizers) may coexist in aggregates and increase the diversity of types of biotreatments, leading to higher treatment efficiencies in one reactor.
- Microbial aggregates may be easily and quickly separated from treated water. Microbial cells immobilized on carrier surfaces such as granulated activated carbon that can adsorb xenobiotics will degrade xenobiotics more effectively than suspended cells.

Chemical Gradients in Microbial Aggregates

However, dense microbial aggregates may encounter problems associated with diffusion limitation, such as slow diffusion both of the nutrients into and the metabolites out of the aggregate. For example, dissolved oxygen levels can drop to zero at some depth below the surface of microbial aggregates. This distance clearly depends on factors such as the specific rate of oxygen consumption and the density of biomass in the microbial aggregate. When the environmental conditions within the aggregate become unfavorable, cell death may occur in zones that do not receive sufficient nutrition or that contain inhibitory metabolites.

Channels and pores in aggregate can facilitate transport of oxygen, nutrients, and metabolites. Channels in microbial spherical granules have been shown to penetrate to depths of 900 μm, and a layer of obligate anaerobic bacteria was detected below the channeled layer. Therefore, there is some optimal size or thickness of microbial aggregates appropriate for application in the treatment of hazardous wastes.

20

Solid Wastes and Soil Biotreatment

Solid Waste

Solid wastes are any discarded solid and semisolid materials resulting from industrial, commercial, and agricultural operations, and from community activities. The following general classification of solid waste sources is commonly used:

- Municipal solid waste (MSW)
- Commercial wastes (from office activities, shops, restaurants, etc.)
- Construction/demolition wastes
- Sludge from wastewater treatment plants
- Industrial solid wastes (remaining after manufacturing of a product)
- Agricultural wastes (crop residues, manure from farms)
- Mining wastes (mine tailings)
- Hazardous solid wastes, which are wastes that pose a danger

Treatment of Solid Wastes

The treatment of solid wastes includes

- Landfilling
- Incineration
- Recovery and reuse

Landfills

The major portion of solid wastes is disposed of in landfills. There are different types of landfills:

- Anaerobic landfills, where solid wastes are filled with water in valley and the layers are either covered or not covered.
- Aerobic landfill, where air is supplied inside landfill.

- Anaerobic or aerobic sanitary landfills, where solid waste is deposited in the trench or area as compacted layers. Each layer is covered with earth; the landfill has an impermeable liner as well as leachate and gas collection systems.

Biological Reactions of Decay in the Landfills

The major reactions are as follows:

$RCONH_2 + H_2O \rightarrow NH_3 + RCOOH$	Deamination: produces toxic ammonia causing eutrophication of reservoirs
$RCH(NH_2)COOH \rightarrow RCH_2NH_2 + CO_2$	Decarboxylation: produces toxic and bad-smelling (the foul odor of putrefying flesh) biogenic amines, for example, cadaverine (the decarboxylation product of the amino acid lysine) and putrescine (the decarboxylation product of the amino acid ornithine)
$3CH_2O \rightarrow 2CH_2O_{0.5} + CO_2$	Fermentation: produces bad-smelling volatile fatty acids used by acetogens, sulfate-, and iron-reducing bacteria
$2CH_2O \rightarrow CH_4 + CO_2$	Methanogenesis: produces methane, which can be used as fuel
$CH_2O + SO_4^{2-} \rightarrow H_2S$	Sulfate reduction: produces toxic, corrosive, and bad-smelling dihydrogen sulfide
$CH_2O + Fe^{3+} \rightarrow Fe^{2+}$	Ferric reduction: produces soluble toxic ferrous ions
$RCH_2O\text{-}PO(OH)_2 + H_2O \rightarrow HPO_4^{2-} + 2H^+ + RCH_2OH$	Dephosphorylation: produces soluble phosphate ions, causing eutrophication of reservoirs

All these toxic products are accumulated in the landfill leachate, which must be treated chemically and/or biologically as hazardous industrial wastewater. Aerobic biotreatment of the landfill leachate can remove a significant portion of toxic substances but the complete removal of ammonia, phosphate, and xenobiotics from the landfill leachate still remains as an unsolved biotechnological problem.

Sewage Sludge Processing

Sewage sludge (biosolids) is composed of biomass generated during anaerobic digestion of activated sludge and settled solids on municipal wastewater treatment plants. The worldwide annual production of sewage sludge is several million tons. Sewage sludge dewatering is performed by

- Vacuum filtration
- Filter press
- Drying bed
- Centrifugation
- Settling by gravity

In many cases, aluminum, ferrous or ferric salts, lime, and synthetic polyelectrolytes are added to improve sludge-settling ability and filterability.

After dewatering, sewage sludge can be disposed of directly or after mixing with chemicals or cement, incinerated, or used as soil conditioner or fertilizer because of its 2% nitrogen and 1% phosphorus content. The obstacles for the use of sewage sludge as a fertilizer are

- The high content of heavy metals
- The presence of pathogens
- The presence of persistent organic compounds such as endocrine disruptors, pharmaceutical residuals, and pesticides

If the contents of heavy metals, pathogens, and persistent organic compounds in sewage sludge are below permitted levels, sewage sludge can be used as organic fertilizer directly or composted in mixture with other organic wastes, such as food processing or agricultural wastes.

Composting

Composting of domestic refuse is a well-known method since ancient times. In medieval times, a process similar to composting was used widely for the production of saltpeter. The primary objective of modern composting is to convert an unstable and potentially offensive material into a stable end-product accompanied by odor reduction, the destruction of pathogens and parasites, and the retention of nutrients. A range of materials can be composted, including municipal refuse, paper, food waste, sewage sludge, and mixtures of these wastes.

Composting is one of the main environmental technologies that can be applied for bioconversion of large quantities of organic waste. Composting is an ancient technology that has been used by farmers for centuries to convert biodegradable horticultural, agricultural, gardening, and kitchen waste into nutrient-rich material for further use as fertilizer. From the microbiological point of view, it is an aerobic mesophilic and thermophilic microbial decomposition of organic substances to a humus-rich, safe, and relatively stable product, called compost and consists of cellulose, lignin, inorganic compounds, and biomass of composting microorganisms.

Microbiology of Composting

The organic component of solid waste is microbiologically decomposed under controlled aerobic, moist, and warm conditions. The process transforms organic forms of nitrogen and phosphorus into inorganic forms that are more bioavailable for uptake by agricultural crops. Heat produced during biooxidation kills human pathogens.

Composting is a complex microbiological process where different groups of microorganisms biodegrade organic waste. Microbial succession during the composting process consists of

- A latent phase (ambient temperature 20°C) of microorganisms adapting to composting conditions, such as temperature, moisture content, aeration, etc.
- A mesophilic phase (20°C–40°C) of intensive microbial growth, resulting in an increase of temperature due to the oxidation of organic compounds by mesophilic bacteria and thermotolerant fungi.
- An initial thermophilic phase (40°C–60°C) with growth of thermophilic bacteria, actinomycetes, and fungi.
- A thermophilic phase (60°C–80°C) due to the growth of thermophilic and spore-forming bacteria, sulfur- and hydrogen-oxidizing autotrophs, and aerobic non-spore-forming bacteria. At the end of this phase, the temperature drops to 40°C.
- A cooling and maturation phase (40°C to ambient temperature), in which bacteria mesophilic/thermotolerant actinomycetes, and fungi are involved in nutrient cycling and degradation of microbial metabolites toxic for plants.

Bulking Agents

Bulking agents include wood chips, leaves, corncobs, bark, peanut and rice husk, dried sludge, or plastic items. Wood chips are the most commonly used bulking agents in composting as they have high carbon content and a high C/N ratio. Functions of bulking agent include

- Offering structural support
- Facilitating aeration during composting
- Diminishing heat transfer between self-heating zones in compost and air due to porosity

Types of Composting Systems

There are typically three main types of composting systems: windrow, aerated pile, and in-vessel composting.

Windrow composting involves stacking the organic waste into piles for natural biodegradation without turning. Windrows are long, narrow, parallel rows with mixed organic waste to a height of 1–2 m, which are periodically turned to provide aeration. Turning is provided more frequently at the beginning of composting when more oxygen is demanded for biodegradation of organics. Due to turning, temperatures above 60°C–70°C cannot be reached in composted wastes. The mesophilic phase lasts in static pile composting for a few days, the thermophilic phases last from a few days to several months, and the duration of cooling and maturation phase consists of several months. Composting in windrows lasts 50–80 days. Windrows usually are used for large volumes of wastes; thus they are situated under covered outdoors and require a lot of space. While windrow composting has low capital costs and produces good quality compost, the disadvantages include odor and leachate problems, cost of turning, loss of ammonia, and potential spreading of allergic fungal spores in the air during turning.

In aerated static piles, air is supplied through perforated piles by blowers. It is the least expensive method of composting, which can be used for small-scale processes. The biodegradation rate depends on weather conditions and does not ensure the reduction of pathogens due to poor mixing.

In-vessel composting, also known as mechanical or enclosed-reactor composting, uses a closed reactor for the bioconversion of organic waste under temperature control and proper air supply. They require little space, minimize odor problems, and are not weather sensitive, but their cost is higher than that of open systems. The duration of this process is 14–19 days. Maturation of the product is provided in piles outside the reactor. The cost of in-vessel composting is higher than composting in piles and windrows and is not always suitable for large volumes of organic wastes.

Optimal Parameters for Composting

The essential parameters for composting are moisture content, C/N ratio, aeration, temperature, pH, particle size, additives, and processing time. The optimal moisture content of the composting material must be 60% but the process can be performed in the 40%–70% range. Moisture content higher than 70% decreases the rate of organic decomposition, and creates anaerobic conditions and odor problems. The moisture content in rice, fruit, and vegetable food waste is approximately 90%; in orange peels, it is 76%; and in sawdust, it is 25%.

The C/N ratio of the material must be between 25:1 and 35:1. A higher C/N ratio reduces the rate of process, but a lower C/N ratio leads to nitrogen loss. Different wastes have different C/N ratios, for example, C/N ratios are as follows: sewage sludge, from 6:1 to 8:1; food waste, 15:1; fruit waste, 35:1; green vegetable wastes and weeds, from 11:1 to 20:1; and sawdust, 500:1. To receive a desirable C/N ratio, a mixture of wastes can be used.

Additional aeration can improve the process. Aeration is provided by turning, mixing, and the use of fans, blowers, and compressors. The optimal air supply is considered to be from 0.6 to 1.8 m^3 air/(d·kg of volatile solids) during the thermophilic phase of composting. Bulking agents, sawdust, and wood chips can be added to improve the mass transfer of oxygen and carbon dioxide between air and material. A temperature of at least 55°C–60°C must be maintained for several days to inactivate the pathogens, parasites, and weed seeds. However, temperatures higher than 70°C may cause inactivation of microorganisms and slow or stop the composting process.

Optimal pH levels of composting are from 7.0 to 8.0. Particle size greatly influences the biodegradation rate. Before composting begins, wastes are shredded to particles less than 5 cm in size. A smaller particle size ensures the greatest surface area and enhances mass transfer between biodegradable material and microorganisms. However, if the particles are too small, the oxygen transfer can be negatively impacted. Additives such as another waste can be added for successful composting to provide optimal C/N ratio (sewage sludge to decrease, and sawdust to increase C/N ratio), water to maintain optimal moisture content, additional sources of nitrogen, phosphorus, and bioessential mineral elements if compost materials lack these elements, lime to provide optimal pH.

Composting of the Mixture of Wastes

Processing time depends on the nature of material and conditions of composting. Food waste can be composted within a few months, but horticultural-waste composting may last from 9 to 12 months. Composting of food waste may take a year in static piles, several months in windrows, and several weeks for in-vessel composting. The final compost must be stable, rich in available plant nutrients, relatively free from pathogens, weed seeds, plant inhibitors, dark brown or black in color, and similar to humus in smell and by touch. The pH must be around 7.0, the preferable C/N ratio approximately 20:1, the moisture content between 35% and 50%, and organic matter content between 40% and 65%. Solid food waste can be mixed with sewage sludge in the ratio of total solids 1:1 to improve both the chemical composition and the texture of the composting material.

C:N ratios are as follows:

Sewage sludge	7:1
Food waste	21:1
Mixture of these wastes	13:1
Matured compost	20:1

To minimize the process duration and the space required for large-scale bioconversion, intensive in-vessel bioconversion of organic waste can be used. An aerobic thermophilic treatment to convert a mixture of sewage sludge and food waste or food waste into fertilizer can be performed within several days. To maintain a neutral pH at the beginning of the bioconversion, $CaCO_3$ is added at 5% to the total solids of organic waste.

The resulting compost can be applied as a soil amendment to improve physical properties such as soil structure, water holding capacity, and porosity. It can be used as a fertilizer to provide plants with essential nutrients such as nitrogen, phosphorus, potassium, and microelements. Compost can also be used as mulch for trees, landscapes, and gardens.

Vermicomposting

Recycling small quantities of organic waste such as household food waste and backyard wastes could be done by vermicomposting. Vermicomposting is based on the use of earthworms to consume the fragments of vegetable and fruit (not meat products) food waste. The activity of the earthworms ensures aeration and mixing of substrate, microbial decomposition of substrate in the intestine of the worm, which leads to stabilization of organic matter and the production of high quality compost. It has been observed that 1 kg of worms can eat 4 kg of waste per week. The process takes place in containers or bins. The addition of excess waste can produce anaerobic conditions.

Composting Scale

It is estimated that around 60 million tons of recoverable organic waste is produced in Europe each year. Approximately 9 million tons are currently recovered through home composting or source separation and centralized composting throughout the EU countries. Composting of municipal organic waste could be economically reasonable in countries where the cost of municipal waste landfilling is high.

Public Health Aspects of Composting

Pathogenic microorganisms and cysts of helminths are present in the composted waste and can be completely or partially removed during composting. If sewage sludge is used for composting, $Ca(OH)_2$ or CaO can be added for 2 h to increase the pH to 12 to kill pathogens and helminth eggs.

However, secondary pathogens such as toxigenic fungi *Aspergillus fumigates*, *Aspergillus flavus*, and some actinomycetes can grow in compost and release spores in air, which could negatively affect the health of compost workers.

Odor Control in Composting

The production of odors due to fermentation and formation of volatile fatty acids can be prevented by the effective mixing and aeration of compost and maintenance of optimal moisture and temperature. In in-vessel composting, exhausted air carrying odors can be treated in biofilters.

Composting of Hazardous Organic Wastes

Hazardous wastes can be treated by composting but long durations are usually needed to reach permitted levels of pollution. Successful applications of all the types of composting are available for the treatment of crude-oil-impacted soil, petrochemicals-polluted soil, and explosives-polluted soil.

Soil

Soil is the material on the surface of the earth that serves as a natural medium for the growth of land plants. The formation of soil has been influenced by parent material (type of the rocks), climate (water and temperature effects), activity of microorganisms and macroorganisms, and topography of the place.

Soil texture depends on the quantity of sand, silt, clay, and colloid fraction. Sand is defined as the portion of soil particles that are between 0.05 and 2.0 mm in diameter; typical content in soil is about 50% by weight. Silt is defined as soil particles with a diameter between 0.002 and 0.05 mm; typical content in soil is about 20%. Clay consists of soil particles <0.002 mm in diameter; typical content in soil is about 20%. The colloidal fraction of soil includes organic matter (typical content is about 3%) and inorganic matter (typical content is about 5%).

Organic Matter in Soil

Peat is unconsolidated soil material consisting of slightly decomposed organic matter accumulated under conditions of excessive moisture. Humus is the organic matter in soil exclusive of peat and soil biomass. The term is often used synonymously with soil organic matter. Humic acid is dark-colored organic material extracted from soil by diluted alkali and that is precipitated by acid. Fulvic acid is the yellow organic material that remains in solution after removal of humic acid by acidification.

Microbial Activity and Soil

The formation of soil from exposed rock may take hundreds of years. The formation of soil involves physical, chemical, and biological processes. Microbes contribute by the production of carbonic acid or organic acids that dissolve rock minerals. Microbes accumulate organic matter by the assimilation of light energy and assimilate nitrogen from air.

The microbes in soil control the availability of many important plant nutrients. Therefore, microbial activity in soil is a key part of soil productivity.

Soil is a habitat for viruses, bacteria and archaea, fungi, algae, protozoa, and nematodes. Indigenous microorganisms live permanently in soil. Nonindigenous microorganisms enter soil by precipitation from air, or from diseased tissues, manure, sewage, and may persist in soil for some time, but they do not perform ecologically important transformations in soil.

Soil and Biodegradation of Xenobiotics

Soil is the main source of microorganisms capable of biodegrading xenobiotic chemicals. Soil is often polluted with xenobiotic chemicals during their manufacturing, transportation, and storage. The secondary effects of this pollution are the pollution of air, food, groundwater, and surface water. Very often, the best way to clean up the polluted site is by the bioremediation of contaminated soil.

Solubility is related to the bioremediation availability of compounds in the liquid phase of soil. The contaminant must be in solution or suspension for bioremediation. Nonpolar, hydrophobic contaminants are concentrated into organic soil material. Nonpolar pollutants are less mobile in soil and groundwater and spread more slowly there than polar pollutants.

Toxic Pollutants of Soil

The key factor driving the need for remediation of contaminated soil and groundwater is the toxicity of pollutants. Disposal or discharge of toxic chemicals to soils presents a difficult problem for bioremediation because of following reasons:

1. Toxic materials are often resistant to biodegradation.
2. Once the materials are in the soil environment, less control exists with respect to their transport and fate.
3. The risk to water supplies is very high because maximum concentration limits (MCLs) for toxic materials in water supplies are often extremely low.

Contaminated soil causes pollution of

- Groundwater
- Air
- Surface water
- Food

Sources of Soil Pollution

Serious soil and groundwater contamination problems also result from spills and improper disposal of toxic materials. For example, accidents during transport of chemicals may result in spillage of large quantities of pure products on small areas of ground. Left untreated, the chemicals may percolate into the soil and eventually contaminate local groundwater.

Improper disposal of chemicals in homes and commercial enterprises can contaminate both soil and groundwater. It could be due to

- Waste oil and cleaning solvents from garages
- Agricultural chemical residues in farms

- Crop dusters at airfields
- Paints and cleaning supplies in enterprises

Leakages from Underground Storage Tank Facilities

In the world, over a million leakages happen annually from underground storage tank facilities. Plumes originating from underground storage tanks (USTs) may be free product, while plumes from waste-disposal ponds are usually composed of a mixture of contaminants. Plumes tend to travel slowly in the unsaturated (vadose) zone, but will eventually reach groundwater if leakage continues. Mixing with water in the aquifer is a function of relative density, with lighter plumes tending to float and dense plumes tending to sink.

Landfill Leachate

Contamination of soil could be from landfill leachate, which is the liquid collecting on the bottom of the landfill and migrating through the landfill liners. The flow is generally downward and spreads through the available channels and porous media, following the path of least resistance.

Bioremediation

Bioremediation is an application of microbial biodegradation capability to clean up contaminated sites. In the majority of cases, petroleum products are involved in the pollution of sites.

The major problems of bioremediation are as follows:

- Absence of indigenous soil microorganisms capable of degrading soil pollutant
- Oxygen requirements for microbial growth and oxidation
- Soluble inorganic nutrient requirements for microbial growth
- Plugging of soil pores with microbial biomass

Soil and Groundwater Bioremediation

Soil and groundwater bioremediation can be classified as

- Landfarming, land treatment, and plant bioremediation
- On-site treatment
- In situ treatment
- Excavation and composting treatment
- Slurry-phase bioremediation

Landfarming, Land Treatment, and Plant Bioremediation

Landfarming, land treatment, and plant bioremediation are performed using soil tilling at a frequency that allows for aeration and activation of biodegradation activity of soil microorganisms. It can be enhanced also by the additions of nutrients, control of soil moisture content by irrigation, control of soil, and cultivation of specific plants. This option is used when the pollution is not strong, when time required for the treatment is not a limiting factor, and when there is no pollution of groundwater.

On-Site and In Situ Soil and Groundwater Bioremediation

These options are used when the level of pollution is high and there is secondary pollution of groundwater. During on-site bioremediation, groundwater is pumped to the surface, treated in the aerobic bioreactor with addition of nutrients and microbial culture if needed, and then returned to the aquifer through polluted soil. Surface treatment could also include other technologies like air stripping or adsorption onto granulated activated carbon.

During in situ (in its original place) treatment, the activity of the degrading microorganisms in the soil and groundwater aquifer is stimulated by the introduction of oxygen and nutrient through injection wells.

Excavation of Contaminated Soil and Its Composting Treatment

The reason for using this option might be that the conditions in situ (pH, salinity, dense texture or high permeability of soil, high toxicity of substance, and safe distance from public place) are not favorable for biodegradation. Contaminated soil can be excavated and treated at a site where conditions and migration of the pollutants can be controlled by compacted clay or plastic liner barriers. In composting, the contaminated soil is mixed with organic bulking agents, such as manure, straw, and sawdust and formed into piles or windrows. Bulking agents help to increase porosity to facilitate airflow, and energy released during biodegradation results in elevated temperatures in the pile. Optimal moisture content is maintained, and the piles or windrows are mechanically mixed for aeration and homogenation.

Slurry-Phase Bioremediation in Bioreactors

This option is used when the level of pollution is so high that it diminishes the biodegradation rate due to toxicity of substances or low mass transfer rate. In slurry-phase bioremediation, the heavily polluted soil is placed in an aerated bioreactor, where water and nutrients are also added to allow continuous mixing. The choice of slurry density depends on the soil quantities to be treated and the contamination concentration. When large quantities of soil are treated or contaminant concentrations are high, lower slurry densities are usually preferable to enhance oxygen transfer rate. Biodegradation is also enhanced by addition of nutrients, microbial cultures, and control pH and temperature.

Slurry-phase bioremediation is operated in a batch or semi-continuous mode. After treatment, the solids are separated from the fluid by sedimentation. The clean solids are redeposited in the original site, while the liquid may be used in the next soil treatment as an inoculum.

Gas Cleaning

Microbial gas, released from on-site, composting, and slurry-phase treatments, is treated from pollutants using biofilter. Packing material for the biofilter is usually a mixture of compost and an inert bulking agent or porous ceramic pellets. Microbial communities grow attached to the packing surface. Microbial gas treatment is used for removing volatile organic compounds (VOCs) released from the bioremediation process.

Pretreatment of Contaminated Soil

Pretreatment is often necessary to remove uncontaminated or less contaminated materials and to remove materials not compatible with slurry-phase bioremediation. Soil fractionation and soil washing are two common methods used to pre-treat contaminated soil. Contaminants are usually hydrophobic and associated with the finer fraction of soil, which has a larger surface-to-volume ratio. Rocks, gravel, and sand are generally relatively clean and can be separated from the soil that is to be treated. Non-soil debris such as sticks, plastic materials, metal parts, and construction materials should be removed prior to treatment.

Steps in Microbial-Culture Development for Slurry-Phase Bioremediation

1. *Selection of microbial sources*: The aim is to find microbial groups or species capable of degrading the target contaminants; more than one source should be used. Soil and activated sludge should be included.
2. *Growth of cultures in the laboratory*: The aim is to select the best sources of microorganisms for future use; mixing sources may be a good strategy.

3. *Determination of kinetic and stoichiometric parameters*: The aim is to develop design and operating parameters; the design of oxygen transfer and nutrient-addition requirements will depend on this step.
4. *Determination of toxicity limitations*: The aim is to determine concentration limits on operation; toxicity to microorganisms is a common problem.
5. *Growth of dense cultures*: The aim is to provide microorganisms for treatment processes; in general, the higher the population density, the better.

Advantages of Soil Bioremediation

Bioremediation offers several advantages over physical and chemical treatment processes used to treat contaminated water and soil:

1. Cleanup costs using bioremediation typically range from $100 to $250 per cubic meter, while more conventional technologies such as incineration or secure landfilling may cost between $250 and $1000 per cubic meter.
2. Bioremediation is aimed at biodegrading and detoxifying hazardous contaminants, whereas other technologies such as venting, adsorption onto activated carbon, solidification and stabilization, soil washing, and disposal into landfills simply transfer the contaminants to a different medium or location.
3. Bioremediation is also a relatively simple technology compared to most others. In situ bioremediation can be carried out with minimal site disruption, minimal VOC emission, and minimal health risk to residents or occupants.

Disadvantages of Soil Bioremediation

1. The most important disadvantages are the difficulty of predicting performance and the difficulty of scaling up from laboratory or pilot-plant tests. The success of a bioremediation project depends on the ability of the process operator to create and maintain environmental conditions necessary for microbial growth.
2. Microorganisms are sensitive to temperatures, pH, contaminant toxicity, contaminant concentration, moisture content, nutrient concentration, and oxygen concentration. A decrease in microbial activity will slow down the biodegradation and extend the treatment period. If microbial activity stops (e.g., due to accumulation of toxic metabolites), restarting the process may be difficult.

3. Cleanup goals may not be achievable during bioremediation because some contaminants are nonbiodegradable or only partially biodegradable or because the levels of contaminant removal cannot be attained microbially. As the contaminant levels are diminished, the biological degradation slows down, and the microorganisms may switch to other energy sources or stop growing.
4. Bioremediation may be relatively time-consuming. The time required generally depends on the rate at which the contaminants are degraded.

Artificial Formation of Geochemical Barrier

One aim of using biotechnology is to prevent the dispersion of hazardous substances from the accident site into the environment. This can be achieved by creating physical barriers on the migration pathway using microorganisms capable of biotransformation of the intercepted hazardous substances, for example, in polysaccharide (slime) viscous barriers in the contaminated subsurface. There may be chemically reacting barriers, like the Fe^0 barrier, which are used for dechlorination of chlorinated methanes and ethanes.

Another approach, which can be used to immobilize heavy metals in soil after pollution accidents, is the creation of biogeochemical barriers. These geochemical barriers could comprise gradients of H_2S, H_2, or Fe^{2+} concentrations, created by anaerobic sulfate-reducing bacteria (in the absence of oxygen and in the presence of sulfate and organic matter), fermenting bacteria (after the addition of organic matter and in the absence of oxygen), or iron-reducing bacteria (in the presence of Fe^{3+} and organic matter), respectively.

Other bacteria can form a geochemical barrier for the migration of heavy metals at the boundary between aerobic and anaerobic zones. For example, iron-oxidizing bacteria will oxidize Fe^{2+} in this barrier and produce iron hydroxides that can diminish the penetration of ammonia, phosphate, organic acids, cyanides, phenols, heavy metals, and radionuclides through the barrier.

21

Microbial Geotechnics

Biogeotechnics

Biogeotechnics is a branch of geotechnical engineering that deals with the applications of biological methods to geotechnical engineering problems. At present, biogeotechnological methods are related mainly to the applications of plants or vegetative soil cover for soil erosion control and slope protection, prevention of slope failure, and reduction of water infiltration into slopes.

A microbial improvement of soil could be one of the most cost-effective methods. The major factors that affect the applications of microorganisms to geotechnical engineering include

- Screening and identification of suitable microorganisms for different applications
- Optimization of microbial activity in situ
- Biosafety of the treatment
- Cost effectiveness
- Stability of soil properties after biotreatment

Among all the factors, cost effectiveness is the most important factor for large-scale application.

Biogeotechnology has advantages in low investment and maintenance costs. It also offers benefits to the environment and aesthetics. The most promising applications are bioclogging and biocementation.

Bioclogging

Bioclogging deals with reducing the hydraulic conductivity of soil and porous rocks using microbial activity or products. It could be used to reduce drain-channel erosion, form grout curtains to reduce the migration of heavy metals and organic pollutants, and prevent piping of earth dams and dikes.

Biocementation

Biocementation deals with enhancing the strength and stiffness properties of soil and rocks though microbial activity or products. It could be used to prevent soil avalanching, reduce the swelling potential of clayey soil, mitigate the liquefaction potential of sand, and compact soil on reclaimed land sites.

Grouting

Grouting is a process used to fill soil voids with fluid grouts. It is often used to control water flow. Common chemical grouts are solutions or suspensions of sodium silicate, acrylates, acrylamides, and polyurethanes.

Industrially produced water-insoluble gel-forming biopolymers of microbial origin such as xanthan, chitosan, polyglutamic acid, sodium alginate, and polyhydroxybutyrate can also be used as grouts for soil erosion control, enclosing of bioremediation zone, and mitigating soil liquefaction.

Suitable microorganisms could be applied to soil to serve the same purpose through microbial growth and biosynthesis of extracellular biopolymers.

Microbial Processes of Bioclogging

Microbial processes that can potentially lead to bioclogging are as follows:

- Formation of impermeable layer of algal and cyanobacterial biomass
- Production of slime in soil by aerobic and facultative anaerobic heterotrophic bacteria, oligotrophic microaerophilic bacteria, and nitrifying bacteria
- Production of undissolved sulfides of metals by sulfate-reducing bacteria
- Formation of undissolved carbonates of metals by ammonifying bacteria

Clogging with Microbial Polysaccharides

An accumulation of bacterial biomass, insoluble bacterial slime, and poorly soluble biogenic gas bubbles in soil in situ will make the soil more impermeable for water. Therefore, bioclogging can be used to seal a leaking construction pit, landfill, or dike. This can be performed by the addition of carbon source and enriched or pure culture of microorganisms to soil. Although a lot of gel-forming water-insoluble microbial polysaccharides are produced in industry, these materials cannot be used for soil grouting because of the high cost involved. Only the growth of microorganisms, which are added to soil, and the accumulation of water-insoluble microbial slime from cheap raw materials in situ can be considered as an economically reasonable option for bioclogging. An addition of microorganisms must be accompanied by the addition of medium that initiates bioclogging.

Bacteria Producing Clogging Polysaccharides

The microorganisms that produce insoluble extracellular polysaccharides to bind the soil particles and fill in the soil pores are oligotrophic bacteria from the genus *Caulobacter*; aerobic Gram-negative bacteria from genera *Acinetobacter*, *Agrobacterium*, *Alcaligenes*, *Arcobacter*, *Cytophaga*, *Flavobacterium*, *Pseudomonas*, and *Rhizobium*; nitrifying bacteria; cellulose-degrading bacteria *Cellulomonas flavigena*; and many species of Gram-positive facultative anaerobic and aerobic bacteria, such as *Leuconostoc mesenteroides* that is used for the production of a water-insoluble exopolymer dextran. Cellulose-containing agricultural and horticultural wastes, and saw dust can be hypothetically used for the propagation of bacteria in soil and formation of the pore-clogging polysaccharide.

Almost all bacteria produce exopolysaccharides under excess of carbohydrates or other water soluble sources of carbon over source of nitrogen. Therefore, such food-processing wastes or subproducts as corn glucose syrup, cassava glucose syrup, and molasses with C:N ratio >20 can be used for the production of bacterial water-insoluble polysaccharides.

Geotechnical Applications of Bioclogging

Geotechnical applications of bioclogging could be in harbor and dam control, erosion potential minimization, earthquake liquefaction mitigation, construction of reactive barriers, seepage control in irrigation channels, sealing of unforeseen leaks in the sheet piling screens around construction wells, and long-term stabilization of contaminated soils. In some cases, biological clogging of porous media and wells is a significant geotechnical problem due to the negative effect of bioclogging on soil bioremediation, sand filtration, and aquifer recharge.

Clogging with Microbially Induced Precipitates

Stable bioclogging could be due to precipitation of inorganic substances in the soil pores, for example, precipitation of calcium carbonate at increased pH. For example, urease-producing bacteria in a medium containing calcium chloride hydrolyze urea to ammonium and carbonate and precipitate calcium carbonate:

$$(NH_2)_2CO + 3H_2O \rightarrow 2NH_4^+ + HCO_3^- + OH^-$$

$$CaCl_2 + HCO_3^- + OH^- \rightarrow CaCO_3 \downarrow + H_2O + 2Cl^-$$

Due to this reaction, the pH is increased and hydrocarbonate is produced. It initiates the precipitation of calcium carbonate, which clogs pores and binds soil particles.

Biobinding

Biobinding is the formation of particle-binding cellular chains to bind sand grains and other soil particles, thus increasing slope stability. Biobinding can be performed by mycelial fungi, actinomycetes, and filamentous phototrophic and heterotrophic bacteria. However, biological binding is unstable and can be degraded by other microorganisms.

Biocementation

Chemical cementation is widely used in geotechnical engineering to fill in the sand voids with fluid grouts to produce sandstone-like masses to carry loads. The chemicals that are used to bind soil particles include sodium silicate, calcium chloride, calcium hydroxide (lime), cement, acrylates, acrylamides, and polyurethanes.

Microbial cementation (or structural microbial grouting) aims at forming soil-particle-binding material after introduction of microbes and specific additives in soil. Microorganisms are often associated with the cemented sediments containing calcium, magnesium, iron, manganese, and aluminum, which are crystallized as carbonates, silicates, phosphates, sulfides, and hydroxides.

The well-known technology of microbial biocementation is a combination of urease-producing microorganisms and soluble calcium salts. Microorganisms provide fast urea hydrolysis, increase the pH during hydrolysis of urea to ammonia, and form calcite in soil or rocks. The urease-producing microorganisms are usually from genera *Bacillus*, *Sporosarcina*, *Sporolactobacillus*, *Clostridium*, and *Desulfotomaculum*.

Microbial Processes of Biocementation

Microbial processes that can potentially lead to biocementation are as follows:

- Binding of soil particles with metal sulfides produced by sulfate-reducing bacteria
- Binding of the particles with metal carbonates produced due to hydrolysis of urea
- Binding of the particles with iron precipitates produced due to activity of iron-reducing bacteria
- Binding of the particles due to precipitation of silica dioxide, caused by microbial acidification

Geotechnical Applications of Biocementation

Microbial cementation could be used for numerous civil and environmental engineering applications, for example:

1. Enhancing stability for retaining walls, embankments, and dams
2. Reinforcing or stabilizing soil to facilitate the stability of underground constructions
3. Increasing the bearing capacity of piled or non-piled foundations
4. Reducing the liquefaction potential of soil
5. Treating pavement surface
6. Strengthening tailings dams to prevent erosion and slope failure
7. Constructing permeable reactive barriers in mining and environmental engineering
8. Binding of dust particles on exposed surfaces to reduce dust levels
9. Increasing the resistance to petroleum-borehole degradation during drilling
10. Increasing the resistance of offshore structures to erosion of sediment beneath foundations and pipelines
11. Stabilizing pollutants in soil by binding
12. Controlling erosion in coastal areas and rivers
13. Creating water filters and borehole filters
14. Immobilizing bacterial cells into a cemented active biofilter

Problems of Microbial Cementation and Clogging

A disadvantage of microbial cementation is that the microbial activity depends on many environmental factors such as temperature, pH, concentrations of donors and acceptors of electrons, and concentrations and diffusion rates of nutrients and metabolites. The design of microbial applications in bioclogging and biocementation must take into account not only soil conditions and grouting-medium content, but also microbiological, ecological, and geotechnical engineering aspects of the process. Design of bioclogging and biocementation requires data of the biological processes (growth, biosynthesis, biodegradation, bioreduction, biooxidation, and specific enzymatic activities); chemical reactions accompanied with formation of insoluble compounds; and physicochemical processes such as precipitation, crystallization, and adhesion.

Screening of Microorganisms for Bioclogging and Biocementation

The group of chemotrophic prokaryotes is most suitable for soil bioclogging and biocementation because of their smallest cell size, typically from 0.5 to 2 μm, ability to grow in soil, and large physiological diversity.

Phototrophic prokaryotes, mainly cyanobacteria, grow only on the soil surface because light penetrates a few millimeters into the soil. These bacteria can form a rigid crust on the surface of soil or sediment, which diminishes soil infiltration rate and improves slope stability. Cyanobacteria can also create millimeter-scale laminated carbonate buildups called stromatolites, which are formed in shallow marine environments, due to the sequence of sedimentation, growth of biofilm, production of a layer of exopolymers, and lithification of sediments by the precipitation of microcrystalline carbonate.

Archaea could be excluded from being considered as bioagents for soil bioclogging and biocementation, because all of them live in extreme environments that are not compatible with the majority of construction or land reclamation site conditions.

Application of Anaerobic Fermenting Bacteria in Biogeotechnics

Anaerobic fermenting bacteria may be involved in cementation of soil particles under the presence of calcium, magnesium, or ferrous ions. This cementation can be due to the increase in pH caused by ammonification (release of ammonia) and carbon dioxide production in soil added with urea or waste protein. The insoluble carbonates and hydroxides of metals will be precipitated at high pH, thus binding soil particles and clogging soil.

If carbohydrates are added to soil, fermenting anaerobic bacteria can diminish the pH due to the formation of organic acids during fermentation of carbohydrates. This could be potentially used in bioclogging and biocementation to precipitate silicates from colloidal silica suspension. It is known that the stability of colloidal silica suspension is reduced at acidic pH. Thus, inorganic acids are added in this type of chemical grouting.

Organic acids produced in fermentation can dissolve carbonates and hydroxides, binding soil particles or plugging soil pores. Anaerobic bacteria cannot clog soil pores by the synthesis of extracellular polymers because they are not able to produce a large quantity of slime due to their low efficiency of biological energy production in fermentation.

Application of Anoxic Bacteria in Biogeotechnics

Organic acids, hydrogen, and alcohols, which are produced by anaerobic fermenting bacteria from polysaccharides and monosaccharides, can be used as donors of electrons by anaerobic respiring bacteria. Anoxic iron-reducing, sulfate-reducing, and nitrate-reducing bacteria can be used in geotechnical applications due to the formation of dissolved ferrous ions, precipitation of sulfides, and increase of pH. Denitrifying bacteria form a large quantity of nitrogen gas, which can be used to diminish the saturation of sand with water and to prevent sand liquefaction.

Application of Facultative Anaerobic Bacteria in Biogeotechnics

Facultative anaerobic bacteria could be considered as the most suitable bioagents for soil bioclogging and biocementation because many species are able to produce large quantities of exopolysaccharides, which usually promote the formation of cell aggregates, and can grow under either aerobic or anaerobic conditions. The last property of facultative anaerobic bacteria is the most essential for biotreatment of soil in situ, where supply of oxygen is limited by the soil porosity and both aerobic and anaerobic microzones co-exist in soil. There are, for example, bacteria from the genera *Alcaligenes*, *Enterobacter*, *Staphylococcus*, *Streptococcus*, *Rhodococcus*, corynebacteria (*Gordonia*, *Nocardioides*), gliding bacteria (*Myxococcus*, *Flexibacter*, *Cytophaga*), and oligotrophic bacteria (*Caulobacter*).

Application of Microaerophilic Bacteria in Biogeotechnics

Microaerophilic bacteria could be used for the biobinding of soil particles because many strains of microaerophilic bacteria are combined in filaments or joined by sheaths, and these filamentous structures can also bind soil particles. The filamentous bacteria from the genera *Beggiatoa*, *Haliscomenobacter*, *Microthrix*, *Nocardia*, *Sphaerotilus*, and *Thiothrix* are common in aerobic tanks of wastewater treatment plants and can be probably used for biobinding of soil particles.

Application of Aerobic Bacteria in Biogeotechnics

Aerobic bacteria could be suitable for soil bioclogging, biocementation, and biobinding of soil particles because many species are able to produce a large quantity of slime, form chains and filaments, increase pH, and oxidize different organic and inorganic substances. Cells of many *Actinomycetes*, a group of Gram-positive bacteria, typical soil inhabitants, form particles, binding mycelium and producing particle-binding slime in soil. These bacteria are most prospective for aerobic soil bioclogging, biocementation, and biobinding.

General Considerations on Bacterial Applicability in Geotechnics

If aerobic bacteria are used for soil clogging or cementation, a major technological problem is the air supply into soil. If the rate of oxygen supply into soil by aeration and diffusion is not sufficient, there will be formation of anaerobic layer or zones, where aerobic bacteria will not be active.

Therefore, from the technological and biological points of view, the most suitable physiological groups for soil bioclogging and biocementation in situ are facultative anaerobic bacteria, which are active under both aerobic and anaerobic conditions.

Another general consideration is that the most suitable bacteria for soil bioclogging or biocementation are bacteria with Gram-positive type of cell wall because these bacteria are most resistant to the changes of osmotic pressure, which is typically seen in soil on construction or reclamation sites.

Biosafety in Biogeotechnics

An important aspect in microbial clogging and cementation is biosafety. To diminish the risk of accumulation and release of pathogenic bacteria during bioclogging and biocementation, the following selective conditions can be used:

- Application of carbon sources, which are used in nature by saprophytic microorganisms, such as cellulose, cellulose-containing agricultural waste, vegetable-processing waste, and molasses.
- Conditions, which are suitable for the growth of autolithotrophic bacteria. Carbon dioxide is used as a carbon source and inorganic substances (NH_4^+, Fe^{2+}, S) are used as electron donors.
- Conditions that are suitable for application of bacteria able to use anaerobic respiration with SO_4^{2-} or Fe^{3+} as electron acceptors.
- Application of solution with low concentration of carbon source for preferable growth of oligotrophic microorganisms in soil.

The problem of biosafety can be solved also by the selection of a safe bacterial strain. The biomass of this safe strain can be produced in a bioreactor and used as a starter culture for bioclogging or biocementation.

22

Microbiology of Air and Air Treatment

Bioaerosols

A bioaerosol is defined as a collection of airborne biological particles: viruses, bacteria, spores of fungi, algae, and other microorganisms. The size of bioaerosols ranges from 0.1 to 30 mm. Microorganisms may also be attached to airborne dust particles. Human exposure to bioaerosols may result in infections and allergic reactions, depending on the size of bioaerosol particles and the depth of penetration into the lungs, types of microorganisms, and their pathogenicity.

Viable bioaerosol particles are living organisms that can be detected by growing individual microorganisms in a suitable medium. The most common viable bioaerosols are bacterial and fungal spores. Nonviable bioaerosols include pollen and dead microorganisms.

Sources of Bioaerosols

Bioaerosols come from two sources:

1. Natural sources, including coughing, sneezing, shedding of infected human skin, and dispersion of soil dust and droplets from aquatic environment by wind.
2. Production sources, including agricultural activity, for example, composting, biolaboratory activity, biotechnological facilities, and aerobic wastewater treatment. Aeration of wastewater may be responsible for the generation of bioaerosols, which could infect humans and cause diseases.

Viral Aerosols

Viruses are intracellular parasites that can reproduce only inside a host cell. They consist usually of RNA or DNA covered with protein. Although naked viruses range from 0.02 to 0.3 μm, most airborne viruses are found as part of droplet nuclei or attached to other airborne particles. Aerosolization can occur by coughing, sneezing, or talking. Viral bioaerosols spread infectious airborne diseases such as colds, flu, chicken pox, and SARS.

Bacterial Aerosols

Bacteria are single-celled organisms with sizes ranging from 0.3 to 10 μm. Some bacteria form dormant spores with high survivability in dry aerosol particles, which are easily carried by air flow. There are many bacterial pathogens spread by air and causing airborne infectious diseases such as tuberculosis and legionellosis. Aerosolization can occur from the surface of soil containing up to 10^9 bacterial cells/g or water containing up to 10^6 bacterial cells/mL. In indoor environments, bacteria can colonize wet sites in ventilation system and become aerosolized by air flow.

Control strategies are intended to prevent exposure of susceptible individuals to aerosols containing infectious agent. For example, to prevent the outbreak of legionellosis, cooling towers should be located so that cooling tower exhaust would not be carried directly into the ventilation systems of buildings.

Fungal Aerosols

The majority of the fungi are saprophytes, which are organisms feeding on dead organic matter. They live in soil and in sites having decaying vegetation. Many species of fungi release spores, which are dispersed in air. These fungal spores are particles of 0.5–30 μm in size, are resistant to environmental stresses, and are adapted to airborne transport. Inhalation of fungi or their spores can cause an infection or allergic disease. Fungi and their spores can be a major problem in buildings where moisture control is poor. Some fungi produce mycotoxins, for example, aflatoxin and the trichothecenes.

Pollen Grain Aerosols

Pollen grains are relatively large, nearly spherical particles produced by plants to transmit genetic material to the flowers of other plants of the same species. Pollen grains range in size from 10 to 100 μm. The release of pollen into air is seasonal and causes seasonal allergic diseases, for example, hay fever.

Concentration of Aerosols

The concentration of biological particles in air is variable and influenced by wind, weather, and the source of the aerosol. Typical concentrations of live bacteria in bioaerosols are 100–1000 cfu/m^3 for outdoor air and same or less for indoor air of ventilated space. However, it could be less than 10 cfu/m^3 in mountain areas and less than 100 cfu/m^3 in clean rooms, or more than 10,000 cfu/m^3 in farms, near aeration tanks of municipal wastewater-treatment plants, or in the indoor air of biotechnological laboratories due to improper microbiological practice (Figure 13.1).

Bioaerosol Sampling

Bioaerosol sampling is the removal and collection of biological particles from the air so that viable organisms could be detected. There should be no change in the culturability of the microorganisms or their biological integrity. The methods of bioaerosol sampling are impaction, impingement, filtration, and settling by gravity.

Impaction separates particles from the airstream by using the inertia of the particles to force their deposition onto the surface of a semisolid medium. Liquid impingement is similar to impaction, but the medium is a liquid. Filtration of air retains cells on the membrane surface that is later exposed to the semisolid medium.

Bioaerosols and Indoor Air Quality

The occurrence of bioaerosols in indoor air can negatively affect human health. Sources of bacterial and fungal aerosols include building materials, furnishings, pets, plants, and air-conditioning systems. Bacteria grow in areas where water is stagnant or condenses within air conditioning or ventilation systems. Fungi have a lower requirement for water than bacteria and grow at the surface of different building materials.

Occupants of enclosed spaces are a major source of bacterial and viral bioaerosols. Human-to-human transmission can occur in high-density indoor environments such as classrooms and military or correctional facilities. A sneeze can release about a million small droplets, forming a bioaerosol after drying.

Fate of Bioaerosols in Environment

A significant number of released microorganisms quickly die after aerosolization due to drying in air. The persistence of vegetative cells of microorganisms is increased at high humidity, low temperature, and low or absence of solar radiation. The endospores of bacteria and spores of actinomycetes and fungi have a high survival rate in the environment. The death of microorganisms in an aerosol can be described conventionally as exponential decay:

$$N = N_o e^{-kt}$$

where

N is the viable organism concentration in air at time t

N_o is the viable organism concentration in air at time zero

k is the decay rate constant

Bioaerosols from Wastewater Treatment, Spray Irrigation, and Biotechnological Plants

Bioaerosols are produced in aeration systems of wastewater-treatment plants, industrial bioreactors, or in aquacultural ponds using intensive aeration. These bioaerosols, if not treated, can spread for at least several hundred meters from the point of aeration. A high level of pollution with bioaerosol can also be caused if wastewater-treatment effluent is used for spray irrigation.

The dispersion of microorganisms in environment (the downwind concentration of microorganisms) depends on wind speed and direction, microbial concentration in the source, intensity of aeration, and microbial decay rate. To remove airborne microbes from air supplied for cultivation or in aseptic transfers of microorganism, the air is filtrated through membrane filters or through depth filters filled with fibers, where the microorganism are adhered.

Odors Generated by Wastewater-Treatment Plants

Odor-producing compounds, mainly dihydrogen sulfide, mercaptans, and volatile organic acids, are formed during aerobic and anaerobic treatment of wastewater. H_2S is produced by sulfate-reducing bacteria and during decay of sulfur-containing amino acids such as methionine, cysteine, and cystine. H_2S is corrosive to many materials and toxic to humans. To prevent odors, frequent cleaning and washing of equipment, as well as maintenance of aerobic conditions in the wastewater-treatment system is required. The best practice is to cover odor-producing units and to treat odorous air prior to release.

Treatment of Odorous and Exhaust Gases

The gases can be treated in a wet scrubber using chemical oxidation by chlorine, hydrogen peroxide, or ozone. The gases can be combusted at temperatures ranging from 500°C to 800°C in flame or oxidized by catalysts (palladium or platinum) at temperatures between 300°C and 500°C. The gases can be adsorbed on granulated activated carbon or other adsorbents, for example, iron oxide. Activated carbon is particularly effective at removing sulfur-containing odorous compounds.

The gases can be oxidized biologically, using biofiltration. In the simplest biofilters, soil or compost beds can be used for odor or toxicity removal. In other cases, more sophisticated metal or plastic carriers for attachment of microbial biofilm and for enhanced mass transfer between biofilm and air are used. Gases flow upward through pores in carrier and are oxidized by microorganisms of the biofilm at the surface. Aerated gas biofilter with fixed microbial biofilm can remove almost all odorous, exhaust, or toxic compounds.

Biofiltration of volatile organic compounds (VOCs), in air emissions of painting facilities, chemical reactors, and wood processing facilities, is often used in practice. Chemical and physical technologies for VOCs destruction are expensive, when VOCs concentrations are low and air volume is large. Industrial ventilation air containing formaldehyde, ammonia, and other substances can also be effectively treated in the bioscrubber or biofilter.

An example of biofiltration of odorous gases, used in human practice in ancient times, is a tomb covered with earth. Soil microorganisms over the tomb degrade odorous volatile compounds of the decaying body.

Aerobic Biotechnological Treatment of Hazardous Waste Gas

The CERCLA Priority List of Hazardous Substances contains many substances released in industry as gaseous hazards and which can be treated biotechnologically, including the following: chloroform, trichloroethylene, 1,2-dibromoethane, 1,2-dibromo-3-chloropropane, carbon tetrachloride, xylenes, dibromochloropropane, toluene, methane, methylene chloride, 1,1-dichloroethene, bis(2-chloroethyl) ether, 1,2-dichloroethane, chlorine, 1,1-trichloroethane, ethylbenzene, 1,1,2,2-tetrachloroethane, bromine, methylmercury, trichlorofluoroethane, 1,1-dichloroethane, 1,1,2-trichloroethane, ammonia, trichloroethane, 1,2-dichloroethene, carbon disulfide, chloroethane, *p*-xylene, hydrogen sulfide, chloromethane, 2-butanone, bromoform, acrolein, bromodichloroethane, nitrogen dioxide, ozone, formaldehyde, chlorodibromomethane, ethyl ether, and 1,2-dichloropropane.

Biotreatment of Toxic Gas

The common way to remove vaporous or gaseous pollutants from gas- or airstreams is to pass contaminated gases through bioscrubbers containing suspensions of biodegrading microorganisms or through a biofilter packed with porous carriers covered by biofilms of degrading microorganisms. Depending on the nature and volume of polluted gas, the biofilm carriers in a biofilter may be cheap porous substrates, such as peat, wood chips, compost, or regular artificial carriers such as plastic or metal rings, porous cylinders and spheres, and fibers and fiber nets.

The bioscrubber also can be used for the treatment. It must be stirred to ensure a high mass transfer rate between gas and microbial suspension. The liquid that has interacted with the polluted gas is collected at the bottom of the biofilter and recycled to the top part of the biofilter to ensure an adequate contact of polluted gas and liquid and optimal humidity of biofilter. Addition of nutrients and fresh water to bioscrubber or biofilter must be made regularly or continuously. Fresh water can be used to replace water that has evaporated in the bioreactor. If the mass transfer rate is higher than the biodegradation rate, the absorbed pollutants must be biodegraded in an additional suspended bioreactor or biofilter connected in series to the bioscrubber or absorbing biofilter.

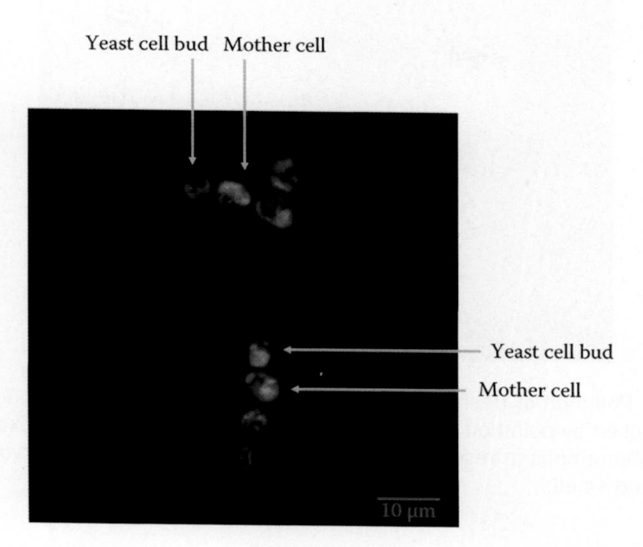

Figure 1.6 Yeast cells.

Figure 1.7 Cultivation of edible mushrooms on saw dust.

Figure 1.11 Pollution of fresh water due to excessive growth of algae and cyanobacteria caused by pollution of water with nitrogen and phosphorous. Excessive growth of phototrophs in reservoirs can clog water filters and give reservoir water a bad taste and smell.

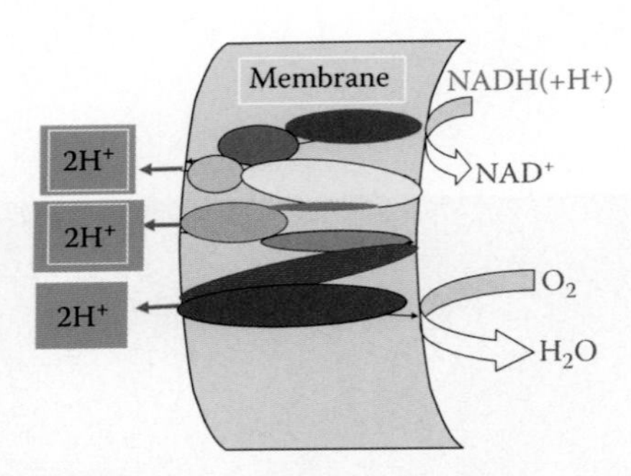

Figure 4.3 Formation of proton-motive force on membrane due to electron transfer from donor to acceptor and vectorial transfer of protons to membrane outside.

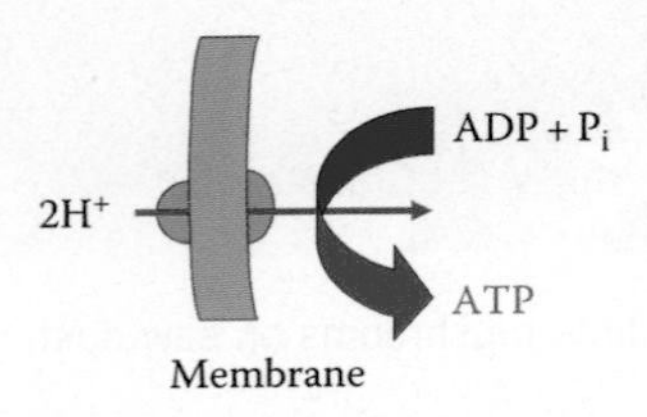

Figure 4.4 Oxidative phosphorylation. ADP, adenosine diphosphate nucleotide; ATP, adenosine triphosphate nucleotide; P_i, inorganic phosphate.

Figure 5.2 Edible mushrooms containing lignin-degrading enzymes. They can be used for biodegradation.

Figure 7.1 Sample ecosystem in New Zealand with different plants and animals as well as viruses and microorganisms: bacteria, fungi, and algae. The surrounding mountains form the physical boundary of the ecosystem. Component ecosystems, for example a single deer farm in New Zealand, (pictured) can be considered part of a bigger ecosystem, despite being separated from the surrounding environment by a fence.

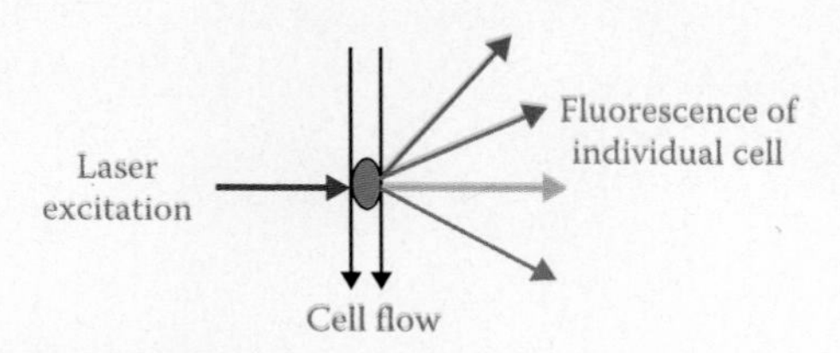

Figure 8.5 Flow cytometry of individual cells.

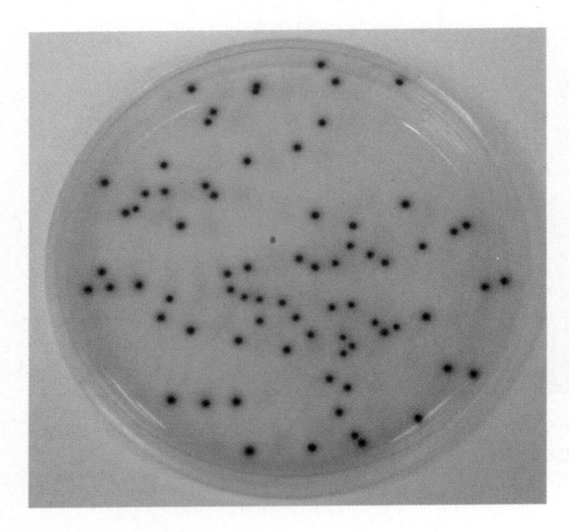

Figure 8.7 Colonies on Petri dish.

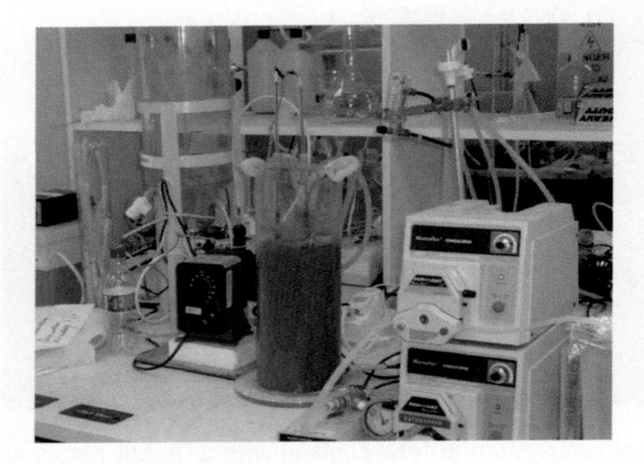

Figure 13.1 This figure depicts a potentially dangerous environmental engineering experiment. Air is passed through an aeration tank containing bacterial biomass. The resulting contaminated air contains aerosols of pathogenic and opportunistic microorganisms and must be filtered before being released into the atmosphere.

Figure 14.5 The industrial-scale reactor used in environmental biotechnology.

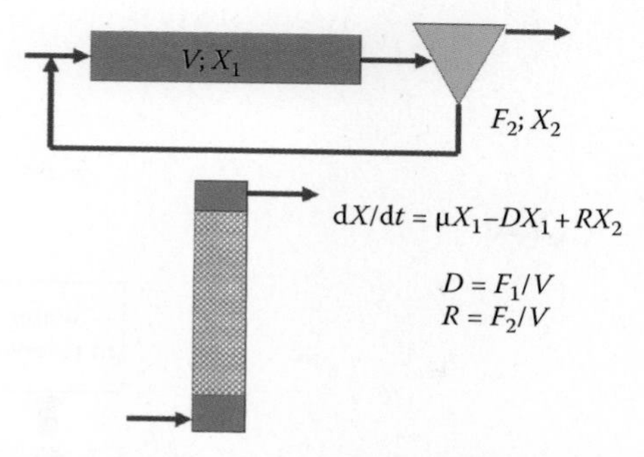

Figure 14.6 Continuous systems with internal recycling or retention of biomass.

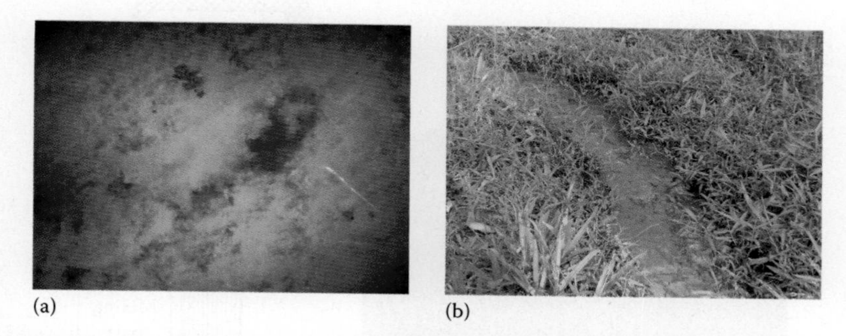

Figure 15.1 Iron hydroxide precipitates occur naturally in a stream (b) due to oxidation of ferrous chelates by neutrophilic iron-oxidizing bacteria; (a) shows a closeup of the precipitates.

Figure 15.2 Eutrophication of urban reservoirs.

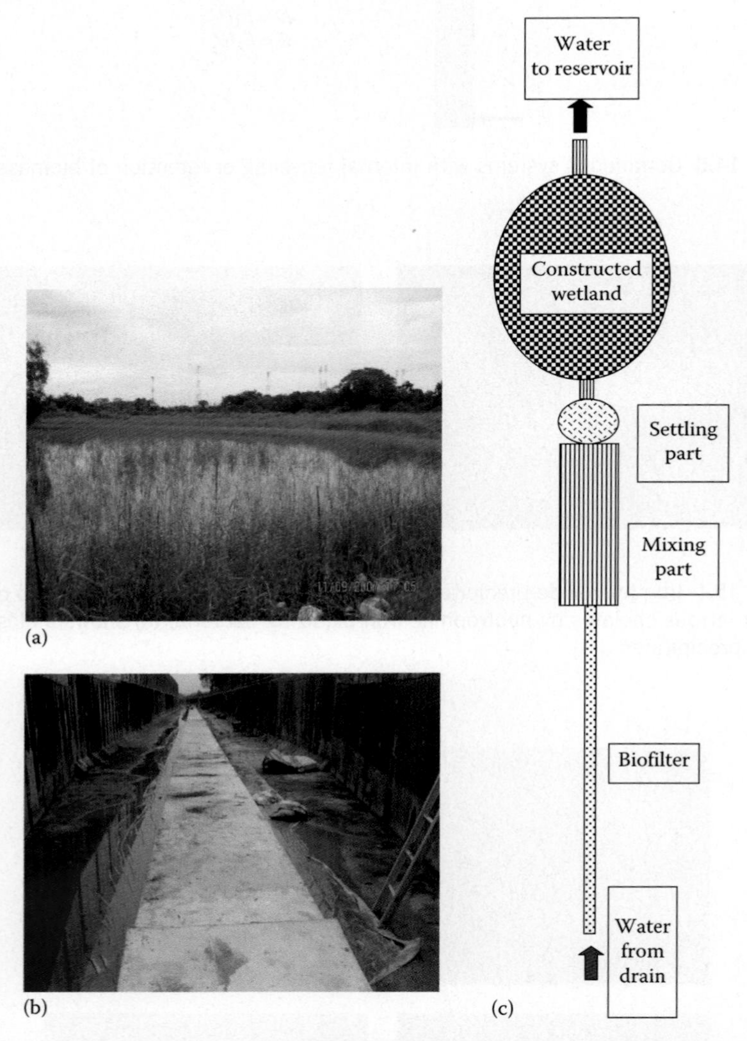

Figure 15.3 Combination of (a) constructed wetland with (b) biofilter inside drain. Schematic of this combination is shown in (c).

Figure 16.1 Anaerobic digesters on municipal wastewater treatment plant.

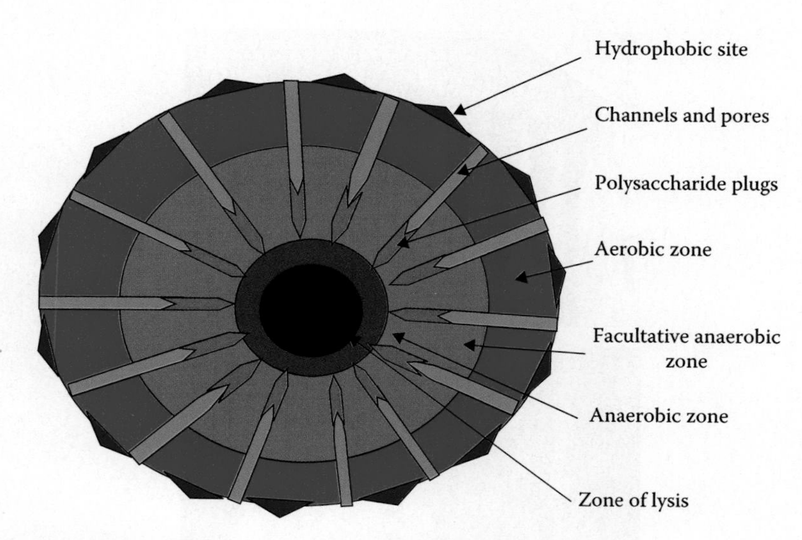

Figure 17.6 Schematic of microbial granule structure.

Figure 23.1 Biodeterioration caused by sulfate-reducing bacteria on a marble statue (black plumes of CaS).

Figure 23.2 Biodeterioration of the columns of an ancient Roman temple caused probably by nitrifying bacteria.

23

Biodeterioration, Biocorrosion, and Biofouling

Prevention of biodeterioration, biocorrosion, and biofouling of materials used in civil and environmental engineering is an important task of environmental microbiology.

Microbial Biodeterioration

Almost all materials can deteriorate due to microbial oxidation/reduction, hydrolysis, production of acids, alkali, and oxygen radicals.

Timber in buildings and other wooden structures decay due to fungi and bacteria that hydrolyze cellulose, hemicelluloses, and degrade lignin. Dry rot, brown rot, and white rot of the timber are due to the different species of fungi specializing on biodegradation of cellulose or lignin.

Mineral acids or alkali excreted by the microorganisms cause a microbial deterioration of cultural heritage materials.

For example, corrosion of steel and cement and deterioration of marble constructions can be caused by the formation of sulfuric or nitric acids:

$$S + 1.5O_2^{+} \rightarrow 2H^{+} + SO_4^{2-}$$

(sulfur-oxidizing bacteria from genus *Acidithiobacillus*)

$$NH_4^{+} + 1.5O_2 \rightarrow 2H^{+} + NO_2^{-} + H_2O$$

(ammonium-oxidizing bacteria from genus *Nitrosomonas*)

Deterioration of Cultural Heritage Objects

Wood and paper can be easily biodegraded by microorganisms under humid conditions. Objects of art, sculptures, and historical buildings are also deteriorated by microorganisms in humid and polluted atmospheres. For example, the marble statue shown in Figure 23.1 has black plums due to the formation of black CaS, which can be described by the following reactions:

$$SO_2 \text{ (released into air by vehicles)} + H_2O + 0.5O_2 \rightarrow H_2SO_4$$

$$H_2SO_4 + 2CH_2O \rightarrow H_2S + 2CO_2 + 2H_2O \text{ (bacterial sulfate reduction)}$$

$$CaCO_3 \text{ (white marble)} + H_2S \rightarrow CaS \text{ (black matter)} + H_2O + CO_2$$

Another example is the biodeterioration of the column in ancient Roman Minerva temple in Assisi caused probably by ammonifying and nitrifying

Figure 23.1 (See color insert following page 292.) Biodeterioration caused by sulfate-reducing bacteria on a marble statue (black plumes of CaS).

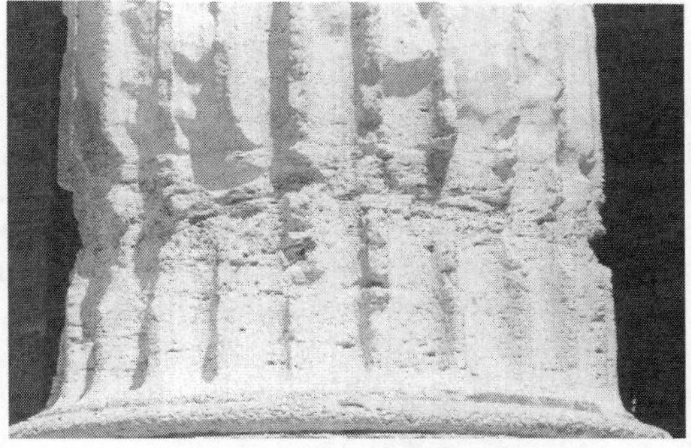

Figure 23.2 (See color insert following page 292.) Biodeterioration of the columns of an ancient Roman temple caused probably by nitrifying bacteria.

bacteria (Figure 23.2) in the upper layer of soil where the bottom part of the column was embedded at some period of time:

$$CO(NH_2)_2 + 3H_2O \rightarrow 2NH_4^+ + CO_2 + 2OH^-$$

$$R(COO)NH_2 + \text{chemical energy} + [2H = \text{bioreductant}] \rightarrow RCOOH + NH_3$$

$$NH_4^+ + 2O_2 \rightarrow NO_3^- + 2H^+ + H_2O$$

$$CaCO_3 + 2H^+ \rightarrow Ca^{2+} + CO_2 \uparrow + H_2O$$

Microbially Influenced Corrosion

Microbially influenced corrosion (MIC) is corrosion that is caused by the presence and activities of microbes.

In all biochemical reactions, corrosion reactions occur with an exchange of electrons. The electrons are released in one area, the anode, travel through a metallic path and are consumed through a different biochemical reaction in another area, the cathode. The major reasons of microbially induced biocorrosion are formations of nitric, sulfuric, and organic acids; H_2S; and H_2.

Microorganisms can induce corrosion:

- By stimulation of the anodic reaction by formation of acids
- By stimulation of the cathodic reaction by microbial production of H_2S
- The biodegradation of protective films
- The increase in conductivity of the liquid environment

Microbial Formation of Acids

Organic acids, produced by bacteria and fungi, are also active corrosive agents. Biocorrosion results in pitting, crevice corrosion, selective de-alloying, stress-corrosion cracking, and under-deposit corrosion. The formation of a microbial biofilm at the surface is an essential stage of the corrosion process. A biofilm provides a localized decrease of pH, concentration of oxygen, and ORP and an increased concentration of microbial metabolites, initiating corrosion.

Very common bioagents of biocorrosion are anaerobic sulfate-reducing bacteria, anaerobic fermenting bacteria, aerobic nitrifying bacteria, aerobic sulfur-oxidizing bacteria, and acid-producing fungi. Stress-corrosion cracks in structures can be initiated by the action of nitrifying bacteria at the concrete surface. MIC can be a serious problem in stagnant-water systems such as the fire-protection system because of fermentation and production of organic acids.

MIC can be controlled by biocides or by conventional corrosion-control methods. Iron-reducing bacteria can protect steel from corrosion, probably due to the formation of a protective biofilm.

A bridge collapse in 2007 in Minnesota, which killed 13 people, has been attributed to a buildup of pigeons' poo. Some experts said that the birds' droppings deposited over the bridge's framework caused the steel beams to rust faster because of the microbial formation of nitric acid from nitrogen compounds of pigeons' poo.

Biofouling of Membranes

Biofouling of membranes, which are used in microfiltration, nanofiltration, reverse osmosis, desalination process, and in membrane bioreactors, is caused by aggregation of microorganisms and deposition of their products at the membrane surface and inside membrane pores. Biofouling of membranes leads to a loss in efficiency of the separation process. Biofouling can be diminished using membrane coatings and technological operations.

Flux through membrane and transmembrane pressure are the best indicators of membrane fouling. Under constant flux operation, transmembrane pressure increases to compensate for the fouling. Under constant transmembrane pressure, flux declines due to the occurrence of membrane fouling.

Membrane Treatments of Water and Wastewater

The types of membrane treatments are as follows:

- Reverse osmosis removes cations and anions, as well as organics from water.
- Nanofiltration removes Ca and Mg compounds from water.
- Ultrafiltration removes colloidal particles from water.
- Microfiltration removes microorganisms from water.
- Membrane bioreactors are used in wastewater treatment for the cultivation of microorganisms with simultaneous separation of cells and liquid.

The applications of membranes in water treatment are as follows:

- Removal of color, odor, and organic contaminants from water
- Seawater desalination
- Treatment of brackish groundwater
- Water softening
- Wastewater recovery

Mechanisms of Membrane Fouling

The mechanisms of membrane fouling are as follows:

- Blockage of pore entrance
- Constriction of pore entrance
- Gel/cake formation due to concentration polarization
- Internal plugging of pore
- Pore narrowing

Types of Foulants

Foulants can be

- Colloidal matter (clays, flocs)
- Biological matter (bacteria, fungi)
- Organic matter (oils, polyelectrolytes, humic acids)
- Scaling matter (mineral precipitates)

Fouling Control

Fouling control can be performed by

- Physical cleaning using back flush
- Biological cleaning using biocides and enzymes to remove microorganisms
- Chemical cleaning using acids, bases, and surfactants to remove foulants
- Selecting a membrane that is less prone to fouling
- Optimal choice of operating conditions

24

Outline and Summary of the Course "Environmental Microbiology for Engineers"

Course Outline

Grading criteria include two quizzes (30%) and a final exam (70% of total marks).

Lecture No.	Content
1	Microorganisms Viruses, prokaryotes, fungi, algae, and protozoa. Cell shape and structure: capsule, wall, membrane, inclusions, spores, and organelles.
2	Static Biochemistry Monomers, oligomers, polymers, proteins, polysaccharides, and nucleic acids.
3	Dynamic Biochemistry Biochemical reactions, enzymes, and metabolic blocks.
4	Biooxidation and Bioreduction Oxidation–reduction, balance of biological energy, and metabolic diversity.
5	Biodegradation

(continued)

Lecture No.	Content
	Biodegradation of natural substances: polysaccharides, proteins, nucleic acids, lipids, and lignin.
	Biodegradation of aliphatic, aromatic, and chlorinated hydrocarbons.
	Co-metabolism.
	Biodeterioration.
	Biocorrosion.
6	Molecular Biology and Genetics
	Replication, transcription, and translation.
	Gene, chromosome, recombination, and mutation.
7	Bioagents of Environmental and Engineering Bioprocesses
	Prokaryotic and eukaryotic cells.
	Cell aggregates.
	Populations.
	Biofilms.
8	Reproduction, Proliferation, and Growth
	Nutrition and growth: nutrients, media, growth, kinetics, and stoichiometry of growth.
9	Microbial Ecology
	Positive and negative interactions in microbial communities.
	Interactions of microorganisms with humans, animals, and plants.
	Environmental factors, ecological behavior, relation to oxygen, temperature, pH, and water activity.
10	Classification of Viruses and Microorganisms
	Phenotypic and genotypic classifications.
	Groups of viruses, prokaryotes, and eukaryotes.
11	Physiological Classification of Prokaryotes
	Classification of prokaryotes according to relation to oxygen and generation of biological energy.
	Periodic table of prokaryotes.
12	The Groups of Prokaryotes
	Properties of different groups of Gram-negative, Gram-positive bacteria, and archaea.
	Functions of prokaryotic groups in environmental applications.
	The periodic table of prokaryotes and isolation and identification of microbial strains.
	Anaerobic microorganisms: anoxygenic phototrophic bacteria and fermenting bacteria.
	Archaea: methanogens and anaerobic digestion of organic substances.
	Anaerobically respiring prokaryotes: sulfate-reducing, iron-reducing, and denitrifying prokaryotes.
	Facultative anaerobic and microaerophilic prokaryotes.
	Aerobic prokaryotes.

Lecture No.	Content
13	Public Health and Water Disinfection Saprophytic, opportunistic, and pathogenic microorganisms. Factors of pathogenicity. Agents of infection. Waterborne diseases and agents. Indicator microorganisms. Bacteriological quality of water.
14	Quiz 1 (qualitative questions)
15	Biotechnological Processes Upstream, core, and downstream processes. Selection, inoculation, batch and continuous cultivation, bioreactors, and separation of biomass.
16	Disinfection of Water Disinfectants. Kinetics of disinfection.
17	Aquatic Systems and Water Biotreatment Removal of biological instability and water pollutants and biofilm processes.
18	Anaerobic and Anoxic Treatment of Wastewater Oxygen and Energy Generation Anaerobic digestion of organic matter: digesters, microbiology, biochemistry, ecology, septic tanks, anaerobic biofilters. Anoxic biotechnological methods including denitrification, iron reduction, and sulfate reduction.
19	Aerobic Treatment of Wastewater Microbiology and biotechnology of aerobic wastewater treatment, activated sludge process, nutrients removal, filamentous bulking and foaming, granulation, and fixed-biofilm treatment.
20	Value-Added By-Products of Environmental Engineering Enzymes, bioplastic, pigments, fertilizers, and biofuel from wastes.
21	Biotreatment of Industrial Hazardous Wastes Composting of solid wastes: windrow, static pile, and in-vessel systems; stages of composting; microbiology; and process control.
22	Solid Wastes and Soil Biotreatment Removal of pathogens from soil and solid wastes. Soil pollutants, sources of pollution, groundwater remediation, soil venting, slurry-phase remediation.
23	Microbial Geotechnics Geotechnical applications of microorganisms using biocementation, bioclogging, and biodesaturation.

(*continued*)

Lecture No.	Content
24	Microbiology of Air and Air Treatment Types and sources of bioaerosols. Sampling. Indoor and outdoor bioaerosols. Biotreatment of polluted air.
25	Biodeterioration, Biocorrosion, and Biofouling Agents and mechanisms of biodeterioration and biocorrosion. Membrane biofouling.
26	Quiz 2 (quantitative questions)

Summary of the Course "Environmental Microbiology for Engineers"

Microorganisms

The objects of study in environmental microbiology are bacteria, microscopic fungi, microscopic algae, and protozoa. Other microscopic organisms such as viruses, metazoa, cysts of the helminthes are also important for environmental microbiology.

Viruses are particles assembled from biopolymers, which are capable of multiplying and assembling as new virus particles inside living cells. Their typical size is about 0.1 μm.

Bacteria are microorganisms with prokaryotic type of cells. The typical cell size is 1–2 μm. Bacteria are most active in the degradation of organic matter and are used in wastewater treatment and soil bioremediation. The removal of pathogenic bacteria from water, wastewater, soil, and solid waste is an important task of environmental biotechnology.

Fungi are eukaryotic microorganisms that assimilate organic substances by adsorption. The typical cell size is 10 μm. Fungi are active degraders of polymers and are actively used in the composting and biodegradation of toxic organic substances.

Algae are floating eukaryotic microorganisms that assimilate light energy. The typical cell size is 10–20 μm. Algae are used in environmental biotechnology for the removal of nitrogen and phosphorus from water and as indicators of water quality.

Protozoa are unicellular animals that digest organic food intracellularly. The typical cell size is 10–50 μm. Some protozoa are pathogenic and have to be removed from water and wastewater.

The helminthes are parasites of humans and animals. The presence of their cysts in water and wastewater is a public health concern.

Static Biochemistry

Monomers combine and form polymers.

The biopolymer units are connected by hydrogen bonds and hydrophobic bonds.

Polysaccharides are composed of carbohydrate (sugar) units joined by a covalent bond.

Proteins consist of chains that are made up of the 20 amino acids, which are joined by covalent bond. The primary structure of proteins is the sequence of amino acids. The secondary structure of proteins refers to the formation of a helix or sheet by the polypeptide chain. The tertiary structure of proteins is due to the folding of the chain. The quaternary structure of proteins is an association of several protein molecules into one functional protein.

Denaturation is a change of the secondary, tertiary, and quaternary structures of proteins (enzymes), following their inactivation.

Typical lipids (fats) contain hydrophilic glycerol and hydrophobic long-chain fatty acids. Lipids are essential components of the cell membrane.

DNA and RNA are chains of nucleic acids composed of nucleotides. Each nucleotide contains a sugar (ribose in RNA and deoxyribose in DNA), phosphate, and a nitrogen-containing base: adenine (A), guanine (G), cytosine (C), and thymine (T, in DNA only) and uracil (U, in RNA only).

The nucleotides are usually represented by the letters A, G, C, and T. Given the sequence of letters on one DNA strand, we know instantly what the sequence of letters must be on the other strand, because of complementary base pairing. G of one strand always pairs with C of another strand. A always pairs with T.

DNA consists of two sugar–phosphate backbones, which are held together by hydrogen bonds between the bases on two strands.

RNA differs from DNA in that it is usually single stranded. RNA molecules contain a few thousand nucleotides. RNA occurs as messenger, transfer, and ribosomal RNAs.

Dynamic Biochemistry

All chemical reactions in cells are catalyzed by protein molecules called enzymes. More than 3000 different protein molecules (enzymes) must be synthesized for self-replication of a bacterial cell.

Metabolism is the system of biochemical reactions in a living organism or microbial community.

The elementary unit of metabolism is the biochemical reaction: substrate (S) transforms into product (P) through the catalytic action of enzyme (E):

$$S + E(+C) \rightarrow [SEC] \rightarrow P + E(+C)$$

The main metabolic blocks are catabolism (oxidation/reduction coupled with energy generation) and anabolism (biosynthetic reactions coupled with energy utilization).

Biooxidation and Bioreduction

The main forms of biologically available and convertible forms of energy are $NAD(P)H_2$, $\Delta\mu_{H^+}$, and ATP.

Oxidation–reduction is the loss-gain of electrons followed by the transformation of protons, H^+. The energy is released from an electron donor (energy source) in coupled oxidation–reduction reactions.

Fermentation is an intramolecular oxidation–reduction reaction. There is no external electron acceptor in fermentation.

Respiration is the oxidation of chemical substances by an electron acceptor. Aerobic respiration is the oxidation of the substances with oxygen as the terminal electron acceptor.

Anaerobic respiration is the oxidation of the substances with other terminal acceptors of electrons, such as NO^{3-}, Fe^{3+}, SO_4^{2-}, and CO_2.

The result of respiration is a proton-motive force. The outer face of the membrane has a lower pH and a more positive charge.

The controlled passage of protons back across the membrane through specific membrane proteins is used to drive ion transport, cell motility, or ATP synthesis.

The integrity of the cell membrane, where a proton-motive force is created and ATP is synthesized, is very important for the microbial cell viability. Organic solvents, heavy metals, and oxidants can easily destroy integrity of cell membrane and production of energy in cell.

Biodegradation

Monosaccharides and storage polysaccharides, as well as proteins and nucleic acids are degraded quickly. Cellulose and other structural polysaccharides, lipids, and lignin are degraded slowly.

Aliphatic hydrocarbons are degraded quickly under aerobic conditions.

Aromatic monocyclic and polycyclic hydrocarbons are degraded more slowly under aerobic conditions. Phenols and aromatic acids can be degraded anaerobically.

Chlorinated aliphatic and aromatic hydrocarbons are degraded very slowly. Usually, the atoms of chlorine must be removed from the molecule prior to the biodegradation.

The initial step of the biodegradation of hydrocarbons involves the introduction of one or two oxygen atoms from O_2 by monooxygenase or dioxygenase. After this initial oxidation, hydrocarbons are converted into organic acids that can be metabolized to CO_2 and water.

The aromatic ring may be cleaved between the atoms of carbon with an attached hydroxylic group (ortho-cleavage) or in the position close to one of these atoms of carbon (meta-cleavage).

The gene(s) of specific biodegradation enzyme(s) is often located not in the bacterial chromosome but on a mobile genetic element such as a plasmid.

Biochemical reactions of aerobic dechlorination include

1. Hydrolytic dechlorination: An atom of chlorine is changed by hydroxylic group of water.
2. Oxygenolytic dechlorination: Oxygenase incorporates both oxygen atoms of O_2 into the substrate.
3. Dechlorination after ring cleavage: The metabolites are chlorinated linear or branched hydrocarbons.

Anaerobic biodegradation of chlorinated hydrocarbons includes the so-called reductive dechlorination step. The products of this process are dechlorinated hydrocarbons.

Co-metabolism is the transformation of an organic compound by microorganisms that are unable to use this compound as a source of energy or carbon but can degrade it due to its similarity with the source of energy or carbon.

Biotransformation is the metabolism of the substance without a significant change of its chemical nature. Biodegradation is a process where the chemical structure of the substance is degraded, usually to carbon dioxide and water.

Biotransformation and biodegradation are used in environmental biotechnology for water and wastewater treatment, soil remediation, and exhaust gas purification.

Molecular Biology and Genetics

Each protein is synthesized using different RNAs, according to information stored in the part of DNA (a gene).

Prokaryotic cells have a single chromosome, a covalently closed circular molecule of DNA, which is twisted. The aggregate of chromosomal DNA in prokaryotic cells, which is microscopically visible is termed the nucleoid.

In addition to the nucleoid, one or more small circular DNA molecules, called plasmids, may be present in the bacterial cell.

The genetic information of eukaryotes is stored in several independent molecules of DNA (chromosomes). The chromosomes in eukaryotic cells are stored in the nucleus, which is separated from the cytoplasm by a membrane.

Bioagents of Environmental and Engineering Bioprocesses

Microorganisms (microbes) are those organisms that are too small to be visible without the aid of microscope. Their size ranges from 1 to 70 μm.

All living organisms are composed of cells.

Prokaryotic cells are relatively simple in structure; they lack a true membrane-covered nucleus. The most common shapes are spherical cell and rod-shaped cell.

Eukaryotic cells are more complex, in that they contain organelles that are compartments for special metabolic functions.

The cytoplasmic membrane is a selective barrier separating the cytoplasm (the inner contents of a cell) from the external environment.

The cell wall protects the cell from changes in osmotic pressure. Bacteria with a thick cell wall stain Gram-positive. Cells that stain Gram-negative have a thin cell wall and an outer membrane.

The glycocalyx (capsule) is an extracellular polysaccharide covering the bacterial cell. It contributes to pathogenicity, helps bacteria to attach to surfaces, and protects the cell against desiccation, heavy metals, antibiotics, and disinfectants.

Lipopolysaccharides of the outer membranes of Gram-negative bacteria are often toxic or allergenic for humans. The variability of lipopolysaccharides among various strains allows pathogenic bacteria to evade the body's immune system.

Intracellular inclusions include granules of glycogen or poly-β-hydroxybutyrate (PHB) in the cytoplasm, elemental sulfur granules, and polyphosphate granules. The vegetative cell can form endospores. The inner dry matter of the endospore is covered by a thick envelope.

Fungi and algae often have a cell wall made up of polysaccharides such as cellulose or chitin. Some algae have inorganic compounds such as calcium carbonate or silica in their cell walls. The cells of animals often have no cell wall.

Three important organelles of eukaryotic cells are mitochondria, chloroplasts, and the nucleus. The nucleus contains the genetic material of the eukaryotic cell. It is a place where DNA for new cells is synthesized. Mitochondria are the organelles responsible for energy generation. Chloroplasts are cell organelles that assimilate light energy.

A microbial biofilm is a layer or aggregates consisting of microbial cells and organic matter attached and growing on the surface. The concentration of dissolved oxygen in biofilms drops to zero after a depth of 100–200 μm. Therefore, deeper layers of the biofilm consist of facultative anaerobic and obligate anaerobic microorganisms.

Reproduction, Proliferation, and Growth

The average content of dry matter of bacterial cells is as follows: protein, 55%; RNA, 15%; polysaccharides, 10%; lipids, 5%; DNA, 5%; and monomers and inorganic ions, 10%. The water content in the cells is 70%–80%.

The elemental composition of biomass can be described approximately by the formula $CH_{1.8}O_{0.5}N_{0.2}$.

Growth is an increase of individual cell mass or the mass of a cell population.

Proliferation is the increase of cell number (N). A balanced growth of microorganisms is followed by a proportional increase of cell number and mass of the population.

The time required to form two new cells from one cell is called the generation time, t_g. Exponential growth is described by the following equations:

$$N = N_0 2^n$$

and

$$\log N = \log N_0 + (t/t_g)\log 2$$

where
 N_0 is the initial cell concentration
 n is the number of generations
 t is the duration of exponential growth

Exponential growth is described by the equations:

$$\frac{dX}{dt} = \mu x$$

and

$$\mu = \frac{(\ln X - \ln X_0)}{t}$$

where
 μ is the specific growth rate
 X is the concentration of the biomass at time t
 X_0 is the initial concentration of biomass

If one nutrient limits the specific growth rate, it can be expressed by Monod's equation:

$$\mu = \mu_{max}\left[\frac{S}{(S + K_s)}\right],$$

where
 μ_{max} is the maximum specific growth rate
 S is the concentration of nutrient (substrate) limiting growth rate
 K_s is a constant (half-saturation constant)

Growth yield ($Y_{X/S}$) is a ratio between quantity of produced biomass and consumed nutrient (substrate). For the batch system, it is determined by the equation

$$Y_{X/S} = \frac{X_t - X_0}{S_0 - S_t},$$

where
 S_t is the substrate concentration in the system at the end of period t
 S_0 is the initial substrate concentration

Growth yield ($Y_{X/S}$) for the continuous system without recycling of the biomass is determined by the following equation:

$$Y_{X/S} = \frac{X}{(S_\mathrm{I} - S_\mathrm{E})}$$

where
X is the biomass concentration
S_0 and S_f are the concentrations of substrate in the influent and effluent, respectively

The semiclosed system of cultivation is called as batch culture and the open system is called as continuous culture.

The following phases can be separated in batch culture: lag-phase (adaptation of the cells), log-phase (exponential growth), stationary phase (absence of growth), and death phase.

The most popular reactors are aerobic and anaerobic fermentors of complete mixing.

The most common type of continuous culture is a chemostat. The dilution rate (d), which is a ratio between flow rate (f) and working volume of the reactor (v), is maintained constant in a chemostat.

The growth of a culture can be monitored as follows:

1. By physical methods (weight of biomass, microscopic enumeration, optical density, turbidity, fluorescence, flow cytometry)
2. By chemical methods (measurement of protein, DNA, the components of cell wall, ATP, photopigments, cytochromes, $NADH_2$, quantitative PCR)
3. By biological methods (plate count, most probable number)
4. By physiological methods (respiration rate, CO_2 production rate, biodegradation rate)

Microbes are often attached to the liquid–solid and liquid–gas interfaces because the nutrient concentrations are higher there.

Stoichiometric limitation of growth means that the dosage of the nutrient in the medium determines the yield of the biomass. Kinetic limitation means that the specific growth rate depends on the concentration of limiting nutrient.

Microbial Ecology

Organisms are combined into a population. Populations are combined into a microbial community. The microbial communities of the flocs of activated sludge, fixed biofilm of biofilter, and soil microbial community are used in environmental biotechnology.

Microorganisms live in natural or engineered habitats in which their growth is affected by the interactions with populations of other microbes. There may be positive interactions (cooperation), negative interactions (competition), or no interactions (neutralism) between the cells of microbial population, between different microbial populations or between microorganisms and macroorganisms.

Competition and cooperation is carried out by the changes in pH of the medium and redox potential, excretion of antibiotics or extracellular digestive enzymes, increased protection from UV or heavy metals, or by the aggregation of the cells.

Types of Interaction between Microbial Populations

	Effect of Interaction	
Type of Interaction	**On Population A**	**On Population B**
Neutralism	0	0
Commensalism	0	+
Mutualism (symbiosis)	+	+
Competition (antagonism)	−	−
Amensalism	0	−
Predation and parasitism	+	−

Bacteria called as R-tactics grow fast in a rich medium but can quickly die under a shortage of nutrients (*Pseudomonas* spp.).

Bacteria called as L-tactics grow fast in a rich medium but form dormant forms like spores under a shortage of nutrients (*Bacillus* spp.).

Bacteria called as K-tactics are adapted to grow slowly in mediums with low concentrations of nutrients (*Hyphomicrobium* spp.).

Aerobes are microorganisms that grow at the atmospheric pressure of oxygen (0.21 atm). Microaerophiles prefer low concentration of oxygen because they have oxygen-sensitive molecules. Aerotolerant anaerobes have no need of oxygen for growth but can tolerate its presence in the medium.

Obligate (strict) anaerobes are very sensitive to oxygen because they have no protection against the toxic products of oxygen reduction: hydrogen peroxide (H_2O_2), superoxide radical (O_2^-), and hydroxyl radical ($OH^\bullet$).

The maximum temperature for growth depends on the sensitivity of protein structure on temperature. The minimum temperature depends on the temperature of the lipid membrane melting.

Psychrophiles have optimal temperatures below 15°C. Mesophiles have optimal growth temperatures in the range between 20°C and 40°C. Thermophiles grow best above 50°C.

Natural biotopes have different pH: pH 1–3 (gastric juice, volcanic soil, mine drainage); pH 3–5 (plant juices, acid soils); pH 7–8 (freshwater and seawater); and pH 9–11 (alkaline soils and lakes). Microbes grow somewhere within the pH range of 5–9 (neutrophiles), at pH lower than 4 (acidophiles), and at pH higher than 9 (alkalophiles). The intracellular pH is an approximately neutral pH.

The extracellular pH influences the dissociation of carboxylic, phosphate, and amino groups of the cell surface, thus changing its charge and adhesive properties. It is important for environmental biotechnology in the sedimentation of activated sludge and in the fixed-biofilm process.

Salts or organic substances cause water to diffuse out of the cell by osmosis. Halophiles require the addition of NaCl in the medium. Extreme halophiles require 15%–30% of NaCl in the medium. Xerophiles are able to live in a dry environment.

Classification of Viruses and Microorganisms

Microorganisms can be isolated as a strain or clone.

The primary classification unit is a species, which is defined as a collection of similar strains. The higher taxonomic units are genus, family, order, and kingdom. The reference strains are stored in culture collections.

Microorganisms are classified (grouped) by their functional properties (phenotypic classification) or their genetic properties (phylogenetic classification).

The major object for phylogenetic classification of prokaryotes is at present the 16S rRNA gene.

All eukaryotes are aerobic organisms. Only a few unicellular fungi (yeasts) can grow under anaerobic conditions.

Fungi are used in the composting, soil bioremediation, and biodegradation of xenobiotics in the soil.

Protozoa are unicellular animals obtaining nutrients by ingesting other microbes, or by ingesting macromolecules. The cells form cysts under adverse environmental conditions.

The cysts are resistant to desiccation, starvation, high temperature and, a condition that is important from the public health point of view, disinfection in wastewater-treatment plants.

Changes in the protozoan community reflect the operating conditions of an aeration tank.

Physiological Classification of Prokaryotes

Prokaryotes can be classified according to relation to oxygen and generation of biological energy as shown in the periodic table of prokaryotes. The periodic table of prokaryotic classification was proposed to give predictive power to prokaryotic classification, clarify the physiological and evolutionary connections between microbial groups, and give a logical basis for students to understand microbial diversity.

Three major types of biological energy generation are the results of evolution of earth's atmosphere from an anaerobic to an aerobic one. Therefore, microbial physiological diversity can be shown as created in three evolutionary periods related to fermenting, anaerobically respiring, and aerobically respiring microorganisms.

There are also intermediate groups, for example, microaerophilic or facultative anaerobic microorganisms, between these groups.

These groups exist in three parallel, semi-independent but coordinated phylogenetic lines:

1. Line of aquatic organisms
2. Line of terrestrial organisms
3. Line of organisms in extreme environment

These lines are semiseparated because of the low frequency of genetic exchanges between organisms in aquatic, terrestrial, and extreme environments.

Prokaryotes of aquatic, terrestrial, and extreme environments are in the following lines: (1) Gram-negative bacteria (*Gracilicutes*), cells with thin wall that originated from environments with a stable osmotic pressure (water, tissues of macroorganisms); (2) Gram-positive bacteria (*Firmicutes*), cells with a rigid cell wall that originated from environments with changeable osmotic pressure (soil); and (3) line of archaea (*Mendosicutes*), cells without the conventional peptidoglycan, originated from environments with some extreme conditions, usually temperature or oxidation–reduction potential.

The physiological diversity of chemotrophic prokaryotes can be shown in the periodic table of prokaryotes as three evolutionary periods of three parallel lines of aquatic, terrestrial, and extreme environment evolution.

Groups of Prokaryotes

Anaerobic Gram-negative and Gram-positive fermenting bacteria, for example, bacteria of genus *Clostridium*, are performing such stages of anaerobic digestion of organic wastes as the hydrolysis of biopolymers and fermentation of monomers.

Anaerobic Archaea are prokaryotes that are distinctive from Gram-positive and Gram-negative bacteria in terms of molecular-biological properties.

Methanogens are a group of Archaea. They are strict anaerobes that convert CO_2, methyl compounds, or acetate to methane gas by anaerobic respiration. This group is extremely important in the anaerobic digestion of organic matter.

Sulfate-reducing bacteria are obligate anaerobes that use organic compounds as energy sources. Sulfate or elemental sulfur serves as the terminal electron acceptor in anaerobic respiration and hydrogen sulfide is the toxic product of this process.

Denitrifying bacteria, for example, *Pseudomonas denitrificans* or *Paracoccus denitrificans*, transform nitrate to nitrogen gas and are used for the removal of nitrate from water and wastewater under anoxic conditions.

Iron-reducing bacteria can reduce different Fe(III) compounds and are important in the anaerobic biodegradation of organics in the aquifers.

Facultatively aerobic bacteria can produce energy by respiration or fermentation. Many species are enterobacteria. There are many pathogens from the genera *Salmonella*, *Shigella*, and *Vibrio*.

The number of indicator species *Escherichia coli* or physiologically similar coliforms is a common indicator of water pollution by feces or sewage.

The growth of microaerophilic filamentous bacteria in poorly aerated or overloaded aerobic tank may be a reason of bulk foaming of activated sludge.

Aerobic heterotrophic Gram-negative bacteria are most active in the aerobic wastewater treatment and biodegradation of xenobiotics. The genera important in aerobic treatment of wastewater are *Pseudomonas*, *Acinetobacter*, *Zoogloea*, *Flavobacterium*, and *Alcaligenes*. Some of them are opportunistic bacteria, that is, they can cause diseases in immunosuppressive, young or old people. Therefore, all treatments of water and soil infected by these bacteria must be performed with precautions against the dispersion of these bacteria in the environment.

Aerobic heterotrophic Gram-positive bacteria, for example, from the genus *Bacillus*, are important in environmental biotechnology. Some of them form spores. *Bacillus* spp. are dominant bacteria in the aerobic treatment of wastewater or solid waste, which are rich in such polymers as starch or protein. Some species of genus *Bacillus*, for example, *Bacillus anthrax*, are pathogenic.

Actinomycetes are aerobic heterotrophic Gram-positive filamentous bacteria which are active degraders of natural biopolymers and are used in the aerobic composting of solid wastes.

Nitrifying bacteria comprise two groups of aerobic bacteria: ammonia-oxidizers and nitrite-oxidizers.

Sulfur-oxidizing chemilithotrophic bacteria oxidize reduced sulfur compounds and are used for the bioremoval of toxic H_2S from water and wastewater.

Iron-oxidizing bacteria are capable of oxidizing Fe(II) in acid or neutral environments. They remove iron from drinking water. The precipitation of biogenic iron hydroxide can clog the pipelines.

Cyanobacteria are oxygenic prokaryotic phototrophs and algae are eukaryotic phototrophic organisms that carry out oxygenic photosynthesis with water serving as the electron donor. Phototrophs are used for the removal of nitrogen and phosphorus from food-processing wastewater. However, blooming of cyanobacteria or algae in water polluted by ammonium or phosphate is a reason for bad odor and taste of water.

Disinfection of Water

Control of unwanted microbial growth can be performed by inhibition of growth, killing the microorganisms or removing them from an environment.

Antimicrobial (cidal) agents kill microbes, thus the terms bactericidal, fungicidal, and viricidal agents are used.

The most sensitive targets of the microbial cell are the integrity of cytoplasmic membrane, active centers and structures of enzymes, and structures of nucleic acids.

Physical sterilization is the process of killing by heat, radiation, or filtration of all living organisms and viruses present in the sample.

Heat sterilization kills all cells. Pasteurization kills the vegetative cells of bacteria, fungi, and protozoa in the liquid.

Electromagnetic irradiation such as microwaves, ultraviolet (UV) radiation, x-rays, gamma rays, and electrons are also used to sterilize the material.

Disinfectants (chlorine gas, chloramine, ozone, quaternary ammonium compounds) are chemical antimicrobial agents that are used on inanimate objects.

Antiseptics (iodine, 3% solution of hydrogen peroxide, 70% solution of ethanol) are chemical antimicrobial agents used on living tissue.

Antibiotics are microbial or synthetic substances that are used to treat infectious diseases because of specific inhibition of microbial species in low concentration.

Filtration, coagulation, and settling, which are used in the treatment of drinking water, help in reducing pathogens due to their removal with the sediments.

Disinfection is the chemical or physical treatment of water or wastewater-treatment plant effluent by strong oxidants (chlorine, chloramines, chlorine dioxide, ozone) or UV.

In general, the resistance of different groups of microbes increases in the following order: vegetative bacteria (the most sensitive), enteric viruses, spore-forming bacteria, and protozoan cysts (the most resistant).

Ozone is more effective against viruses and protozoa than chlorine.

UV radiation at 260 nm wavelength damages microbial DNA. The cell may be reactivated under visible light by the cell repair system.

Ferrous ions, nitrites, hydrogen sulfide, and various organic molecules exert a demand for oxidizing disinfectants such as chlorine. Pre-chlorination is practiced in the treatment sequence as one method to alter taste- and odor-producing compounds, to remove iron and manganese.

Suspended solids (clay, sludge) significantly reduce inactivation of microorganisms during disinfection.

Public Health and Water Disinfection

Saprophytic microorganisms feed on dead organic matter. The pathogenic (infectious) microorganisms grow in animal tissue and can cause diseases in macroorganisms.

Opportunistic pathogens are normally harmless but have the potential to be pathogens in debilitated or immunocompromised organisms.

Infectious microorganisms can enter humans due to direct contact between individuals or by reservoir-to-person contact. The diseases may be conventionally distinguished as airborne, waterborne, soilborne, and foodborne infectious diseases.

When an infectious agent is spread by an insect, such as a mosquito, flea, lice, biting fly, or tick, they are referred to as vectors.

Infectious diseases still account for 30%–50% of deaths in developing countries.

Transmission of waterborne diseases is directly related with the bacteriological quality of water and an effluent of wastewater treatment. Sources of pathogens other than sewage outlets are wildlife watersheds, farm lots, garbage dumps, and septic tank systems.

Pathogens can produce toxins as extracellular proteins called exotoxins, host damage can occur at sites far removed from a localized focus of infection.

Enterotoxins are exotoxins that act in the small intestine. These cause diarrhea, the secretion of fluid into the intestinal passage. Almost all Gram-negative bacteria contain endotoxin. It is the lipopolysaccharide of the outer membrane.

Waterborne pathogenic bacteria are pathogenic strains of *E. coli*, *Leptospira* sp., *Vibrio cholera*, *Shigella* spp., *Salmonella* spp., and *Campylobacter* spp.

The indicator microorganisms are detected in water to indicate the possible presence of disease-causing constituents. Their presence is evidence that pollution associated with fecal contamination from man or other warm-blooded animals has occurred.

Two protozoans of major concern as waterborne pathogens are *Giardia intestinalis* and *Entamoebia histolytica*. There are over 100 known waterborne human enteric viruses such as infectious hepatitis A, poliovirus, and rotaviruses (the major cause of acute gastroenteritis).

Biotechnological Processes

Upstream processes

1. Medium preparation and storage: washing/sonication/thermal pre-treatment/disinfection/sterilization and conservation/cooling/freezing/drying/heating/pasteurization/thermostabilization pre-oxidation
2. Equipment preparation: cleaning/washing/sterilization and disinfection/refilling/checking/testing
3. Creation of the conditions: temperature/aeration/stirring
4. Selection of inoculum: specific selection of cultures/pure culture/enrichment culture/recombinant strain/toxicity and pathogenicity tests/stability tests
5. Cultivation of inoculum: batch, continuous, specific selection
6. Aseptic conditions: air for aeration, indoor air, aseptic sampling

Core processes (biotreatment)

1. Inoculation (start-up) and adaptation
2. Growth of microorganisms
3. Performance of useful functions: degradation, oxidation, reduction, synthesis
4. Selection/succession
5. Monitoring and control

Downstream processes

1. Concentration
2. Separation

3. Recycling
4. Conservation
5. Packing
6. Disinfection
7. Disposal

Aquatic Systems and Water Biotreatment

Aquatic environment often develops vertical density gradients due to thermal stratification. The bottom may become anaerobic because organic matter produced through photosynthesis in the surface layer sinks to the bottom, where heterotrophic bacteria degrade it and consume oxygen.

The sources of water pollution are the following: agriculture (fertilizers, pesticides, solid wastes), farms (wastewater, manure, insects), soil (chemical and biological pollutants), wildlife (biowaste, dead biomass), domestic wastewater (organic and inorganics pollutants, viruses, and microorganisms), industrial effluent (organic and inorganic pollutants), solid wastes (organic and inorganic pollutants if not properly managed and disposed of), and air (sulfur and nitrogen oxides, dust).

Steps of eutrophication include the following: increase of water turbidity; concentration of microorganisms; concentration of toxic, allergic, or bad smelling microbial products; decrease of dissolved oxygen concentration; and decrease of biodiversity in aquatic system.

Raw water from various sources is conveyed by pipelines to the waterworks, where it is chemically treated, filtered, and disinfected. Treatment frees the water of harmful bacteria, makes it clear, sparkling, odorless, colorless, and safe for consumption.

Biological treatment is used to remove biodegradable electron donors (biological instability) from water. These electron donors include biodegradable organic matter (BOM), ammonium, ferrous iron, manganese (II), and sulfides.

Donor of electrons for denitrification can be organic substances (methanol, ethanol, acetate) and inorganic substances such as H_2, S, H_2S, or Fe^{2+} (autotrophic denitrification).

Chemical oxygen demand (COD) is the amount of oxygen required to oxidize organic matter in water chemically. Biological oxygen demand (BOD) is the amount of oxygen required to oxidize organic matter in water biologically. Theoretical oxygen demand (TCOD) for a substance of known content can be estimated from the stoichiometry of the complete oxidation of carbon to CO_2 and nitrogen to nitrate.

Rapid filters, slow sand filters, and soil biofiltration are active in the removal of biological instability of water as well as micropollutants of water such as hydrocarbons, surfactants, endocrine disruptors, bad taste-and-odor compounds, and cyanobacterial toxins.

Anaerobic and Anoxic Treatment of Wastewater Oxygen and Energy Generation

The advantages of anaerobic treatment include the following: (1) it requires no oxygen, (2) it produces lower amounts of well-stabilized sludge (3–20 times less than aerobic processes), and (3) it produces fuel and methane.

Five groups of prokaryotes are involved in the anaerobic digestion of organic wastes: (1) hydrolytic bacteria, (2) fermentative acidogenic bacteria, (3) acetogenic bacteria, (4) hydrogenotrophic methanogens, and (5) acetotrophic methanogens.

Sulfate reducers and methanogens are competitive at COD/SO_4^{2-} ratios of 1.7–2.7. An increase of this ratio is favorable to methanogens, while a decrease of this ratio is favorable to sulfate reducers.

The upflow anaerobic sludge blanket (UASB) reactor uses immobilized biomass to retain sludge in the treatment system.

Anoxic processes and biotechnological methods use nitrate bioreduction, ferric bioreduction, sulfate bioreduction, manganese, selenium and arsenic bioreductions, anoxic oxidation of ammonia, and bioreduction of perchlorate.

Aerobic Treatment of Wastewater

The major contaminants of wastewater, which must be removed, are biodegradable organic compounds, volatile organic compounds, recalcitrant xenobiotics, toxic metals, suspended solids, nutrients, microbial pathogens, and parasites.

The domestic wastewater typically contains 800 mg of total solids/L, 240 mg of suspended solids/L, and 200 mg BOD/L.

The BOD/N/P weight ratio required for biological treatment is 100/5/1.

A waste sludge equivalent to 20%–40% of the applied BOD is generated in aerobic biological treatment of wastewater.

The conventional activated sludge system include an aeration tank for growth of aerobic microorganisms and biodegradation of pollutants, a settling tank for separation of biomass and effluent, and selection of fast-settling microorganisms for aeration tank.

Mixed liquor suspended solids (MLSS) is the amount of organic and mineral suspended solids, including microorganisms, in the mixed liquor. MLSS in an aeration tank ranges typically from 2500 to 3500 mg/L.

Mixed liquor volatile suspended solids is the organic portion of MLSS including non-microbial organic matter, dead and live cells, and cellular debris. MLVSS ranges typically between 65% and 75% of MLS.

Food-to-microorganism (F/M) ratio is a measure of the organic load into the activated sludge system. It is expressed in terms of mass of substrate (BOD) per unit mass of biomass per day, or in kg BOD per kg MLSS per day.

HRT is the average time spent by the influent liquid in the aeration tank. It is the reciprocal of the dilution rate, which is a ratio of influent flow rate to the working volume of aeration tank.

Sludge age (or solids retention time) is the mean cell residence time of microorganisms in the system.

Removal of nitrogen: nitrification followed by denitrification or particle nitrification to nitrite followed by anammox process.

Removal of phosphate: chemically using Ca or Fe salts and biologically using alternation of anaerobic and aerobic processes.

The overgrowth of filamentous bacteria in activated sludge may be due to low F/M ratio, low concentration of dissolved oxygen, and presence of H_2S.

Foaming in aerobic tank is due to undegraded surfactants/detergents, rising denitrifying sludge, excessive growth of actinomycetes producing biosurfactants. Foam control: (1) chlorination; (2) reducing sludge age; (3) use of a biological selector; (4) reduce aeration rate; (5) reduction in pH, and oil and grease levels; and (6) application of antifoam agent.

Biofilm reactors for wastewater treatment include trickling filters, rotating biological contactors (RBCs) and submerged filters. Microbial biomass is attached to a medium/carrier.

Value-Added By-Products of Environmental Engineering

Enzymes, polysaccharides, edible mushrooms, bioplastic, pigments, fertilizers, and biofuel can be produced from liquid or solid organic wastes.

The recovery of metals can be used in the treatment of industrial wastewater.

The transformation of sewage sludge into organic fertilizer, recovery of organic acids, phosphate, and ammonium from reject water could be value-added by-products of municipal wastewater treatment.

Biotreatment of Industrial Hazardous Wastes

Sources of solid waste are as follows: municipal solid waste (MSW), commercial solid wastes, construction/demolition solid wastes, sludge from wastewater-treatment plants, industrial solid wastes, agricultural solid wastes, and mining solid wastes.

Approximately 60% of solid wastes go to sanitary landfills, which are engineered systems, where the waste is deposited in compacted layers and covered with earth at the end of each day. To avoid negative effects of the landfill leakage and gas emission, the landfill has an impermeable liner, and leachate and gas collection systems.

Composting is an engineered biological system which performs thermophilic and mesophilic aerobic decomposition of organic wastes. Windrow systems, aerated pile systems, and in-vessel systems of composting are used.

Microbial succession during composting includes mesophilic phase (20°C–40°C), thermophilic phase (60°C–80°C), and cooling and maturation phase. The most active roles are played by thermophilic bacillus and actinomycetes. The major factors in composting are C:N ratio, intensity of aeration, temperature, and moisture content.

Solid Wastes and Soil Biotreatment

The formation of soil is influenced by parent material, climate, activity of microorganisms and macroorganisms, and topography of the place.

Soil texture is defined as the relative proportions of the sand, silt, clay, and colloid fraction. Humus is defined as the total of the deeply transformed organic compounds in soil.

Microbial activity in soil is limited by water activity, dissolved oxygen, and nutrients concentrations. Aerobic or anaerobic microzones are small volumes of soil, where biological processes differ from those of the soil as a whole.

Soil is a habitat for viruses, bacteria and archaea, fungi, algae, protozoa, and nematodes. Microorganisms colonize soil and form an indigenous community.

Microbial function in soil is the biogeochemical cycling of carbon and other elements following accumulation of metabolites and evolution of biosphere.

The biogeochemical cycle of carbon includes assimilation of CO_2 from atmosphere and mineralization of organic matter. The biodegradation of organic matter by fermentation and by anaerobic and aerobic respiration is the most important process in the wastewater and solid waste biotreatment.

The anaerobic decomposition of organic matter requires the participation of four metabolic groups of bacteria: (1) hydrolytic bacteria, (2) fermentative acidogenic bacteria, (3) acetogenic bacteria (fatty acid oxidizing bacteria), and (4) methanogens.

The biogeochemical cycle of nitrogen includes nitrogen fixation by free-living nitrogen-fixing bacteria and symbiotic nitrogen-fixing bacteria, ammonification, nitrification by ammonium-oxidizing and nitrite-oxidizing bacteria, and denitrification.

The electron donors for denitrification can be such organic substances as methanol and ethanol and inorganic substances such as H_2, S, H_2S, or Fe^{2+}.

The biogeochemical cycle of sulfur includes oxidation of organic sulfur, H_2S or S, and sulfate reduction.

The biogeochemical cycle of phosphorus includes mineralization of organic phosphorus, assimilation of phosphate, precipitation or solubilization of phosphate, and the accumulation of polyphosphate under aerobic conditions.

The biogeochemical cycle of iron includes oxidation of Fe(II) and reduction of Fe(III). Fe(II) is stable under neutral pH and anaerobic conditions. The solubility of Fe(III) and Fe(II) ions depends on the formation of their chelates with organic acids.

The problems of soil bioremediation could be (1) absence of indigenous soil microorganisms capable of degrading soil pollutant, (2) providing oxygen for microbial growth and oxidation, (3) providing soluble inorganic nutrients for microbial growth, and (4) preventing plugging of soil pores with microbial biomass.

There are numerous methods for in situ (air sparging, bioventing), on-site (landfarming, composting), and off-site (slurry-phase) bioremediation of soil.

Microbial Geotechnics

The geotechnical applications of microorganisms, using biocementation, bioclogging, and biodesaturation, are aimed at improving the mechanical properties of soil so that it will be more suitable for construction or environmental purposes.

Bioclogging is the production of pore-filling materials through microbial means so that the porosity and hydraulic conductivity of soil can be reduced.

Biocementation is the generation of particle-binding materials through microbial processes in situ, so that the shear strength of soil can be increased.

The most suitable microorganisms for soil bioclogging or biocementation are facultative anaerobic and microaerophilic bacteria, although anaerobic fermenting bacteria, anaerobic respiring bacteria, and obligate aerobic bacteria may also be suitable for use in geotechnical engineering.

Due to complexity, the applications of microbial geotechnology would require an integration of microbiology, ecology, geochemistry, and geotechnical engineering knowledge.

Microbiology of Air and Air Treatment

The content of modern atmosphere, especially the content of such gases as CH_4, H_2, and H_2S, significantly depends on the microbial activity of corresponding bacteria on the earth.

A bioaerosol is a collection of airborne biological particles including viruses, bacteria, and spores of fungi.

The sources of bioaerosols are (1) natural sources include coughing, sneezing, shedding of human skin, and disturbance of soil and aquatic environments by wind action; (2) production sources such as agricultural practices (e.g., cleaning of silos), processing of diseased animals in abattoirs, laboratory operations (blending and sonicating), and wastewater-treatment operations.

Bioaerosols in outdoor environments are dominated by fungal spores, bacterial cells and pollen. In clean rooms, concentrations of culturable agents are from 100 to 1000 cfu/m^3, but in certain industrial or agricultural sites the concentration may be from 10^5 to 10^{10} cfu/m^3.

There are many airborne infectious diseases. Legionnaires' disease and tuberculosis are examples of illnesses resulting from exposure to aerosol with specific microorganisms. The control of Legionellae in cooling towers is an established code of practice for in Singapore.

Steps for the prevention of odor in wastewater-treatment plants includes mechanical cleaning of collection systems, frequent washing to remove grit and organic debris, frequent scraping to remove scum and grease, keeping vents clear in trickling filters to keep biofilms aerobic, and keeping tank walls, air pipes, and diffusers clean in activated sludge systems.

Odorous gases can be treated in wet scrubbers, by combustion, by adsorption, chemical oxidation, and biofiltration.

Biodeterioration, Biocorrosion, and Biofouling

The prevention of biodeterioration and biocorrosion of materials is an important task of environmental microbiology. The main reason for microbially induced biocorrosion is the formation of nitric, sulfuric, organic acids, or H_2S.

The biofouling of membranes used in microfiltration, nanofiltration, reverse osmosis, desalination process, and cultivation in membrane bioreactor is caused by the aggregation of microorganisms and deposition of their products on the membrane surface and in membrane pores, leading to the loss of efficiency of the separation process. Biofouling can be diminished using membrane coatings and technological operations.

25

Quiz Bank

Microorganisms

$(2 \times 10^{-11}) \times 4 \times 10^{6} \times 2 \times 10^{-3} =$

A. 1.6×10^{-8}
B. 16×10^{-8}
C. 1.6×10^{-7}
D. 16×10^{-7}

Virus is

A. Particle
B. Bioparticle
C. Cellular particle
D. Cellular organism

Typical size of virus particle is

A. 10^{-6} m
B. 0.001 μm
C. 1000 nm
D. 10^{-5} cm

Bacteria

A. Are prokaryotes
B. Are cell aggregates
C. Are viral particles
D. Are eukaryotic organisms

What is a suitable diameter for pores in a filter concentrating viruses from 100 mL of water sample?

A. 5 nm
B. 50 nm
C. 0.5 μm
D. 5.0 μm

Volume of bacterial cells could be

A. $2.5\,mm^3$
B. $2 \times 10^{-8}\,m^3$
C. $3.5\,\mu m^3$
D. $1 \times 10^{-5}\,L$

A microbe is

A. A unicellular organism
B. A multicellular microorganism
C. A unicellular macroorganism
D. A unicellular or multicellular microorganism

The mycelium of fungi is

A. A cell aggregate
B. A cellular filament
C. Filaments of cells
D. A part of hyphae

Fungi are important bioagents in

A. Photosynthesis
B. Degradation of polymers
C. Removal of nutrients from water
D. Biodegradation of viruses

Which eukaryotes are able to use light energy?

A. Autotrophic chemolithotrophes
B. Brown algae
C. Fungi
D. Cyanobacteria

Molds are

A. Bacteria
B. Algae
C. Fungi
D. Protozoa

Protozoa are

A. Prokaryotes
B. Microplants
C. Microanimals
D. Microfungi

Surface-to-volume ratio (S/V) for a spherical cell with diameter D can be determined by following equation: $S/V = (\pi D^2)/(\pi D^3/6)$.

Assume that the rate of biodegradation is proportional to the *S*/*V* ratio of cells. Therefore, the ratio of biodegradation rates for two types of cells with diameters 1 and 50 μm, respectively, is

A. 50
B. 5
C. 0.2
D. 0.02

The mass of a dry bacterial cell is approximately

A. 1 mg
B. 1 μg
C. 1 ng
D. 1 pg

Static Biochemistry

The presence of a capsule

A. Increases bacterial-cell sensitivity to heavy metals
B. Decreases bacterial-cell sensitivity to disinfection
C. Protects cell from high temperature
D. Enhances cell growth

Glycogen is

A. The source of lipids
B. A storage polymer
C. A structural biopolymer
D. A component of cell wall

The tertiary structure of a protein is

A. An association with other polypeptides
B. A sequence of amino acids connected by peptide bonds
C. A final folded shape of a molecule
D. The sequences of nucleotides

A hydrogen bond is formed between

A. O, H, and P atoms
B. N, H, and O atoms
C. N, H, and P atoms
D. O, H, and H atoms

In its double-stranded form, the strands of DNA in the chromosome are said to

A. Be held together by covalent bonds
B. Be held together by hydrogen bonds

C. Contain the same amount of guanine as adenine
D. Contain the same amount of adenine as cytosine

The force of dipole–dipole interaction between molecules is approximately

A. 0.1% of covalent bond force
B. 1% of covalent bond force
C. 10% of covalent bond force
D. 100% of covalent bond force

The function of lipids in a cell is to

A. Strengthen the cell wall
B. Prevent free transport of ions into and from cell
C. Neutralize cellular ions
D. Control cell surface charge

All proteins are composed of

A. 10 amino acids
B. 20 amino acids
C. 4 nucleotides
D. 15 monosaccharides

Dynamic Biochemistry

Enzyme is a catalyst that

A. Lowers the activation energy of reaction
B. Increases the specificity of reaction
C. Lowers the rate of reaction
D. Is composed of DNA

Hydrolase

A. Is an enzyme hydrolyzing C–C bonds
B. Adds water across a bond, hydrolyzing it
C. Hydrates a surface
D. Is an enzyme useful for transporting water across a membrane

The carbon source for chemolithotrophs is

A. Organic compounds
B. Inorganic compounds
C. Carbon dioxide
D. All of the above

Which of the following is an enzyme?

A. Cellulose
B. Cellobiose

C. Cellulitis
D. Cellulase

The products of fat hydrolysis are

A. Acetic acid and formaldehyde
B. Lypase and lipoids
C. Glycerol and fatty acid
D. Glucose and energy

All the following characteristics apply to enzymes except that

A. They act on substances known as substrates
B. They are highly specific in their catalytic action
C. They increase the rate of chemical reactions in the cell
D. They are composed solely of carbohydrate molecules

Biooxidation and Bioreduction

Energy in life systems

A. Is released into the environment
B. Is utilized from the environment
C. Is transformed from one form to another
D. Is characterized by all of the above

The source of energy for chemolithotrophic microorganisms is

A. Anaerobic oxidation of organic compounds
B. Oxidation of inorganic compounds
C. Reduction of organic compounds
D. Fermentation of inorganic compounds

Bacteria are able to use such energy sources as

A. Chemical energy of organic compounds
B. Chemical energy of inorganic compounds
C. Energy of light
D. All of the above

In aerobic respiration, the final electron acceptor is

A. Hydrogen
B. Oxygen
C. Water
D. ATP

The enzymatic reaction depicted by substrate + electron → product is

A. Oxidation
B. Reduction

C. Hydrolysis
D. Oxygenation

The oxidation number of S in SO_3^- is

A. +3
B. +4
C. +5
D. +6

Oxidation of an organic substance coupled with the reduction of Fe^{3+} is

A. Fermentation
B. Aerobic respiration
C. Anaerobic respiration
D. Lithotrophy

A characteristic of fermentation is that

A. Oxygen can be used for fermentation
B. There is no external acceptor of electrons
C. Nitrate can be used as the terminal electron acceptor
D. Sulfate can be reduced

During a reduction bioprocess

A. Electrons are lost from a substrate molecule
B. Large amounts of biologically available energy are usually obtained
C. Electrons are added to a substrate molecule
D. The substrate molecule is oxidized

Which of the following is true about anaerobic respiration?

A. It involves the reduction of oxygen
B. It involves the oxidation of sulfate
C. It involves the reduction of nitrate
D. It involves the oxidation of oxygen

How do substances change when they are reduced?

A. By the addition of an electron
B. By the loss of an electron
C. By the addition of protons
D. By the release of protons

Oxidation without an external electron acceptor is

A. Anaerobic respiration
B. Fermentation
C. Aerobic respiration
D. Microbial respiration

The oxidation number of carbon in lactic acid ($CH_3COCOOH$) is

A. +3
B. +2/3
C. −3
D. −2/3

The number of moles of electrons donated by one mole of lactic acid ($CH_3COCOOH$) during its oxidation to CO_2 coupled with reduction of ferric hydroxide $Fe(OH)_3$ to ferrous ions Fe^{2+} is

A. 2
B. 4
C. 6
D. 8

Which of the following substances does not participate in the production of biological energy?

A. Adenosine
B. Adenosine triphosphate
C. Adenosine diphosphate
D. Nicotineamide adenine dinucleotide

Which substance can be a product of fermentation?

A. C_2H_6
B. H_2S
C. CO_2
D. NH_3

Biodegradation

The enzyme catalyzing starch hydrolysis is

A. Amylose
B. Amylopectine
C. Amylizin
D. Amylase

Lignin is biodegraded due to

A. Regular depolymerization
B. Radical-based random depolymerization
C. Hydrolysis of C–O–C bonds between aromatic units
D. Reduction of oxygen in C–O–C bonds

The primary step in biodegradation of chlorophenol is

A. Dechlorination
B. Ortho-cleavage of aromatic ring
C. Meta-cleavage of aromatic ring
D. Oxygenation

The plasmid of degradation

A. Is a specific enzyme that is used by cell in the degradation of xenobiotics
B. Is the genes of specific enzymes of biodegradation
C. Is the sequence of nucleotides that codes for the specific enzymes of biodegradation
D. Is a part of bacterial chromosome that codes for the specific enzymes of biodegradation

The first step in microbial degradation of pentachlorophenol would be

A. Cleavage of aromatic ring
B. Oxidation of carbon atoms in aromatic ring
C. Dechlorination of aromatic ring
D. Incorporation of oxygen in aromatic ring

Co-metabolism is

A. The biodegradation of xenobiotics mediated by the biodegradation of another substance
B. Not the complete aerobic biodegradation of xenobiotic
C. The utilization of xenobiotic for microbial growth
D. The assimilation of xenobiotic followed by their biodegradation

Ortho-cleavage of aromatic ring after initial oxygenation is

A. Cleavage of aromatic ring in the position opposite to one of the carbon atoms with an attached hydroxylic group
B. Cleavage of aromatic ring in the position close to the most reduced carbon atoms
C. Cleavage of aromatic ring in the position close to one of the carbon atoms with attached hydroxylic group
D. Cleavage of aromatic ring between the carbon atoms with attached hydroxylic group

Which enzyme can be used for co-metabolism?

A. Amylase
B. Protease
C. Monooxygenase
D. DNAase

Biosynthesis

A gene is

A. A sequence of nucleotides that codes for a specific protein
B. A sequence of nucleotides in rRNA coded in DNA
C. A bacterial chromosome that codes genome
D. A sequence of amino acids in a specific protein

Translation is

A. Correspondence between the trinucleotides in mRNA and tRNA, which is specific for every amino acid
B. Transfer of genetic information from chromosome to mRNA
C. Transformation of genetic information from the sequence of mRNA into the sequence of protein
D. Interaction between codon and anticodon

Genetic information of cell is coded by the sequence of

A. Amino acids
B. Ribose and deoxyribose residuals
C. Nucleotides
D. Phosphate residuals

Mutation is

A. Inheritable change in the base sequence of DNA
B. Inheritable change in the base sequence of mRNA
C. Non-inheritable change in the sequence of nucleotides
D. Spontaneous change in the nucleotide sequence

Genetic information of cell is coded by the sequence of

A. Amino acids
B. Ribose and deoxyribose residuals
C. Nucleotides
D. Phosphate residuals

Complementation of DNA strands is due to

A. Negative charge of phosphate groups
B. Similarity of strand structure
C. Stereospecificity of hydrogen bonds
D. Specificity of covalent bonds between deoxyribose residuals

What is not true about plasmid?

A. It is a small part of chromosomal DNA
B. It is a small DNA molecule
C. It reproduces autonomously
D. It is a circular DNA molecule

Translation is

A. The transfer of information from DNA to mRNA to protein
B. The transfer of information from protein to rRNA
C. The transfer of information from protein to DNA
D. The transfer of information from mRNA to protein

Bioagents of Environmental Bioprocesses

Volume of 10^9 spherical cells with diameter 1000 nm is

A. 520 μm^3
B. 0.52 mm^3
C. 5.2 cm^3
D. $5.2 \times 10^{-6} m^3$

The cellular function of cytoplasmic membrane is

A. Protection of cell from disinfection
B. Regulation of cell growth
C. Control of the transport of ions
D. Cell resistance to high temperature

Shape of filamentous bacteria demonstrates an adaptation to

A. Nonviscous environment
B. Soil environment
C. Viscous environment
D. Freshwater environment

Function of cytoplasmic membrane is

A. To create viscous environment for cell
B. To control flows of nutrients
C. To produce cell capsule
D. To protect cell from high temperature

Gram-negative prokaryotic cells

A. Stain Gram-negatively because of their structure
B. Stain Gram-negatively because of negative charge of cell wall
C. Stain Gram-negatively because of negative effect of cell membrane
D. Stain Gram-negatively because of negative interactions between cells

Which of these methods can be used to determine the number of viable microorganisms in a sample?

A. Filtration and determination of biomass weight
B. Counting cells under light microscope without specific staining
C. Measuring content of microbial DNA
D. Measuring number of colony forming units

The shape of spirilla demonstrates an adaptation to

A. Nonviscous environment
B. Soil environment
C. Viscous environment
D. Freshwater environment

Bacterial spore .

A. Is produced to improve cell survival
B. Is sensitive to temperature
C. Is an intracellular enzyme
D. Is produced during fast growth

Cytoplasmic membrane separates

A. Nucleus from cytoplasm
B. Ribosomes from nucleus
C. Cytoplasm from environment
D. Cell wall from environment

Organelle is

A. An inclusion in cytoplasm
B. A cell component with specific function
C. An inclusion in cytoplasm separated with membrane
D. Visible in microscope cell component

Reproduction and Growth

Microorganisms undergo rapid cell division during the

A. Lag phase of growth
B. Exponential phase of growth
C. Stationary phase of growth
D. Death phase of growth

Equation describing growth yield is

(Y=yield, X=biomass concentration, S=substrate concentration, t=time)

A. $Y = \Delta \ln X / S\Delta t$
B. $Y = \Delta X / \Delta S$
C. $Y = \Delta S / X\Delta t$
D. $Y = \Delta S / \Delta X$

What is Monod's equation?

A. $\mu = \mu_{max} [S + K_s]$
B. $\mu = \mu_{max} [1/(1 + S/K_s)]$
C. $\mu = \mu_{max} [S/(S + K_s)]$
D. $\mu = \mu_{max} [S/K_s]$

When the population doubles during each given unit of time, the growth is said to be

A. Linear
B. Slow
C. Exponential
D. Fast

Which equation describes the growth rate?

(X = biomass concentration, t = time, S = substrate concentration, Y = yield)

A. $dS/dt = 0$
B. $dX/dt = k$
C. $dX/dS = Y$
D. $X = YS$

A microbial cultivation began with 4×10^6 cells/mL and ended with 1.28×10^8 cells/mL. Assuming binary fission of cell and no cell death, how many generations did the cells go through?

A. 5
B. 6
C. 32
D. 64

Microbial Ecology

A number of one-directional interactions in a system with two biotic components and two factors of environment is

A. 4
B. 8
C. 12
D. 16

What is a boundary for a lake considered as a system?

A. Shoreline on the map
B. Water mirror
C. Bottom sediment in lake
D. All of the above

Microorganisms capable of living under either aerobic or anaerobic conditions are described as

A. Obligate anaerobes
B. Facultative anaerobes
C. Oxygen-tolerant anaerobes
D. Microaerophiles

Which bacterial group is the most suitable for water biotreatment?

A. R-tactics
B. L-tactics
C. K-tactics
D. All of the above

Which is the interaction between two microorganisms which positively affects the survival and activity of both microorganisms?

A. Mutualism
B. Amensalism
C. Commensalism
D. Competition

Which microbial species cannot dominate in an ecosystem with a high concentration of nutrients?

A. Oligotrophic species
B. Species with the highest maximum of growth rate
C. Acidophilic species
D. Species with the highest efficiency of biomass synthesis

Which is the specific interaction between two microorganisms, which is negative for the activity of only one microorganism?

A. Mutualism
B. Amensalism
C. Commensalism
D. Competition

An example of commensalism is

A. Phototrophs producing oxygen and inhibiting anaerobes
B. Facultative anaerobes using oxygen and creating favorable conditions for anaerobes
C. Coexisting of two anaerobic species in biofilms
D. Simultaneous biodegradation of organics by two species

The sensitivity of a microbial strain to oxygen depends on

A. Its actual habitat
B. Its ability to reduce toxic oxygen radicals
C. Its ability for anaerobic respiration
D. Its ability to reduce oxygen

Classifications of Viruses, Prokaryotes, and Eukaryotes

Microorganisms capable of living in a number of environments are described as

A. Obligate
B. Facultative

C. Anaerobic
D. Mesophilic

The name of a bacterial species must be written as

A. *Pseudomonas*
B. *pseudomonas putida*
C. *Pseudomonas putida*
D. *Pseudomonas Putida*

Bacteria differ from Archaea by

A. Cell shapes
B. Cell sizes
C. Cell wall components
D. Cell structures

A point in rRNA-based phylogenetic classification is that

A. Relatedness of two compared rRNA sequences reflects the similarity in the structure of nucleotides
B. Difference between two compared rRNA sequences reflects the evolutionary time after their origin
C. Relatedness of compared rRNA sequences reflects their sequencing technique
D. Difference of compared rRNA sequences reflects their physiological difference

Gram-positive prokaryotic cells are Gram-positive because of

A. Positive charge between cell wall and cell membrane
B. Positive correlation between cell wall thickness and cell wall staining
C. Positive effect of cell membrane on cellular activity during staining
D. Positive interactions between staining cells

Unicellular fungi are

A. Molds
B. Yeasts
C. Mushrooms
D. Fruit bodies

Protozoa are differentiated by

A. Means of digestion
B. Means of locomotion
C. Activity of photosynthesis
D. Activity of biodegradation

Flagellates are

A. Fungi
B. Protozoa
C. Algae
D. Bacteria

Gram-negative prokaryotic cells

A. Are stained Gram-negatively because of their structure of cell wall
B. Are stained Gram-negatively because of negative charge of cell wall
C. Are stained Gram-negatively because of negative effect of cell membrane
D. Are stained Gram-negatively because of negative interactions between cells

The ratio of consumed CO_2 to produced O_2 in oxygenic photosynthesis is

A. 4
B. 2
C. 1
D. 0.5

Physiological Classification of Prokaryotes

Which electron acceptor is not used in anoxic energy generation?

A. Nitrate (NO_3^-) used by denitrifying bacteria
B. Methane (CH_4) used by methylotrophic bacteria
C. Sulfate (SO_4^{2-}) used by sulfate-reducing bacteria
D. Ferric ions (Fe^{3+}) used by iron-reducing bacteria

Gram-positive prokaryotic cells will dominate in

A. Oceans
B. Lakes
C. Soil
D. Rich liquid medium

Gram-negative prokaryotic cells

A. Stain Gram-negatively because of their structure
B. Stain Gram-negatively because of the negative charge of cell wall
C. Stain Gram-negatively because of the negative effect of cell membrane
D. Stain Gram-negatively because of the negative interactions between cells

Gram-positive cells are adapted to

A. Life in extreme environment
B. High salinity of medium
C. Low salinity of medium
D. Changes of salinity

The parallelism in the evolution of the genes in the evolutionary lines of prokaryotes of aquatic, terrestrial, and extreme environment origins is due to

A. Horizontal gene exchange between the prokaryotes of aquatic, terrestrial, and extreme environment origins
B. Transfer of the genes between the representatives of one phylogenetic line
C. Directional mutations in the DNA sequences of the prokaryotes of aquatic, terrestrial, and extreme environment origins
D. Random mutations in the DNA sequences of the prokaryotes of aquatic, terrestrial, and extreme environment origins

Groups of Prokaryotes

Bacteria from genus *Clostridium* perform

A. Fermentation of ammonia
B. Fermentation of methane
C. Fermentation of cellulose
D. Fermentation of lignin

Methanogens are

A. Obligate anaerobic bacteria
B. Able to reduce CO_2
C. Bioagents of metal reduction and precipitation
D. Bioagents of metal oxidation and dissolution

Anoxygenic phototrophic bacteria can be used in environmental engineering for

A. Oxidation of Fe^{2+}
B. Oxidation of SO_4^{2-}
C. Oxidation of H_2O
D. Reduction of H_2S

Bacteria from genus *Clostridium* are

A. Aerobes
B. Anaerobes
C. Microaerophiles
D. Facultative anaerobes

Which of the following bacteria produce acetate?

A. *Nitrosomonas* spp.
B. *Syntrophobacter* spp.
C. *Desulfobacterium* spp.
D. *Pseudomonas* spp.

Which of the following is a specific feature of methanogens?

A. They can oxidize hydrogen
B. They can reduce sulfate
C. They can degrade biopolymers
D. They can oxidize methane

Public Health and Water Disinfection

Which prokaryotes cannot be used as indicators of fecal pollution of water?

A. Nitrifying bacteria
B. Sulfate-reducing bacteria
C. Aerobic bacteria from genus *Pseudomonas*
D. Facultative anaerobes

Which species is not related to enterobacteria?

A. *Escherichia coli*
B. *Enterococcus faecalis*
C. *Nitrosomonas europea*
D. *Vibrio cholera*

To measure the concentration of heterotrophic bacteria in a sewage sample, it was diluted by 10^7 times with sterile medium, then 0.1 mL of diluted suspension was spread on surface of a semisolid medium in a Petri dish and incubated for 24 h. A picture of grown colonies (dark dots) is shown. What was the concentration of heterotrophic bacteria in the sewage sample?

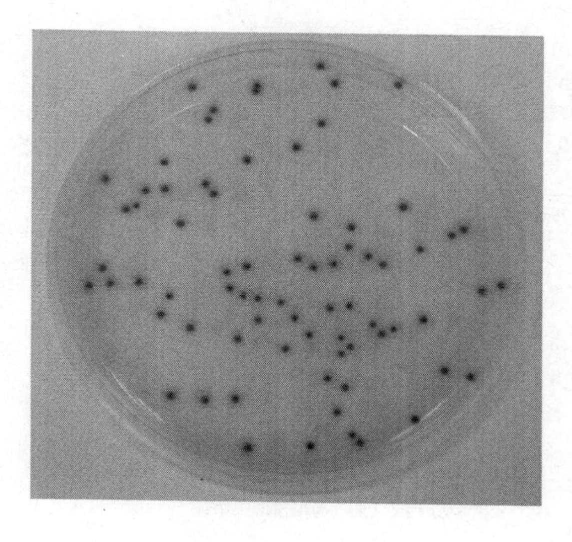

A. 7×10^6 cells/mL
B. 7×10^7 cells/mL
C. 7×10^8 cells/mL
D. 7×10^9 cells/mL

Which prokaryotes dominate in human feces and can be used as indicators of fecal pollution of water?

A. *Bacteroides* spp.
B. *Ruminobacter* spp.
C. *Escherichia coli*
D. *Desulfovibrio* spp.

A saprophytic microbe

A. Grows in or on animal tissue
B. Feeds by dead organic matter
C. Feeds on dead organic matter but can grow in or on animal tissue
D. Derives its nutritional requirements from the host organism

The concentration of indicator bacteria in water indicates

A. The presence of pathogenic bacteria in water
B. The risk of waterborne infectious diseases
C. The level of biological pollution of water
D. The presence of bacterial cells in water

Biotechnological Processes

Genetic engineering is

A. Microbial engineering
B. Adaptation of microbial genome to industrial needs
C. Engineering of genetic material
D. Selection of useful microbial strains

The cloning vector

A. Introduces recombinant RNA into cells
B. Introduces recombinant DNA into cells
C. Clones a recombinant molecule inside a vector
D. Transfers a recombinant molecule from a cell into the environment

One of the stages in the polymerase chain reaction is

A. Heating DNA to destroy single strands and adding primers
B. Cooling single-stranded DNA and adding primers
C. DNA hydrolysis by restrictases
D. Cyclic transfer of genetic material from mRNA to tRNA

Restriction fragment length polymorphism (RFLP) is used in environmental biotechnology for the following purpose

A. To identify relatedness between different genes of similar organisms
B. To identify differences between different genes of similar organisms
C. To identify relatedness between similar genes of different organisms
D. To identify differences between chromosomes of same genome

The time between inoculation and the beginning of growth is usually called as

A. Lag phase
B. Log phase
C. Stationary phase
D. Death phase

The advantage of biotechnological treatment of wastes is

A. Cheap nutrients which must be present in the medium
B. Biodegradation of a wide spectrum of hazardous substances
C. Selection and testing of microorganisms for biotechnological treatment
D. Control of optimal conditions to maintain microbial activity in the system

Chemostat is

A. A reactor with fixed biomass
B. A reactor of complete mixing
C. A reactor with plug flow
D. A reactor with control pH, temperature, and chemical components of a medium

A sequencing batch reactor is

A. A system of batch cultivation with the periodical addition of nutrients
B. A system of continuous cultivation with the periodical addition of nutrients and removal of microbial suspension
C. A system of continuous cultivation with continuous addition of nutrients and removal of microbial suspension
D. A system of batch cultivation with the removal of microbial suspension when it is needed

The primary step in the development of biotechnology is

A. Isolation of pure or enrichment culture
B. Cultivation of pure or enrichment culture
C. Selection of pure or enrichment culture
D. Identification of pure or enrichment culture

Selection pressure for the production of enrichment culture is

A. Isolation and cultivation of enrichment culture
B. Selective conditions for cultivation of enrichment culture

C. Disinfection of medium
D. Tolerable pressure during cultivation

Which of the following operations is not related to DNA recombinant technology and creation of recombinant bacterial strain?

A. DNA is extracted from bacterial cell
B. Small circular DNA sequences are bound together by specific enzymes
C. Small sequences of DNA are introduced into DNA vector
D. Vector is transferred into the cell and produces multiple copies of the introduced gene

Aquatic Ecosystems and Water Treatment

What is the concentration of microorganisms in water if after diluting 10^5 times and a spread of 0.05 mL on the surface of a Petri dish, the number of colonies was 18?

A. 1.8×10^5 cfu/mL
B. 3.6×10^6 cfu/mL
C. 3.6×10^7 cfu/mL
D. 1.8×10^7 cfu/mL

Biotechnology can be most applicable for waste treatment in the following cases:

A. Unpleasant smell of wastes
B. Large volume of wastes
C. Toxic and viscous wastes
D. Chemical industry wastes

Which of the following is not related to eutrophication?

A. Increase of phosphorous concentration in water
B. Fast growth of phototrophs
C. Pollution of water with xenobiotic
D. Low concentration of dissolved oxygen in water

Which of the following reactions is ammonification?

A. $N_2 + [2H] + x\text{ATP} \rightarrow RNH_2$
B. $R(COO)NH_2 + [2H] \rightarrow RCOOH + NH_3$
C. $NH_4^+ + 2O_2 \rightarrow NO_3^- + 2H^+ + H_2O$
D. $2NO_3^- + 5[2H] \rightarrow N_2\uparrow + 2OH^- + 4H_2O$

Microbiological processes of the phosphorus cycle do not include

A. Mineralization of organic phosphorus
B. Bioassimilation of phosphate
C. Precipitation of phosphates as salts of Ca, Mg, Fe, and Al
D. Biosolubilization of phosphates due to the production of organic acids

The electron acceptor in bioreduction of sulfur compounds is

A. S^{2-}
B. RSH
C. S^0
D. SO_4^{2-}

The source of non-point pollution of reservoir with nutrients is

A. Agricultural drainage
B. Air dust
C. Golf course surrounding the reservoir
D. Farm

A feature of constructed wetland for removal of nutrients from stormwater

A. Round wetland
B. Shallow pond
C. Pond with the gravel on the bottom
D. Wetland without trees

The biological instability of water is caused by

A. Ammonium
B. Nitrate
C. Sulfate
D. Ferric ions

Chemical oxygen demand (COD) is a parameter to measure

A. Concentration of oxygen in water and wastewater
B. Concentration of organic substances in water and wastewater
C. Quantity of oxygen needed for microbial growth
D. Concentration of all oxidized compounds in water and wastewater

Bioremoval of unstable organic matter in water is performed by

A. Oxidation
B. Reduction
C. Hydrolysis of organics
D. Incorporation of oxygen into molecules

Which of the following substances cannot be used for denitrification of water and wastewater?

A. SO_4^{2-}
B. SO_2
C. S^0
D. S^{2-}

Anaerobic and Anoxic Treatment of Wastewater

An important application of anaerobic Gram-positive fermenting bacteria in environmental engineering is

A. Production of biofuel
B. Anoxic treatment of wastewater
C. Methanogenic fermentation
D. Anaerobic nitrogen fixation

The molar ratio of reduced sulfate to oxidized glucose (x_1 stoichiometrical coefficient) in bacterial sulfate reduction process: $C_6H_{12}O_6 + x_1SO_4^{2-} + x_2H^+ \rightarrow x_3CO_2 + x_4H_2S + x_5H_2O$ is

A. 12
B. 6
C. 3
D. 1

An application of anoxic bacteria in environmental biotechnology is

A. Oxidation of sulfate
B. Reduction of CO_2
C. Precipitation of heavy metals
D. Nitrification in wastewater

What is not true about prevention of eutrophication in water reservoir?

A. Eutrophication can be prevented using bioremoval of nutrients from the streams
B. The rate of eutrophication can be diminished using aeration of water in reservoir
C. Eutrophication can be prevented by collection and reuse of runoff
D. Eutrophication can be prevented by the proper use of pesticides

The number of moles of electrons donated by a mole of lactic acid ($CH_3COCOOH$) during its oxidation to CO_2 coupled with reduction of ferric hydroxide $Fe(OH)_3$ to ferrous ions Fe^{2+} is

A. 3
B. 7
C. 10
D. 14

The equation for the "anammox" process is

A. $NH_4^+ + 2O_2 \rightarrow NO_3^- + 2H^+ + H_2O$
B. $NH_4^+ + NO_2^- \rightarrow N_2 + 2H_2O$
C. $NH_4^+ + NO_2^- + H_2O \rightarrow NO_3^- + 1/2N_2 + 3H_2O$
D. $NH_4^+ + 0.5O_2 \rightarrow NO_2^- + 2H^+ + H_2O$

The molar ratio (*b*/*a* ratio) of nitrate (NO_3^-) to ethanol (C_2H_5OH) in the bacterial denitrification process ($aC_2H_5OH + bNO_3^- + cH^+ \rightarrow dN_2 + eCO_2 + fH_2O$) is

A. 12.0
B. 5.0
C. 2.4
D. 0.4

Which bacteria could be used for removal of nitrogen from wastewater?

A. Nitrate-oxidizing bacteria
B. Nitrate-fermenting bacteria
C. Nitrate-reducing bacteria
D. Nitrate-hydrolyzing bacteria

Which of the following depicts methanogenesis performed by *archaea*?

A. $C_2H_5OH \rightarrow CH_4 + CO_2$
B. $CO_2 + 4H_2 \rightarrow CH_4 + 2H_2O$
C. $CH_3COOH + 4H_2 \rightarrow 2CH_4 + 2H_2O$
D. $H_2 + CH_2 \rightarrow CH_4$

Which substance is the product of anaerobic respiration?

A. NH_3
B. H_2SO_4
C. CH_3CH_2OH
D. N_2

Aerobic Treatment of Wastewater

Filamentous microaerophilic bacteria cause a problem of wastewater treatment because of

A. Bulking of activated sludge
B. Foaming in anaerobic digester
C. Formation of dense flocs in aeration tank
D. Sedimentation and clogging in anaerobic digester

Microbial aerobic biotransformation of nitrogen includes formation of

A. NH_3
B. N_2
C. HNO_3
D. NH_4OH

Optimal BOD:N:P ratio for the aerobic biotreatment of wastewater, which satisfies the C, N, and P requirements of a wide variety of microorganisms, is

A. 100:5:1
B. 100:6:3

C. 100:8:5
D. 100:10:7

Which species has a relation to the formation of "sewage fungus?"

A. *Escherichia coli*
B. *Sphaerotilus natans*
C. *Methanobrevibacter* spp.
D. *Bacillus subtilis*

Recycle of a portion of activated sludge after settling into the aeration tanks ensures

A. The maintenance of specific activity of biomass in aeration tank
B. The total absence of filamentous microorganisms in aeration tank
C. The retention of slow-growing organisms in aeration tank
D. The retention of actively respiring microorganisms in aeration tank

Which prokaryotes of the following genus are typical representatives of biocomponents of activated sludge flocs?

A. *Vibrio*
B. *Clostridium*
C. *Methanobacterium*
D. *Pseudomonas*

Which is not an essential factor for domination of filamentous bacteria in activated sludge?

A. Low concentration of dissolved oxygen in aeration tank
B. High phosphate load in aeration tank
C. Low concentration of nutrients
D. Presence of sulfides in aeration tank

The typical specific surface area of the medium (carrier) in a fixed biofilm reactor is

A. $10\,m^2/m^3$
B. $100\,m^2/m^3$
C. $1{,}000\,m^2/m^3$
D. $10{,}000\,m^2/m^3$

Value-Added By-Products

Which is not a value-added by-product of food waste biotreatment?

A. Cellulose
B. Proteolytic enzymes
C. Fuel ethanol
D. Polylactic acid

Organic acids produced during anaerobic treatment of wastes can be used for the production of

A. Biodegradable plastics
B. Proteins
C. Cellulases
D. Microbial polysaccharides

Which is an application of microbial polysaccharides?

A. Drilling fluid in oil industry
B. Raw material for microbial cultivation
C. Biofuels
D. Anticorrosive materials

Which waste cannot be used for cultivation of edible mushrooms?

A. Saw dust
B. Compost from agricultural wastes
C. Excess of activated sludge
D. Cheese whey

Which waste cannot be used for biofuel production?

A. Excess of activated sludge
B. Vegetable oil
C. Food wastes
D. Industrial wastewater

Which is an effective way for recovery of phosphate and ammonia?

A. Precipitation with Mg salts
B. Precipitation with Al salts
C. Phosphatization and denitrification
D. Precipitation with carbonates

Biotreatment of Industrial Hazardous Wastes

Co-metabolism is

A. Simultaneous oxidation of two similar substances while only one substance serves as the energy source
B. Simultaneous oxidation of two substances, which are energy sources
C. Simultaneous reduction of two substances
D. Mineralization of a substance, which serves as a carbon source, and the assimilation of a substance, which serves as an energy source

Which of these substances is a xenobiotic?

A. Protein
B. Pesticide

C. Lignin
D. Lipid

Which pre-treatment cannot be used to intensify the biotreatment of hazardous waste?

A. Mechanical suspension of hazardous hydrophobic substances
B. Pyrolysis (high temperature decay) of solid waste
C. Removal of hazardous substances by adsorption
D. Preliminary oxidation of hazardous wastes by H_2O_2

What is the component of landfilling leachate?

A. Polysaccharides
B. Proteins
C. Amino acids
D. Ammonium

Combinations of anaerobic and aerobic treatments are useful in the following situations

A. Biodegradation of chlorinated aromatic hydrocarbons including anaerobic dechlorination and aerobic ring cleavage
B. Sequential nitrogen removal including aerobic nitrification and anaerobic denitrification
C. Anaerobic reduction of Fe(III) and microaerophilic oxidation of Fe(II) with production of fine particles of iron hydroxide
D. Anaerobic adsorption of phenols followed by their aerobic biodegradation

Biodegradation can be enhanced by such electron acceptor as

A. NO_3^-
B. N_2
C. NH_3
D. NH_4^+

Solid Wastes and Soil Biotreatment

Which prokaryotes could be most suitable for aerobic bioremediation of soil, polluted with organic substances?

A. Gram-negative iron-reducing bacteria
B. Gram-positive fermenting bacteria
C. Gram-positive heterotrophic bacteria
D. Aerobic *Archaea*

Which could be considered as a biomass?

A. Food waste
B. Coal produced from plants

C. Organic matter of wastewater
D. Organic matter of soil

Which physiological group of bacteria is suitable for anaerobic bioremediation of polluted soil?

A. Sulfate-reducing bacteria
B. Nitrifying bacteria
C. Filamentous Gram-positive microaerophilic bacteria
D. Iron-oxidizing bacteria

Which bacteria could be used for anaerobic treatment of pesticide-polluted soil?

A. Iron-oxidizing bacteria
B. Fermenting bacteria
C. Phototrophic bacteria
D. Ammonia-oxidizing bacteria

Microbial Geotechnics

Biocementation is

A. The use of microorganisms to enhance cement
B. Enhancing the strength of soil using microbial activity
C. Enhancing natural soil cementation
D. The use of microorganisms to clog pores in soil

The microbial process that cannot produce a biocementation effect is

A. Production of enzymes degrading lignin
B. Production of sulfides by sulfate-reducing bacteria
C. Production of carbonate ions
D. Precipitation of silica dioxide due to microbial acidification

The bioagents of soil bioclogging and biocementation could be

A. Archaea
B. Methanogens
C. Oligotrophs
D. Aerobic bacteria from genus *Pseudomonas*

To diminish the risk of pathogenic bacteria accumulation and release during the bioclogging and the biocementation, the following selective conditions can be used:

A. Application of a carbon source such as cellulose
B. Application of a carbon source such as glucose
C. Conditions that are suitable for the growth of heterotrophic mesophilic bacteria
D. Conditions that are suitable for the growth of fungi

The microbial processes that can produce a bioclogging effect is

A. The production of enzymes of sulfur oxidation
B. Methanogenic activity
C. The bioreduction of ferric compounds
D. The biooxidation of hydrocarbons

Microbiology of Air and Air Treatment

The particles of nonviable bioaerosols could be

A. Clay particles
B. Dead bacterial cells
C. Viruses
D. Bacterial spores

The source of a fungal aerosol could be

A. Brick masonry
B. Glass window
C. Wooden wall
D. Concrete walls

The bioaerosol with the highest survival rate in environment is

A. The vegetative cells of pathogenic bacteria
B. The spores of chemotrophic bacteria
C. Cells of protozoa
D. Cells of phototrophs

Odor-producing compounds are

A. Volatile organic acids
B. Volatile hydrogen
C. Nitrates
D. Ferric and ferrous ions

Biocorrosion, Biodeterioration, and Biofouling

Microbiologically influenced corrosion (MIC) can be caused by

A. Fermentation
B. Aerobic respiration
C. Anaerobic respiration
D. All of the above

Microbially influenced corrosion (MIC) can be caused by

A. Lactic acid fermentation
B. Anoxygenic iron reduction

C. Glucose assimilation
D. Mineralization of hydrocarbons

Microorganisms cannot induce corrosion by

A. The biodegradation of metal
B. The biooxidation of protective film
C. The production of microbial polysaccharides
D. The enzymatic hydrolysis of cellulose in soil

An important mechanism of membrane biofouling is

A. Biodegradation of membrane material
B. Biochemical modification of membrane surface
C. Growth of bacteria in water
D. Formation of microbial polysaccharides during microbial growth

26

Tutorial Bank

Tutorial 1: Microorganisms: Cell Size and Shape

Self-Examination 1

A bacterial cell has a dry biomass of 3×10^{-12} g. How many cells would be present in 1 metric ton (MT) of dry biomass?

Solution

$N = 1 \times 10^6$ g/MT: 3×10^{-12} g/cell $= 0.33 \times 10^{18}$ cells/MT

Self-Examination 2

What is the maximum of biomass quantity (X_{max}) in 40 m^3 of working volume of fixed biofilm reactor with specific surface of biofilm 300 m^2/m^3, bacterial biofilm thickness 42 μm, and density of biomass 1.04 g/cm^3?

Solution

$$X_{max} = \text{volume of biomass} \times \text{density of biomass}$$

$$= \text{thickness} \times 1\,m^2 \times \text{specific surface} \times \text{volume of reactor} \times \text{density of biomass}$$

$$= 42 \times 10^{-6}\,m \times 1\,m^2 \times 3 \times 10^2\,m^2/m^3 \times 40\,m^3 \times 1.04 \times 10^6\,g/m^3$$

$$= 5241 \times 102\,g = 524,\ 100\,g = 524.1\,kg$$

Q26.1.1

Assume that the average diameter (d) of a bacterial cell of spherical shape is 1 μm, and the average size of bacterial cell of a shape of rod with hemispherical ends is 1 μm in diameter and 3 μm of length (l).

Q26.1.1.1 What is the cell volume of the spherical cell (v_c) and the bacterial rod (v_b)?

Q26.1.1.2 What is the surface-to-volume ratio of a cell with a spherical shape, (S/V)c, or rod, $(S/V)_c$?

Q26.1.1.3 What is the importance of different cell shape for environmental engineering?

Q26.1.2
Assume that the diameter (d) of a spherical algal cell is about 5×10^{-5} m and that the diameter of a spherical bacterial cell is about 2 μm.

Q26.1.2.1 What is the ratio of cell surface-to-cell volume for *algae* $(S/V)_a$ and bacteria $(S/V)_b$?

Q26.1.2.2 What is an important consequence for environmental engineering?

Q26.1.2.3 What should be the diameter of the pore (d_p) in membrane filter to separate algae and bacteria?

Q26.1.3
Assume that the average size of rod-shaped bacterial cells is 1 μm in diameter (d) and 3 μm in length (l), density of the intact cell is 1.04 g/cm^3 (d), and content of water in the intact biomass is 75% (w/w) (w).

Q26.1.3.1 What is the number of the cells in 1 g of intact biomass (n_w)?

Q26.1.3.2 What is the number of the cells in 1 g of dry biomass (n_d)?

Q26.1.3.3 What is the maximal number of the cells in 1 m^3 of working volume of bioreactor (n_{max})?

Q26.1.4
How many rod-shaped bacterial cells (n_{max}) with hemispherical ends with average diameter (d) 1 μm and length (l) 2 μm may be detected theoretically on a 25×75 mm microscope slide? Consider that the magnification of the microscope is 1000.

Q26.1.5
What is the maximal number of bacterial layers (b) in the biogranule with radial and laminar structures of the layers and diameter (d_g) 5×10^{-3} m? Consider that the bacterial community is presented by rods with diameter (d_c), 1 μm, and length (l), 2 μm.

Q26.1.6
How many algal cells with a diameter (d_f), 5 μm, and length (l_f), 10 μm, may be detected visually without microscope, on a 25×75 mm microscope slide? *Important note*: Consider that the algal cells are not aggregated.

Q26.1.7
Wastewater is biologically treated in a fixed biofilm reactor where the specific surface of biofilm (s) is 40 m^2/1 m^3 of reactor volume, colonization of the biofilm carrier (n_s) is 4×10^8 cells/mm^2, bacterial community is presented by rods with diameter (d) 1 μm and length (l) 3 μm, and cell density (d) is 1.04 g/cm^3.

Q26.1.7.1 What is the number of bacterial cells in the reactor (n)?

Q26.1.7.2 What is the wet biomass of the bacteria in the biofilm (x)?

Q26.1.7.3 What is the minimal number of bacterial layers in the biofilm (b)?

Q26.1.8
How many portions of microbial seeds for wastewater treatment (N) may be produced in a 40 m^3 bioreactor containing 10 g of dry biomass/L after cultivation? Assume that one portion of the seeds must contain at least 10^{15} viable bacterial cells; 1 g of dry biomass contains 1.5×10^{12} bacterial cells, and their viability is 18%.

Q26.1.9
What is the concentration of viruses in 1 mL of water if the concentration of viral biomass is 6.3 ng/mL? Consider that the diameter of a viral particle is 100 nm and its specific density is 1.2.

Tutorial 2: Cell Chemistry and Structure

Q26.2.1
What is the content of carbon, hydrogen, oxygen, and nitrogen in biomass represented by the empirical formula $CH_{1.8}O_{0.5}N_{0.12}$? Use the following atomic weights in your calculations: H = 1; C = 12; N = 14; O = 16.

Q26.2.2
What is the empiric formula of microbial product if the contents (w/w) of carbon, hydrogen, oxygen, and nitrogen in this product are 60%, 10%, 20%, and 10%? Use the following atomic weights in your calculations: H = 1; C = 12; N = 14; O = 16.

Q26.2.3
Calculate the molar concentration of a 16S rRNA-targeted oligonucleotide probe, which is sufficient to detect 10^6 bacterial cells of *Paracoccus denitrificans* suspended in 1 mL of water. Consider that the number of ribosomes in a cell is 1.5×10^5. Note that 1 mol of a substance contains 6.023×10^{23} molecules (Avogadro's number).

Q26.2.4
What is the content of lipids (g/g) in the spherical algal cell? The diameter of the cell is 28 μm, and the diameter of the 6 spherical lipid globules is 6 μm. The density of cell cytoplasm and cells wall is 1.04 g/cm^3; the density of lipid globules is 0.8 g/cm^3.

Q26.2.5
Calculate the average number of ribosomes in an *E. coli* cell. Consider that the quantity of rRNA in the cell is 15% of dry cell mass, there are 3500 nucleotides in one ribosome, the average molecular weight of one nucleotide is 348, and that the average dry mass of one cell is 2×10^{-12} g. Note that 1 mol of a substance contains 6.023×10^{23} molecules (Avogadro's number).

Tutorial 3: Metabolism and Biodegradation

Q3.1

Determine substances that can be produced from glucose ($C_6H_{12}O_6$), hydrogen sulfide (H_2S), or hydrogen (H_2) by fermentation, anaerobic respiration, and respiration.

Q3.2

Determine the products of microbiological oxidation/reduction substances, balance the equations of this reaction, calculate the mass ratio (Y) between the quantities of electron donor and electron acceptor, and assume practical application of the process in environmental engineering.

The parameters of microbial oxidation/reduction.

Electron Donor	Electron Acceptor	Products	Y, g/g	Environmental Application(s)
NH_4^+	O_2			
S	NO_3^-			
H_2	CO_2			

Q3.3

Iron-reducing bacteria can be used for the production of ferrous ions (Fe^{2+}) from iron ore (Fe_2O_3) as electron acceptor and cellulose $(C_6H_{10}O_5)_n$ as electron donor. Ferrous ions precipitate phosphate ions (HPO_4^{2-}), forming $FeHPO_4$. The efficiency of iron reduction is 70%.

Q3.3.1 Balance the reaction of phosphate removal with ferrous ions. Determine the quantity of iron ore for precipitation of 100% of phosphate from wastewater containing 20 mg P/L during 30 days. The flow rate of wastewater is 2000 m^3/day.

Q3.3.2 Balance the reaction of ferric reduction with cellulose. Determine the mass ratio between cellulose and added iron ore.

Q3.4

Calculate the parameters of photosynthesis:

Q3.4.1 Balance the reactions of anoxygenic and oxygenic photosynthesis.

Q3.4.2 What may be the percentage of sulfur inclusions in the biomass of anoxygenic phototrophic bacteria?

Q3.4.3 What is the mass ratio of produced biomass and oxygen? What is the ratio of volume of produced air (21 vol.% of oxygen) to 1 g of produced biomass?

Q3.5

Balance the reactions of denitrification with ethanol:

$$?C_2H_6O + ?NO_3^- + ?H^+ = ?CO_2 + ?H_2O + ?N_2$$

Q3.6

Balance the reactions of denitrification with glucose:

$$?C_6H_{12}O_6 + ?NO_3^- \rightarrow ?N_2 + ?CO_2 + ?H_2O + ?OH^-$$

Q3.7

Calculate the daily consumption of methanol (CH_4O) for the removal of nitrate (NO_3^-) from groundwater using bacterial denitrification performed according to equation:

$$?CH_3OH + ?NO_3^- = ?CO_2 + ?H_2O + ?N_2 + ?OH^-.$$

The flow rate of treated groundwater is 800 m^3/h; the concentration of nitrate in groundwater is 25 mg N-NO_3^-/L.

Tutorial 4: Growth and Ecology

Q4.1

Iron-reducing bacteria produce ferrous ions (P) from insoluble Fe(III) compounds. How does one determine the specific growth rate (μ) of iron-reducing bacteria attached to the particles of iron ore using data on ferrous production? What are the other ways of determining the growth rate of bacteria attached to a surface in a fixed biofilm reactor?

Q4.2

Bacteria were grown in an aerobic batch culture. Ethanol (C_2H_5OH) was the sole carbon source, which was oxidized to CO_2 and used to produce the microbial biomass ($CH_2O_{0.5}N_{0.2}$). Experimental data are shown in the table below.

	Time (t), h						
Parameters	**0**	**12**	**24**	**36**	**48**	**60**	**72**
Biomass concentration (X), g/L	0.04	0.06	0.5	3.8	7.2	8.4	8.2
Concentration of ethanol (S), g/L	16.0	16.0	15.0	8.6	2.6	0.1	0.0

Use the experimental data from this table to determine the specific growth rate (m) and growth yield (Y)

Q4.3

Derive the equation connecting specific growth rate (μ) and generation time (g) of microbial population during increase of microbial biomass concentration from X_0 to X and increase of bacterial cell concentration from N_0 to N during period of time t.

Q4.4

Use the experimental data of batch cultivation from the table below to calculate the parameters of microbial growth. Consider that growth is an exponential process.

	Source of Carbon and Energy	
Parameter	**Glucose**	**Methanol**
S_0, g/L	20	5
S, g/L	14	?
X_0, g/L	1	1.2
X, g/L	4	5.6
t, h	?	8
Y, g/g	?	0.3
m, h^{-1}	0.5	?
T (lag phase duration), h	3	2

Q4.5

Calculate the concentration of heterotrophic bacteria in sewage sample assuming that the number of grown colonies of heterotrophic bacteria in Petri dish = 70. To grow the colonies, 0.1 mL of sample diluted 10^7 times was spread on the surface of semisolid medium in Petri dish and incubated for 24 h.

Q4.6

Consider that eutrophication in reservoir occurs by phosphate enrichment due to runoff of storm water from a surrounding golf course, following the growth of phototrophic and heterotrophic prokaryotes, causing lowered levels of oxygen.

a. Draw a schematic of interactions in this ecosystem.
b. Propose a set of technologies to prevent the negative effects of eutrophication due to runoff from the golf course.

Q4.7

Engineered wetlands are used for the treatment of agricultural wastewater, effluent of wastewater-treatment plants, and removal of nutrients from urban stormwater.

a. Describe the reason for constructing engineered wetlands and their principles.
b. Calculate the surface of conventional constructed wetland with hydraulic retention time (HRT) = 5 days and innovative engineered wetland with ability to remove 3.5 g phosphate/m^3/day. The flow rate is 100,000 m^3/day, the concentration of phosphate is 4 mg/L, and the depth of constructed wetland is 1 m.

Q4.8

The biomass of fermenting bacteria and methanogens in an anaerobic digester must be accumulated during the start-up period of anaerobic digestion from

10 mg/L to at least 80 g/L and from 20 mg/L to at least 20 g/L, respectively. The average specific growth rates of fermenting bacteria and methanogens in an anaerobic digester are 0.02 and 0.005 h^{-1}, respectively. What is the duration of the start-up period of an anaerobic digester?

a. 29 days
b. 57 days
c. 75 days
d. 134 days

Q4.9

Calculate the stable ratio between the concentration of spherical bacterial and protozoan cells in an ecosystem, which is isolated from the surrounding environment. Consider that conditions are as follows:

- Protozoa are feeding on bacteria.
- 10% of bacterial biomass is transformed into protozoan biomass.
- The diameters of bacterial and protozoan cells are 2.3 and 38.9 μm, respectively.
- Densities of both types of cells are the same.

Tutorial 5: Anaerobic Processes

Q5.1

Select the group of microorganisms from the list below that is suitable for the bioremoval of heavy metals from wastewater, and describe its (their) function(s) in this treatment:

Lithotrophs
Methanogens
Sulfate-reducing bacteria
Sulfur-oxidizing bacteria
Enterobacteria
Nitrifying bacteria

Q5.2

Draw the balanced equations of fermentations, determine mass ratio (*Y*) between the quantities of product and substrate, and assume practical application of the process in environmental engineering.

TABLE Parameters of Microbial Fermentation

Substrate	Product	Reaction	*Y*, g/g	Application
Glucose $C_6H_{12}O_6$	Ethanol CH_3CH_2OH			
Propionic acid CH_3CH_2COOH	Acetic acid CH_3COOH			

Q5.3

Fermenting bacteria can be used for the reductive dechlorination of pentachlorophenol ($C_6C_{15}OH$) in industrial wastewater at a flow rate of 500 m^3/day and concentration of pentachlorophenol 20 mg/L. Another wastewater stream containing 10 g glucose ($C_6H_{12}O_6$) per 1 L can be used for the production of hydrogen (H_2) by microbial fermentation. In one experiment on fermentation of glycose-containing medium, the molar concentrations of the products of fermentation were 140 mmol/L for hydrogen (H_2), 150 mmol/L for acetate (CH_3COO^-), and 80 mmol/L for ethanol (CH_3CH_2OH).

(a) Balance the reaction of fermentation using experimental data.
(b) Balance the reaction of reductive dechlorination.
(c) Calculate the flow rate of glucose-containing wastewater considering that only 20% of produced hydrogen is used for reductive dechlorination.

Use the following atomic weights in your calculations:
H = 1, C = 12, O = 16, and Cl = 35.5.

Q5.4

Balance the reaction of cellulose fermentation considering that products of fermentation are acetate, ethanol, and CO_2. The molar ratio of acetate to CO_2 is 2:1.

$$C_6H_{10}O_5 + H_2O \rightarrow C_6H_{12}O_6 \rightarrow X_1C_2H_4O_2 + X_2C_2H_6O + X_3CO_2$$

Q5.5

A biotechnological method for converting cellulose-containing waste to fuel ethanol includes acid hydrolysis, fermentation, and recovery of fuel ethanol. Calculate the theoretical production of fuel ethanol from cellulose-containing wastes. The density of ethanol is 0.79 kg/L.

Q5.6

Sulfate-reducing bacteria can be used for the removal of copper from acid mine drainage water at flow rate of 500 m^3/day and concentration of copper at 20 mg/L. Acetate (CH_3COO^-), the main product of cellulose-containing wastes ($C_6H_{10}O_5$) fermentation, can be used as a cheap donor of electrons for sulfate reduction.

a. Explain how the process will be performed and the microorganisms that will participate.
b. Calculate the quantity of cellulose used for the removal of copper for 1 year of wastewater treatment. The molecular weight of cellulose monomer is 162; the atomic weight of copper is 63.5.

Q5.7

Calculate the daily consumption of ethanol (C_2H_5OH) for the removal of nitrate (NO_3^-) from groundwater using bacterial denitrification performed according to the following equation:

$$\underset{46\text{ g}}{5C_2H_5OH} + \underset{14\text{ g}}{12NO_3^-} = 10CO_2 + 9H_2O + 6N_2 + 12OH^-$$

The flow rate of treated groundwater is 1200 m^3/h, the concentration of nitrate in groundwater is 45 mg NO_3^--N/L, and the concentration of dissolved oxygen is 7 mg/L. Note that the essential condition to start-up denitrification is that dissolved oxygen must be removed from groundwater using microbial oxidation of ethanol according to the following equation:

$$\underset{46\text{ g}}{C_2H_5OH} + \underset{96\text{ g}}{3O_2} = 2CO_2 + 3H_2O$$

Q5.8
Calculate the content of methane (%, v/v) in the biogas produced during anaerobic digestion of activated sludge ($CH_2O_{0.5}N_{0.12}$). The material balance of the anaerobic digestion process is shown in the following equation:

$$CH_2O_{0.5}N_{0.12} + 0.34H_2O \rightarrow 0.58CH_4 + 0.42CO_2 + 0.12NH_3$$

Assume that 90% (v/v) of ammonia and 25% (v/v) of carbon dioxide are dissolved in the liquid of anaerobic digester.

A. 16%
B. 32%
C. 64%
D. 76%

Q5.9
What is the mass ratio (g/g) of oxidized glucose to reduced sulfate in sulfate reduction? Note that the molecular weight of glucose = 180 g, and the molecular weight of sulfate = 96 g.

A. 0.21
B. 0.62
C. 1.86
D. 5.58

Tutorial 6: Aerobic Processes

Q6.1
Calculate the theoretical COD (mg O_2/g of matter) for solid waste containing
15% (w/w) of lipids, $C_{50}H_{100}O_6$
35% (w/w) of carbohydrates, CH_2O
60% (w/w) of water, H_2O

Q6.2
Cometabolism of trichloroethylene (TCE) by methylotrophic bacteria is an effective approach for the remediation of polluted aquifers. These reactions are described by the following equations:

Calculate the quantity of the biomass of methanotrophic bacteria (X) needed to purify 250,000 m^3 of groundwater (V) polluted by TCE for 60 days (T). The initial concentration of pollutant (S) is 0.20 mg/L, its final concentration must be 0.00 mg/L, the rate of co-metabolization (Q) of TCE by methane monooxidase (MMO) is 200 mg/g × day, and the content of methane monooxidase in the biomass of methanotrophic bacteria (C) is 4 mg/g.

Q6.3
Calculate the empirical formula of the mixture of glucose ($C_6H_{12}O_6$, 500 mg/L) and formic acid (CH_2O_2, 1800 mg/L).
H = 1; C = 12; O = 16

Q6.4
Wastewater contains 15 g/L of carbohydrates (CH_2O) and 3 g NO_3-N/L.

Calculate the percentage of BOD that can be removed from this wastewater using only anaerobic bacterial denitrification. Consider that all oxidizable matter are carbohydrates.

Q6.5
Calculate the yield of biomass (the empirical formula of microbial biomass is $CH_2O_{0.5}$) during aerobic growth in a medium with ethanol (C_2H_6O) as the source of carbon and energy. The production of CO_2, measured during growth, was 0.8 g CO_2/g of consumed ethanol.

Q6.6
Raw water contains 2 mg NH_4^+-N/L in a flow of 1000 m^3/day. The treatment goal is to create biologically stable drinking water with NH_4^+-N concentration lower than 0.1 mg/L. Assume that treatment could be performed in the fixed biofilm reactor with the design nitrification load 0.25 g NH_4^+-N/m^2-day. Calculated the required surface area of the fixed biofilm reactor.

A. 250 m^2
B. 800 m^2
C. 2800 m^2
D. 7600 m^2

Q6.7
Calculate the ratio of the costs for aeration in nitrification/denitrification process and in partial nitrification/anammox process.

A. 1.33
B. 2.66
C. 5.32
D. 10.64

Q6.8
Calculate the molar ratio of added ethanol (C_2H_5OH) to remove water nitrate (NO_3^-) in the denitrification process.

$$aC_2H_5OH + bNO_3^- + cH^+ \rightarrow dN_2 + eCO_2 + fH_2O$$

A. 4.2
B. 6
C. 0.42
D. 0.6

Q6.9
A student suffering from flu releases 10^3 viral particles into the indoor air of a classroom during every sneeze and has 12 sneezes per hour. How many students will be infected during 1 h of class if the volume of the classroom is 300 m^3, exchange of indoor air by ventilation is 500 m^3/h, and 200 inhaled viral particles can initiate flu?

Tutorial 7: Solid Waste Biotreatment

Q7.1

(a) Analyze the aerobic digestion of activated sludge (shown by the formula $CH_2O_{0.5}N_{0.12}$):

$$CH_2O_{0.5}N_{0.12} + x_1H_2O + X_2O_2 \rightarrow x_3CO_2 + x_4HNO_3$$

(b) Calculate the supply of air in m^3/kg of digested sludge. The concentration of oxygen in air is 21% (v/v). Assume that the utilization efficiency of supplied air is 10%.
(c) Describe which physiological groups of microorganisms can participate in aerobic digestion of activated sludge.

Q7.2
Select the groups of microorganisms from the list below, which are suitable for the aerobic degradation of biopolymers from food-processing waste, and describe their function in this treatment:

Iron-oxidizing bacteria
Actinomycetes
Methanogens
Nitrifying bacteria

Q7.3
Calculate the parameters of anaerobic digestion of activated sludge. Consider that elemental content of activated sludge corresponds to the formula $CH_2O_{0.5}N_{0.12}$

("MW" 23.7). The material balance of anaerobic digestion is shown in the following equation:

$$CH_2O_{0.5}N_{0.12} + 0.34H_2O \rightarrow 0.58CH_4 + 0.42CO_2 + 0.12NH_3$$

(a) Calculate the content of methane in biogas. Assume that the components of biogas are not dissolved in water.
(b) Calculate the production of methane, in terms of m^3/kg of digested activated sludge.

Q7.4

Calculate the required addition of KH_2PO_4 (MW of H = 1, O = 16, P = 31, K = 40) in g/m^3 of polluted soil, which is needed to support the bioremediation of the polluted site with the designed microbial biomass level = 200 g/m^3 of soil and the content of P in the microbial biomass = 1.3% (w/w):

A. 1.2 g/m^3
B. 11.5 g/m^3
C. 60 g/m^3
D. 115 g/m^3

Q7.5

Calculate the content of COD per 1 kg of inorganic solid waste. The empirical formula of organic waste can be presented as $C_{16}H_{24}O_5$. The content of this organic matter in solid waste is 26% (w/w)? MW of H = 1, C = 12, and O = 16.

A. 0.3 kg O_2/kg of waste
B. 0.6 kg O_2/kg of waste
C. 1.2 kg O_2/kg of waste
D. 2.4 kg O_2/kg of waste

Q7.6

Calculate the yield of the microbial biomass (empirical formula is $CH_{1.8}O_{0.5}$) from organic waste (empirical formula is CH_2O) using the following equation describing the aerobic biotechnological process:

$$CH_2O + 0.7O_2 \rightarrow X_1CO_2 + X_2CH_{1.8}O_{0.5} + X_3H_2O$$

MW of H = 1, C = 12, and O = 16.

A. 0.08 g of biomass/g of waste
B. 0.18 g of biomass/g of waste
C. 0.34 g of biomass/g of waste
D. 0.50 g of biomass/g of waste

Q7.7

Calculate the addition of $NH_4H_2PO_4$ (MW of H = 1, N = 14, O = 16, P = 31) in g/m^3 of industrial wastewater containing petroleum hydrocarbons. The initial concentration of hydrocarbons is 8 kg/m^3 of wastewater, the required final concentration is 100 g/m^3. This addition is needed to support the balanced growth of microbial biomass degrading hydrocarbons. The yield of biomass for the balanced growth is 0.60 g of dry biomass/g of consumed hydrocarbons. The contents of N and P in microbial biomass are 10% and 2% (w/w).

A. 38 g/m^3
B. 3.5 g/m^3
C. 3.8 g/m^3
D. 0.35 g/m^3

Q7.8

What is the half-life of a pollutant in soil treated in a slurry-phase bioreactor, if the content of the pollutant decreased from 250 to 56 mg/kg of soil in 17 days. Assume that the kinetics of bioremediation can be expressed as a first-order reaction: $S = S_0 e^{-kt}$, where S_0 is the initial pollutant concentration (mg of pollutant/kg of dry soil), S is the pollutant concentration at time t days after the beginning of composting (mg of pollutant/kg of dry soil), and k is the first-order degradation rate constant depending on degradation activity of microorganisms (day–1):

A. 7.9 days
B. 18.2 days
C. 24.8 days
D. 32.1 days

Q7.9

What is the content of COD per 1 kg of inorganic solid waste if the formula of organic matter of solid waste is $C_{16}H_{24}O_5N_4$ ("molecular weight" is 352 g) and the content of this organic matter in solid waste is 26% (w/w)?

A. 120 g O_2/kg waste
B. 250 g O_2/kg waste
C. 290 g O_2/kg waste
D. 390 g O_2/kg waste

Q7.10

Calculate the needed supply of air in m^3/1 kg of aerobically digested sludge (represented by the formula $CH_2O_{0.5}N_{0.12}$):

$$CH_2O_{0.5}N_{0.12} + X_1H_2O + X_2O_2 \rightarrow X_3CO_2 + X_4HNO_3$$

The concentration of oxygen in air is 21% (v/v). Assume that the utilization efficiency of supplied air is 10%. Note that the volume of 1 mol of gas at these conditions is 22.4 L.

A. 0.6 m^3/kg of dry sludge
B. 6.3 m^3/kg of dry sludge
C. 63.5 m^3/kg of dry sludge
D. 635 m^3/kg of dry sludge

Q7.11

The bioremediation can be described by the equation $S = S_0 e^{-kt}$, where S and S_0 are current and initial content of the pollutant in soil, respectively, t is the time, and k is a constant.

A. What is k if the content of the pollutant decreased from 200 mg/kg of soil to 1 mg/kg of soil for 28 days?
B. What will be the duration of the bioremediation if the required final content of the pollutant in soil is 10 times bigger, 10 mg/kg of soil?
C. What will be the duration of the bioremediation if the degradation activity of soil microorganisms is increased twice?

Q7.12

What is the minimum time for the spread of mycelia of wood-decaying fungi through a wooden construction of 5 m length, if the fungal cell length is 15 μm and the cell reproduction rate on one hyphae is 50 new-formed cells of fungi per day.

27

Exam Question Bank

Question 27.1

The carriers in the reactor are initially colonized at a density (X_0) of 4 g of dry biomass/m^2 of carrier surface. The biofilm density (X, g of dry biomass/m^2 of carrier surface) increases linearly with time (t, days of reactor operation) as follows:

$$X = X_0 + (0.72 \times T)$$

The following information is also provided:

$Q_m = 0.15$ g of COD consumed/m^2 of carrier surface · (h)
$H_{OPT} = 100\,\mu M$
$k = 0.2$
Density of wet biomass (d) = 1.04 g/cm^3
The content of water in an intact microbial cell (w) = 75% by weight of the cell

(a) Calculate the biofilm thickness after 50 days of reactor operation.
(b) Calculate the specific biodegradation activity of the microbial biomass after 50 days of reactor operation.
(c) The reactor must cease operation and excess biomass must be washed off the carriers when the specific biodegradation activity of the biomass decreases to below 30% of its maximum value. When must the reactor be stopped for backwashing to be performed to clean the carrier surfaces?
(d) What is the minimum number of bacterial layers in the biofilm (B) at the time the reactor is stopped for backwashing? Assume that the microbial community is represented by rod-shaped bacterial cells with a diameter (D) of 1 μm and an overall length (L) of 2 μm, and that the rods are arranged in layers with the longitudinal axis of each cell aligned parallel to the carrier surface.

Question 27.2

Batch cultivation is used to grow *Pseudomonas putida* strain EERC285 in the laboratory. Data from this batch cultivation is used to design a large-scale reactor for the industrial cultivation of strain EERC285. The biomass produced in this industrial-scale cultivation is used to bioremediate groundwater polluted by BTEX (benzene, toluene, ethylbenzene, and xylenes). Strain EERC285 is a rod-shaped bacterium. Each cell can be considered as having a cylindrical central portion and two hemispherical ends, with a diameter (D) of 1 μm and an overall length (L) of 3 μm. The water content (W) in each intact cell is 75%, and the density (d) of each intact cell is 1.04 g/cm^3. The parameters associated with batch cultivation in the laboratory are as follows:

Working volume of the reactor (V_r) = 3 L
Initial BTEX concentration (S_0) = 369 mg/L
Final BTEX concentration (S_f) = 12 mg/L
Amount of evaporated BTEX (E) = 425 mg
Overall duration of the process = 73 h
Duration of lag phase = 8 h
Duration of stationary phase = 38 h
Initial concentration of EERC285 cells = 4 × 10^4 cells/mL
Final concentration of EERC285 cells = 8 × 10^7 cells/mL

(a) Calculate the maximum specific growth rate (μ_{max}) of *Pseudomonas putida* strain EERC285 in the laboratory reactor.
(b) Calculate the yield of dry biomass (Y) of *Pseudomonas putida* strain EERC285 in the laboratory reactor.
(c) Calculate the working volume of the industrial reactor (V_{ir}) and time (T_b) for industrial batch cultivation to produce 25 kg of dry microbial biomass for use in enhanced soil bioremediation. Assume the following:
Initial concentration of BTEX in the industrial reactor (S_{oir}) = 600 mg/L.
Final concentration of BTEX in the industrial reactor (S_{fir}) = 1 mg/L.
Duration of lag phase (T_{lag}) = 14 h.
The inoculum represents 1% by weight of the final amount of biomass.
The amount of BTEX evaporated is proportional to the initial BTEX concentration.
(d) Calculate the quantity of biomass of strain EERC285 needed to bioremediate 300,000 m^3 of BTEX-polluted groundwater over a period of 50 days. The initial concentration of BTEX pollutant is 0.3 mg/L, the final concentration of pollutant must be less than 1 μg/L, and the rate of BTEX biodegradation is 15 mg of BTEX/g of biomass · day.

Question 27.3
Consider the anaerobic digestion of activated sludge. The elemental content of activated sludge can be represented by the formula $CH_2O_{0.5}N_{0.12}$. The activated sludge is hydrolyzed and digested by microorganisms, and the final products of the anaerobic digestion are methane, carbon dioxide, and ammonia.

(a) Describe the microbial groups participating in the anaerobic digestion of activated sludge.
(b) Write down the complete and balanced chemical equation for the anaerobic digestion of activated sludge.
(c) Assume that the ammonia produced is dissolved in water. Methane and carbon dioxide will constitute the biogas that is produced in this

anaerobic digestion. This biogas may be used as a fuel. Calculate the content of methane in the biogas.

(d) The activated sludge is mixed with agricultural waste in the ratio 1:1 on the basis of dry weight. The agricultural waste consists mainly of cellulose and is represented by the formula $C_6H_{10}O_5$. Write down the complete and balanced chemical equation for the anaerobic digestion of the mixture of activated sludge and agricultural waste.

(e) Calculate the content of methane in the biogas that is produced from the anaerobic digestion of the mixture of activated sludge and agricultural waste.

Question 27.4

The 16S ribosomal RNA sequence of *Escherichia coli* is shown below in the $5' \rightarrow 3'$ direction:

1	AAAUUGAAGA	GUUUGAUCAU	GGCUCAGAUU	GAACGCUGGC	GGCAGGCCUA	ACACAUGCAA
61	GUCGAACGGU	AACAGGAAAC	AGCUUGCUGU	UUCGCUGACG	AGUGGCGGAC	GGGUGAGUAA
121	UGUCUGGGAA	ACUGCCUGAU	GGAGGGGGAU	AACUACUGGA	AACGGUAGCU	AAUACCGCAU
181	AACGUCGCAA	GACCAAAGAG	GGGGACCUUC	GGGCCUCUUG	CCAUCAGAUG	UGCCCAGAUG
241	GGAUUAGCUA	GUAGGUGGGG	UAACGGCUCA	CCUAGGCGAC	GAUCCCUAGC	UGGUCUGAGA
301	GGAUGACCAG	CCACACUGGA	ACUGAGACAC	GGUCCAGACU	CCUACGGGAG	GCAGCAGUGG
361	GGAAUAUUGC	ACAAUGGGCG	CAAGCCUGAU	GCAGCCAUGC	CGCGUGUAUG	AAGAAGGCCU
421	UCGGGUUGUA	AAGUACUUUC	AGCGGGGAGG	AAGGGAGUAA	AGUUAAUACC	UUUGCUCAUU
481	GACGUUACCC	GCAGAAGAAG	CACCGGCUAA	CUCCGUGCCA	GCAGCCGCGG	UAAUACGGAG
541	GGUGCAAGCG	UUAAUCGGAA	UUACUGGGCG	UAAAGCGCAC	GCAGGCGGUU	UGUUAAGUCA
601	GAUGUGAAAU	CCCCGGGCUC	AACCUGGGAA	CUGCAUCUGA	UACUGGCAAG	CUUGAGUCUC
661	GUAGAGGGGG	GUAGAAUUCC	AGGUGUAGCG	GUGAAAUGCG	UAGAGAUCUG	GAGGAAUACC
721	GGUGGCGAAG	GCGGCCCCCU	GGACGAAGAC	UGACGCUCAG	GUGCGAAAGC	GUGGGGAGCA
781	AACAGGAUUA	GAUACCCUGG	UAGUCCACGC	CGUAAACGAU	GUCGACUUGG	AGGUUGUGCC
841	CUUGAGGCGU	GGCUUCCGGA	GCUAACGCGU	UAAGUCGACC	GCCUGGGGAG	UACGGCCGCA
901	AGUUAAAACU	CAAAUGAAUU	GACGGGGGCC	CGCACAAGCG	GUGGAGCAUG	UGGUUUAAUU
961	CGAUGCAACG	CGAAGAACCU	UACCUGGUCU	UGACAUCCAC	GGAAGUUUUC	AGAGAUGAGA
102	AUGUGCCUUC	GGGAACCGUG	AGACAGGUGC	UGCAUGGCUG	UCGUCAGCUC	GUGUUGUGAA
108	AUGUUGGGUU	AAGUCCCGCA	ACGAGCGCAA	CCCUUAUCCU	UUGUUGCCAG	CGGUCCGGCC
114	GGGAACUCAA	AGGAGACUGC	CAGUGAUAAA	CUGGAGGAAG	GUGGGGAUGA	CGUCAAGUCA
120	UCAUGGCCCU	UACGACCAGG	GCUACACACG	UGCUACAAUG	GCGCAUACAA	AGAGAAGCGA
126	CCUCGCGAGA	GCAAGCGGAC	CUCAUAAAGU	GCGUCGUAGU	CCGGAUUGGA	GUCUGCAACU
132	CGACUCCAUG	AAGUCGGAAU	CGCUAGUAAU	CGUGGAUCAG	AAUGCCACGG	UGAAUACGUU
138	CCCGGGCCUU	GUACACACCG	CCCGUCACAC	CAUGGGAGUG	GGUUGCAAAA	GAAGUAGGUA
144	GCUUAACCUU	CGGGAGGGCG	CUUACCACUU	UGUGAUUCAU	GACUGGGGUG	AAGUCGUAAC
150	AAGGUAACCG	UAGGGGAACC	UGCGGUUGGA	UCACCUCCUU	A	

Note that uracil (U) in the ribosomal sequence is analogous to thymine (T) in the equivalent DNA sequence.

(a) Design a molecular DNA probe for *Escherichia coli* that targets positions 338–354. Write down the DNA probe sequence in the $5' \rightarrow 3'$ direction.
(b) State several environmental biotechnology applications in which this probe can be used.
(c) What is the minimum concentration of this probe (in pmol/mL of sample) that must be introduced if the concentration of *E. coli* cells in the sample is 5×10^5 cells/mL and the average content of ribosomes is 8×10^5 per cell. Note that one mole of probe contains 6.023×10^{23} probe molecules (Avogadro constant).
(d) Which of the following fluorochromes is most suitable to be used as a label for this molecular probe if the source of excitation is a green laser with a wavelength of 533 nm? The fluorochromes and their wavelengths at the emission maximum (in nm) are as follows: Cy2, 506; CY3, 570; ROX, 607; and CY7, 767.
(e) What is the melting temperature (T_m) for this probe? Assume that

$$T_m = 4(G+C)+2(A+T)-3$$

where A, T, G, or C refer to the number of the corresponding bases in the probe sequence.

Question 27.5

An industrial wastewater system contains ammonium, 46 mg/L; chlorophenol, 16 mg/L; iron (II), 80 mg/L; naphthalene, 12 mg/L; phosphate, 2 mg/L; potassium, 8 mg/L; sodium, 260 mg/L; and sulfate, 25 mg/L.

(a) Which compounds should be removed from this wastewater during biological treatment? Which compounds may remain untreated? Assume that the treated effluent is eventually disposed into a freshwater stream.
(b) Which groups of microorganisms can be deployed to treat this wastewater?
(c) Several of these compounds can undergo biotransformation. Define the term "biotransformation" and describe the possible biotransformation processes that can occur in this industrial wastewater.
(d) Several of these compounds can undergo biodegradation. Define the term "biodegradation" and describe the possible biodegradation processes that can occur in this industrial wastewater.
(e) Design a scheme that can optimally combine the different microbial groups and biodegradation/biotransformation processes to treat this industrial wastewater.

Question 27.6

The growth of microorganisms and biodegradation of wastewater pollutants often take place synchronously.

(a) To facilitate industrial wastewater treatment, a bioreactor containing 500 m^3 of wastewater was inoculated by 10 m^3 of bacterial suspension with a cell concentration of 5×10^9 cell/mL and an average cell mass of 3×10^{-12} g/cell. Calculate the mean specific growth rate of the microbial community treating the industrial wastewater if the biomass concentration increased to 5.3 g/L after 18 h of batch cultivation.

(b) Select and justify one microbial group from the list below, which would be suitable for anaerobic biodegradation of industrial wastewater containing starch.
- Nitrifiers
- Methanotrophs
- Denitrifiers
- Cyanobacteria
- Methanogens
- Clostridia

(c) Select and justify one microbial group from the list below, which would be suitable for aerobic biodegradation of industrial wastewater containing protein.
- Anammox bacteria
- Fermenting bacteria
- Actinomycetes
- Anoxygenic phototrophs
- *Bacteroides* spp.
- Sulfate-reducing bacteria

(d) Balance the reactions of autotrophic denitrification and microbial growth shown below:

$$?NO_3^- + ?S + ?H_2O \rightarrow ?N_2 + ?SO_4^{2-} + ?H^+$$

$$?CO_2 + ?S + ?H_2O \rightarrow ?CH_2O + ?SO_4^{2-} + ?H^+$$

The "?" indicates a missing integer. CH_2O is the empirical formula of biomass.

(e) Wastewater is biologically treated in a rotating biological contactor (RBC), which is a fixed-biofilm reactor that consists of a series of disks mounted on a horizontal shaft rotating slowly in the wastewater. The RBC disks have a surface area of 25.56 m^2 for biofilm attachment. The biomass content in the RBC is 28 kg. The cell number in the biofilm attached to the RBC disks is 6×10^8 cell/mm^2 of disk surface. The microbial community in the biofilm consists of rod-shaped *Bacillus subtilis* A27 with average cell mass 6×10^{-12} g/cell, and spherical cells of *Micrococcus* sp. with average cell mass 1×10^{-12} g/cell. Calculate the biomass of *Bacillus subtilis* A27 in the RBC.

Question 27.7

Return liquor is the effluent from dewatering of anaerobic sludge. This sludge is produced during anaerobic digestion of activated sludge. The composition of return liquor is as follows (in mg/L):

Phosphorus of PO_4^{3-}	128
Nitrogen of NH_4^+	437
Isobutyric acid, $(CH_3)_2CHCOOH$	531
Valeric acid, $CH_3(CH_2)_3COOH$	159
Isocaproic acid, $(CH_3)_2CH(CH_2)_2COOH$	397

(a) Propose a microbial technology combining nitrification and denitrification for the removal of nitrogen from the return liquor. Show the balanced equations and describe the microorganisms performing nitrification and denitrification.

(b) Calculate the quantity of oxygen in terms of g/m^3 needed for oxidation of organic acids and ammonium in return liquor.

(c) Describe the stages of eutrophication in the aquatic environment, the roles of related microorganisms, and the measures to prevent eutrophication.

(d) Describe the microbiology and chemistry of anaerobic digestion of activated sludge in municipal wastewater treatment plants.

Question 27.8

An innovative technology to remove phosphate from stormwater in multistage bioreactor was proposed. The technology includes the following steps:

(i) Formation of anaerobic conditions in bioreactor using microbial reduction of oxygen

(ii) Microbial reduction of Fe^{3+} in iron ore (Fe_2O_3) with carbohydrates (empirical formula CH_2O) as electron donors

(iii) Chemical reaction of ferrous ions and phosphate ions following chemical oxidation of ferrous phosphate ($FeHPO_4$) with oxygen and precipitation of ferric phosphate

(a) The unbalanced reaction of microbial reduction of Fe(III) is shown below:

$$?Fe^{3+} + ?CH_2O + H_2O \rightarrow ?Fe^{2+} + ?CO_2 + ?H^+$$

The "?" indicates a missing integer. CH_2O is the empirical formula of electron donor for reduction of Fe(III).

(i) Write down the balanced chemical equation of microbial reduction of iron ore with carbohydrates as electron donors.

(ii) Describe microorganisms, which are able to reduce Fe^{3+}, and their applications in environmental engineering.

(b) Microbial nitrification and denitrification processes are also performed to remove nitrogen from stormwater. The unbalanced reaction of microbial denitrification is shown below:

$$?NO_3^- + ?CH_2O \rightarrow ?N_2 + ?CO_2 + ?H_2O + ?OH^-$$

The "?" indicates a missing integer. CH_2O is the empirical formula of electron donor for denitrification.

(i) Balance the denitrification reaction shown above.

(ii) Describe microorganisms, which are able to reduce nitrate, and their applications in environmental engineering.

(c) The average concentration of dissolved oxygen in stormwater is $7\,mg\ O_2\ L^{-1}$.

(i) Calculate the concentration of carbohydrates in stormwater, which is sufficient for microbial reduction of oxygen and the creation of anaerobic conditions in a bioreactor. Consider that there is no diffusion of oxygen in the bioreactor.

(ii) Describe microorganisms, which are able to reduce oxygen, and their applications in environmental engineering.

(d) Eutrophication in water bodies deteriorates water quality.

(i) What are the reasons and sources of eutrophication?

(ii) What microorganisms participate in eutrophication?

(iii) What bioprocesses and microorganisms can diminish the eutrophication rate?

Question 27.9

Solid organic wastes such as food-processing wastes, horticultural wastes, and sewage sludge can be transformed using anaerobic and aerobic microorganisms into value-added products.

(a) Food-processing waste with the biodegradable part 80% (w/w) which can be shown by empirical formula CH_2O can be digested to methane and carbon dioxide.

(i) Describe the processes and microorganisms performing anaerobic digestion of organic wastes to methane and carbon dioxide.

(ii) Calculate the content of methane in biogas (v/v) and maximum yield of methane in terms of kg CH_4/kg of food-processing waste with the empirical formula CH_2O. Assume that a fermentation of the food-processing wastes can be described by the following equation:

$$3CH_2O + H_2O \rightarrow CH_3COOH + CO_2 + 2H_2$$

Neglect the solubility of the gases in water.

(b) A slurry-phase bioreactor with a working volume of 20 m^3 was inoculated with 1 m^3 of bacterial suspension with a concentration of biomass at 10.7 g/L. Calculate the mean specific growth rate and growth yield of the microbial community treating suspended solid wastes in slurry-phase bioreactor using the following experimental data obtained during 18 h of log-phase of batch cultivation:
   (i) The increase of biomass concentration was 5.3 g/L.
   (ii) The decrease of organic-matter concentration was 24.5 g of total organic carbon L^{-1}.

(c) Microbial species have different abilities to degrade, oxidize, or reduce chemical substances. The combinations of microbial species are commonly used in environmental engineering processes.
   (i) Select at least one combination of the microbial groups from the list below, which can be suitable for aerobic digestion of activated sludge and describe the functions of selected microbial groups in this process.
   List of microbial groups:
   - Methanogens
   - Methanotrophs
   - Fermenting bacteria
   - Nitrifying bacteria
   - Sulfate-reducing bacteria
   - Algae
   - Fungi
   - Actinomycetes
   - Aerobic Gram-negative bacteria

   (ii) Select at least one combination of the microbial groups from the list below, which can be suitable for removal of heavy metals from wastewater and describe the functions of the selected microbial groups in this process.
   List of microbial groups:
   - Methanogens
   - Fermenting bacteria
   - Nitrifying bacteria
   - Sulfate-reducing bacteria
   - Protozoa
   - Iron-reducing bacteria

Question 27.10

Consider the anaerobic treatment of industrial wastewater containing methanol (CH_3OH, MW 32). The material balance for the anaerobic degradation of methanol is as follows:

$$4CH_3OH \rightarrow 3CH_4 + CO_2 + 2H_2O$$

(a) Write down the equation for chemical oxygen demand (COD) calculation.
(b) Calculate COD per 1 g of dry organic matter.
(c) Gas produced in this anaerobic process is called biogas and may be used as a fuel. Calculate the content of methane (v/v) in the biogas.
(d) Calculate the production of biogas as m^3 per 1 kg of degraded COD. Assume that 1 mol of gas occupies a volume of 22.4 L.
(e) Describe briefly the physiological groups of bacteria that are usually present in a typical anaerobic digester treating municipal wastewater. Which of these groups are expected to be absent in a digester treating industrial wastewater that contains mostly methanol?

Question 27.11

Nitrification refers to the biological oxidation of ammonium to nitrate by two physiological groups of bacteria. Ammonium-oxidizing bacteria convert ammonium to nitrite and nitrite-oxidizing bacteria convert nitrite to nitrate. The specific growth rates (μ) of these bacteria depend on substrate concentrations according to the following equations:

$$\mu_1 = \mu_{1max}\left[\frac{S_{am}}{(K_{am} + S_{am})}\right]\left[\frac{S_{ox}}{(K_{ox1} + S_{ox})}\right],$$

$$\mu_2 = \mu_{2max}\left[\frac{S_{ni}}{(K_{ni} + S_{ni})}\right]\left[\frac{S_{ox}}{(K_{ox2} + S_{ox})}\right],$$

where

μ_1 and μ_{1max} are specific growth rate and maximum specific growth rate for ammonium-oxidizing bacteria, respectively

μ_2 and μ_{2max} are corresponding values for nitrite-oxidizing bacteria, respectively

S_{am}, S_{ni}, and S_{ox} are concentrations of ammonium, nitrate, and dissolved oxygen, respectively

K_{am} and K_{ox1} are Monod half-saturation coefficients for ammonium and dissolved oxygen for ammonium-oxidizing bacteria, respectively

K_{ni} and K_{ox2} are Monod half-saturation coefficients for nitrate and dissolved oxygen for nitrite-oxidizing bacteria, respectively

Both bacterial groups coexist in environmental engineering systems within a nitrifying structure such as a biofilm or a granule. The growth yields of ammonium-oxidizing bacteria (Y_1) and nitrite-oxidizing bacteria (Y_2) are described by the following equations:

$$Y_1 = \frac{X_1}{(S_{0am} - S_{am})}$$

$$Y_2 = \frac{X_2}{\left(S_{0\mathrm{am}} - S_{\mathrm{am}} - S_{\mathrm{ni}}\right)},$$

where

X_1 and X_2 are the biomass concentrations of ammonium-oxidizing and nitrite-oxidizing bacteria, respectively

S_{0am} and S_{am} are concentrations of ammonium outside and inside the nitrifying structure, respectively

S_{ni} is the concentration of nitrate inside the nitrifying structure

Given $\mu_1 = \mu_2$; $S_{am} = 1$ mg NH_4-N/L; $\mu_{1max} = 1.27$ day^{-1}; $\mu_{2max} = 1.57$ day^{-1}; $K_{am} = 0.15$ mg NH_4-N/L; $K_{ni} = 2.25$ mg NO_2-N/L; $K_{ox1} = 0.3$ mg/L; $K_{ox2} = 0.4$ mg/L; $Y_1 = 0.28$ mg dry biomass/mg NH_4-N; and $Y_2 = 0.08$ mg dry biomass/mg NO_2-N.

(a) Write down the chemical equations for oxidation of ammonium to nitrite, oxidation of nitrite to nitrate and overall oxidation of ammonium to nitrate.
(b) Calculate the concentration of nitrate inside the nitrifying structure (S_{ni}) for $S_{ox} = 0.5$ mg/L and $S_{0am} = 10$ mg NH_4-N/L.
(c) Calculate the ratio of the biomass of ammonium-oxidizing bacteria (X_1) and the biomass of nitrite-oxidizing bacteria (X_2) in the nitrifying community of an activated sludge for $S_{ox} = 0.5$ mg/L and $S_{0am} = 10$ mg NH_4-N/L.

Question 27.12

Wastewater is biologically treated in a fixed-biofilm reactor. The specific surface of biofilm (S) is 40 m^2/m^3 of reactor volume and colonization of the biofilm's carrier (N_s) is 4×10^8 cells/mm^2. The bacterial community is represented by spherical cells (cocci) with an average diameter of 2.5 μm (D_c) and bacterial rods with an average diameter (D_r) of 1 μm and an average length (L) of 3 μm. The spherical cells make up 32% (α_c) of the total number of bacterial cells. Cell density (d) is 1.04 g/cm^3.

(a) What is the number of bacterial cells per m^3 of reactor volume (n)?
(b) What is the weight of wet biomass of the biofilm per m^3 of reactor volume (x)?
(c) What is the ratio of the biomass of the spherical cells and the biomass of the bacterial rods (x_c/x_r)?
(d) What is the minimum number of bacterial layers in the biofilm (b)?

Notes:

Volume of spherical cell (V_c) is

$$V_c = \frac{\pi D_c^3}{6}$$

Volume of bacterial rod (V_r) is the sum of the volumes of two hemispherical ends (or the volume of one sphere) and the volume of the cylindrical part of the cell:

$$V_r = \frac{\pi D_r^3}{6} + (L - D_r)\left(\frac{\pi D_r^2}{4}\right)$$

Minimum number of bacterial layers in the biofilm may be calculated by considering the sum of the area of the projections of spherical cells (projection of one cell is D_c^2) and bacterial rods (projection of one cell is $D_r L$) in m^2/m^2 of biofilm surface.

Question 27.13

Co-metabolism of trichloroethylene (TCE) by methylotrophic bacteria is considered an effective mean of bioremediating a polluted aquifer. The bioremediation process includes oxidation of methane (MW 16) and co-metabolism of TCE (MW 131.5):

$$2CH_4 + 2O_2 + 4NAD \rightarrow 2CO_2 + 4NADH_2$$

(catabolism of methane by methylotrophic bacteria)

$$C_2Cl_3H + 4NADH_2 + 3.5O_2 \rightarrow 2CO_2 + 3H_2O + 4NAD + 3HCl$$

(Co-metabolism of trichloroethylene by methylotrophic bacteria)

(a) Calculate the minimum ratio (in mg CH_4/mg TCE) of oxidized methane and TCE during remediation of the aquifer.
(b) Calculate the ratio (in mg O_2/mg TCE) of consumed oxygen and degraded TCE during TCE co-metabolism.
(c) Calculate the quantity of the biomass of methanotrophic bacteria (x) needed to bioremediate 250,000 m^3 of groundwater (v) polluted by TCE for 60 days (t). The average concentration of pollutant (S_0) is 0.2 mg/L, final concentration (S_f) of TCE must be less than 1 μg/L, and rate of co-metabolism of TCE by methylotrophic bacteria (q) is 0.01 g of TCE/g of biomass · day.
(d) Calculate the working volume of the reactor (V_r) needed for production of this quantity of the biomass. The biomass will be produced by batch cultivation over a period of 72 h (T_c). The concentration during inoculation (X_0) is 0.1 g/L, the specific growth rate (μ) is 0.05 h^{-1}, and the lag phase (T_a) is 20 h.

Note:

$$\mu = \frac{\ln[(X/V_r)/X_0]}{(T_c - T_a)}$$

or

$$\frac{(X/V_r)}{X_0} = e^{\mu(T_c - T_a)}$$

Question 27.14

Either chemical substances or microbial biomass may be used to enhance the bioremediation of polluted sediments. Nitrate, as well as the biomass of *Bacillus* sp. 12kt, were used to enhance the bioremediation of marine sediments polluted with aromatic hydrocarbons. These bacteria are able to degrade the hydrocarbons by denitrification (nitrate respiration) under anoxic conditions. The kinetics of bioremediation can be expressed as a first-order reaction:

$$S = S_0 e^{-kt}$$

where

S_0 is initial hydrocarbon content (mg of hydrocarbon/kg of sediment)
S is the hydrocarbon content at time t after beginning of the bioremediation (mg of hydrocarbon/kg of sediment)
k is the rate of degradation (day^{-1})

(a) Calculate the rate of hydrocarbon degradation for the case of stimulation of bioremediation (k_{ab}) by nitrate only where the hydrocarbon content is halved every 29 days ($t_{1/2}$).
(b) Calculate the rate of hydrocarbon degradation for the case of stimulation of bioremediation (k_{bi}) with *Bacillus* sp. 12kt biomass and nitrate added into the sediment. In this case, $t_{1/2}$ is 12 days.
(c) The initial pollution in the marine sediments is 960 ppm and the permissible level of aromatic hydrocarbons is less than 10 ppm. Calculate the time for bioremediation enhanced by the addition of nitrate only (t_{ab}).
(d) The initial pollution in the marine sediments is 960 ppm and the permissible level of aromatic hydrocarbons is less than 10 ppm. Calculate the time required for bioremediation enhanced by the addition of *Bacillus* sp. 12kt biomass and nitrate (t_{bi}).

Question 27.15

Bacteria use oxidation/reduction (redox) reactions to utilize energy. These redox reactions involve changes of oxidation states of chemical reactants with the transfer of electrons from an electron donor to an electron acceptor. The incomplete and unbalanced redox equation is shown below for three different combinations of electron donors and electron acceptors:

(i) $?H_2 + ?NO_3^- + ?H^+ \rightarrow ?N_2 + ???$
(ii) $?Fe^{2+} + ?O_2 + ?H^+ \rightarrow ???$
(iii) $?NH_3 + ?O_2 \rightarrow ?NO_3^- + ???$

The "?" indicates a missing integer and the "???" indicates one or more missing chemical species. For each of the three incomplete reactions above

(a) Write down the complete and balanced redox equation.
(b) Calculate the mass ratio (Y) of electron donor to electron acceptor in each redox reaction.
(c) Suggest a practical application of each redox reaction in environmental biotechnology.

Notes:

Element	Main Oxidation Numbers			
Fe	+2	+3		
H	0	+1		
N	−3	0	+3	+5
O	−2	0		

Question 27.16

The microbial cultures of *Bacillus megaterium* and *Acinetobacter calcoaceticus* were grown simultaneously in batch culture. Inoculum was 0. 5 g/L h for *Bacillus megaterium* and 0.2 g/L for *Acinetobacter calcoaceticus*. The concentration of glucose, the sole source of carbon and energy in the medium, was 31.2 g/L. The duration of lag phase (d) was 3 h for *Bacillus megaterium* and 9 h for *Acinetobacter calcoaceticus*. The maximum of specific growth rate was 0.28 h^{-1} for *Bacillus megaterium* and 0.35 h^{-1} for *Acinetobacter calcoaceticus*. The average yield was 0.43 g dry biomass/g glucose for *Bacillus megaterium* and 0.25 g dry biomass/g glucose for *Acinetobacter calcoaceticus*.

Note that the specific growth rate (μ) over a duration of exponential phase can be determined from the following equation:

$$X_i = X_{i-1}\exp\left[\mu_{max}\left(t_i - t_{i-1}\right)\right]$$

where

X_i and X_{i-1} are current for moment t_1 and previous for the moment t_{i-1} biomass concentration, g/L

μ_{max} is the maximum of specific growth rate, h^{-1}

Assume that the exponential phase changes immediately for the stationary phase when all glucose is consumed. The average yield is calculated as the ratio of biomass produced to substrate consumed from $t = 0$ h to the end of the period of increasing biomass concentration.

(a) Determine what may be the concentration of *Bacillus megaterium* for each 3 h interval.
(b) Determine what may be the concentration of *Bacillus megaterium* and *Acinetobacter calcoaceticus* for after each 3 h interval.
(c) Determine the consumption of glucose for each 3 h interval.
(d) Determine the ratio (R_t) of the biomass of *Bacillus megaterium* and *Acinetobacter calcoaceticus* at the end of batch cultivation, in the stationary phase.

Question 27.17

Methanotrophic bacteria are used to degrade a mixture of trichloroethylene (TCE) and chloroform (CF) by co-metabolism. Here, the enzyme methane monooxygenase (MMO) initiates the oxidation of these two compounds. Since TCE and CF compete with one another for MMO, the rate of their degradation in the mixture is governed by competitive inhibition kinetics.
In competitive inhibition,

$$V = \frac{V_{max} S}{S + K_s[1 + (S_i/K_i)]}$$

where

V (mg substrate/L-day) is the rate of substrate transformation
V_{max} (mg substrate/L-day) is the maximum rate of substrate transformation
S (mg/L) is the substrate concentration
S_i (mg/L) is the inhibitor concentration
K_s (mg/L) is the half-saturation constant
K_i (mg/L) is the inhibition constant

(a) Under the following conditions, calculate the individual rates of transformation of TCE and CF in mg/L-day. Assume K_i for each competitive inhibitor equals its respective K_s value.
Concentration of TCE = 7.0 mg TCE/L
Concentration of CF = 4.0 mg CF/L
V_{max} for TCE transformation = 100 mg TCE/L-day
V_{max} for CF transformation = 30 mg CF/L-day
$K_{s\,TCE}$ = 2.0 mg TCE/L
$K_{s\,CF}$ = 1.0 mg CF/L
(b) Explain the mechanism of co-metabolism in the biodegradation of xenobiotics.
(c) Calculate the minimum biomass of methanotrophic bacteria needed to degrade a mixture of 7000 mg trichloroethylene (TCE) and 3000 mg of chloroform (CF) by co-metabolism for 30 days. Assume that the average rates of transformation of TCE and CF in the mixture are 2.5 mg/g biomass-day and 0.8 mg/g biomass-day, respectively.
(d) Calculate the working volume of the bioreactor required to produce a final 20,000 g biomass of methanotrophic bacteria after 5 days,

assuming that (1) bacterial growth is proportional to the consumption of methane, (2) the average rate of methane consumption = average rate of methane mass transfer from gas to the liquid, (3) the average rate of methane mass transfer is 0.2 g CH_4/L–h, (4) the average yield of biomass produced from consumed methane is 0.6 g biomass/g of methane, and (5) inoculum of methanotrophic bacteria (initial addition of biomass in the reactor) is 10% of final biomass in the reactor.

Question 27.18

A wastewater stream has a flow of 70,000 m^3/day and an average nitrate concentration of 10 mg NO_3^- L^{-1}. The nitrate in the wastewater stream can be removed by denitrification with electron donors such as methanol (CH_3OH) and ethanol (C_2H_5OH), which are oxidized to CO_2, or elemental sulfur (S), which is oxidized to sulfate (SO_4^{2-}). Prices for methanol, ethanol, and elemental sulfur are $260, $460, and $60 per metric ton, respectively.

(a) Write down the balanced equation for denitrification with methanol as electron donor and calculate the daily cost of supplying methanol to completely remove nitrate from the wastewater stream.
(b) Write down the balanced equation for denitrification with ethanol as electron donor and calculate the daily cost of supplying ethanol to completely remove nitrate from the wastewater stream.
(c) Write down the balanced equation for denitrification with sulfur as electron donor and calculate the daily cost of supplying sulfur to completely remove nitrate from the wastewater stream.
(d) What are the essential conditions for microbial denitrification? What other microbial processes can remove nitrogen from wastewater?

Question 27.19

The main cause of eutrophication in a reservoir is stormwater runoff polluted with ammonia and nitrate, which stimulates the growth of microorganisms and diminishes the levels of dissolved oxygen in the reservoir. You are given the following information:

The volume of reservoir = 6×10^6 m^3

The average monthly stormwater runoff = 1.5×10^4 m^3

The average nutrient concentrations in stormwater runoff = 0.2 mg NH_4^+-N L^{-1} and 0.1 mg NO^{3-}-N L^{-1}

(a) Briefly describe the conditions that are characteristic of the final stage of eutrophication in the reservoir. What bacterial groups are expected to dominate the microbial community during this final stage?
(b) What bacterial groups can participate in the removal of nitrogen from the stormwater runoff?
(c) You design a constructed wetland to intercept and remove the nitrogen from the stormwater runoff. The design ammonia load on the aerobic portion of the wetland is 0.1 g NH_4^+-N m^{-3}/day and the design nitrate

load on the anoxic portion of the wetland is 0.03 g NO_3^--N m^{-3}/day. The wetland has an average depth of 1 m. Calculate the surface area of the constructed wetland.

(d) With a diagram, illustrate the microbial ecosystem in the reservoir and how the stormwater runoff causes eutrophication.

(e) With a diagram, illustrate the microbial ecosystem in the constructed wetland and how nitrogen is intercepted and removed from the stormwater runoff.

Question 27.20

Hydrolysis of cellulose and fermentation of produced glucose can be used in environmental engineering for the production of fuel ethanol from organic wastes. Use the following chemical formulae: cellulose, $(C_6H_{10}O_5)_n$, where n is the number of monomers: glucose, $C_6H_{12}O_6$, and ethanol, C_2H_6O.

(a) Describe processes and microorganisms performing hydrolysis of cellulose and fermentation of glucose.

(b) Write down the balanced chemical equations for hydrolysis of cellulose to glucose and fermentation of glucose to ethanol and carbon dioxide.

(c) Calculate the quantity of fuel ethanol which can be produced from 250,000 ton of dry organic waste with cellulose content of 72% (w/w), using hydrolysis of cellulose to glucose and microbial fermentation of this glucose to ethanol.

(d) What are the other applications of fermenting bacteria in environmental engineering?

(e) Cellulose-containing waste can be transformed by anaerobic microorganisms to methane.

  (i) Write down the balanced chemical equation for methanogenesis from cellulose.

  (ii) Describe the steps and microorganisms of anaerobic digestion of organic wastes.

Question 27.21

Microbial anaerobic respiration is used in environmental engineering for different processes.

(a) Describe electron donors and acceptors for iron-reducing bacteria and applications of these bacteria in environmental engineering.

(b) The unbalanced reaction of microbial reduction of sulfate is shown below:

$$?SO_4^{2-} + ?C_6H_{12}O_6 \rightarrow ?H_2S + ?CO_2 + ?OH^-$$

The "?" indicates a missing integer. $C_6H_{12}O_6$ is the formula of glucose, which is used for reduction of sulfate. Balance the reaction of sulfate reduction.

(c) Assume that bacterial sulfate reduction is used to remove copper ions from the waste stream by the following reaction:

$$H_2S + Cu^{2+} \rightarrow CuS \downarrow + 2H^+$$

Concentrations of pollutants in the wastewater stream are as follows: copper ions (Cu^{2+}), 42 mg/L; sulfate ions (SO_4^{2-}), 58 mg/L; and total organic carbon (TOC), 30 mg/L. The empirical formula of organic matter is CH_2O. What is the maximum percentage of the copper removal from the waste stream?

(d) The unbalanced reaction of microbial denitrification is shown below:

$$?NO_3^- + ?CH_3COOH \rightarrow ?N_2 + ?CO_2 + ?H_2O + ?OH^-$$

The "?" indicates a missing integer. CH_3COOH is the formula of acetic acid, which is used for denitrification. Balance this denitrification reaction.

(e) Describe the major microbial transformations of sulfur; write down the related chemical equations, and explain their importance for environmental engineering.

Question 27.22

The growth of microorganisms is an essential part of bioprocessing of wastes.

(a) Calculate the mean specific growth rate and growth yield of bacterial strain *Pseudomonas veronii* EN07 using experimental data obtained during the lag and log phases of batch cultivation:

$$S_0 = 15.2 \text{ g TOC/L}$$

$$S_F = 1.4 \text{ g TOC/L}$$

$$X_0 = 0.2 \text{ g dry biomass/L}$$

$$X_F = 6.9 \text{ g of dry biomass/L}$$

$$t_{lag} = 4 \text{ h}$$

$$t = 14 \text{ h}$$

where

S_0 and S_F are the initial and final concentrations of total organic carbon (TOC), respectively

X_0 and X_F are the initial and final concentrations of biomass, respectively

t_{lag} and t are the durations of lag phase and batch cultivation, respectively

(b) The concentration of microbial biomass of activated sludge during steady-state suspended continuous cultivation in an aeration tank at dilution rate $0.25\,h^{-1}$ is constant. The activated sludge of effluent from the aeration tank is concentrated in a settling tank. A portion of the activated sludge after settling, 15% w/w, is recycled into the aeration tank. What is the specific growth rate of biomass of activated sludge?

Note: Dilution rate = flow rate through aeration tank/working volume of aeration tank.

(c) Nitrification is an important step in the aerobic treatment of wastewater.
  (i) Which microorganisms perform nitrification?
  (ii) Calculate the daily consumption of oxygen for microbial removal of ammonium from wastewater using the nitrification process performed according to the following equation:

$$NH_4^+ + 2O_2 = NO_3^- + H_2O + 2H^+$$

  The flow rate of wastewater is $25{,}000\,m^3$/day; the concentration of ammonium in wastewater is 25 mg NH_4^+-N/L.
  (iii) What will be the daily consumption of oxygen for microbial removal of ammonia in this wastewater, if nitrification will be replaced by the anammox process, performed according to the following equation?

$$NH_4^+ + NO_2^- \rightarrow N_2 + 2H_2O$$

(d) Microbial species have different abilities to degrade, oxidize, or reduce chemical substances. Combinations of microbial species are usually used in environmental engineering. Select the group(s) of microorganisms from the list below, which is (are) suitable for the bioremoval of chloroethanes and chloroethylenes from wastewater, and describe its (their) function(s) in this treatment:
- Methanogens
- Methanotrophs
- Enterobacteria
- Fermenting bacteria
- Nitrifying bacteria

Question 27.23

An innovative technology to remove phosphate from stormwater in multistage bioreactor has been proposed. The technology includes the following steps:

(a) Formation of anaerobic conditions in bioreactor using microbial reduction of oxygen
  (i) Microbial reduction of Fe^{3+} in iron ore (Fe_2O_3) with carbohydrates (empirical formula CH_2O) as electron donors

(ii) Chemical reaction of ferrous ions and phosphate ions following chemical oxidation of ferrous phosphate ($FeHPO_4$) with oxygen and precipitation of ferric phosphate.

(b) The unbalanced reaction of microbial reduction of Fe(III) is shown below:

$$?\,Fe^{3+} + ?\,CH_2O + H_2O \rightarrow ?\,Fe^{2+} + ?\,CO_2 + ?\,H^+$$

The "?" indicates a missing integer. CH_2O is the empirical formula of electron donor for reduction of Fe(III).

(i) Write down the balanced chemical equation of microbial reduction of iron ore with carbohydrates as electron donors.

(ii) Describe microorganisms, which are able to reduce Fe^{3+}, and their applications in environmental engineering.

(c) Microbial nitrification and denitrification processes are also performed to remove nitrogen from stormwater. The unbalanced reaction of microbial denitrification is shown below:

$$?\,NO_3^- + ?\,CH_2O \rightarrow ?\,N_2 + ?\,CO_2 + ?\,H_2O + ?\,OH^-$$

The "?" indicates a missing integer. CH_2O is the empirical formula of electron donor for denitrification.

(i) Balance the denitrification reaction shown above.

(ii) Describe microorganisms, which are able to reduce nitrate, and their applications in environmental engineering.

(d) The average concentration of dissolved oxygen in stormwater is 7 mg $O_2\,L^{-1}$.

(i) Calculate the concentration of carbohydrates in stormwater, which is sufficient for microbial reduction of oxygen and the creation of anaerobic conditions in a bioreactor. Consider that there is no diffusion of oxygen in the bioreactor.

(ii) Describe microorganisms, which are able to reduce oxygen, and their applications in environmental engineering.

(e) Eutrophication in water bodies deteriorates water quality.

(i) What are the reasons and sources of eutrophication?

(ii) Which microorganisms participate in eutrophication?

(iii) What bioprocesses and microorganisms can diminish the eutrophication rate?

Question 27.24

Polylactic acid is a biodegradable thermoplastic, which can be produced from food-processing waste. Calculate the quantity of polylactic acid, which can be produced from 25,000 ton of dry food-processing waste. The parameters and conditions are as follows:

- The biodegradable part of food-processing waste is 80% (w/w).
- The biodegradable part of food-processing waste can be shown with the empirical formula, CH_2O.
- Polylactic acid with the formula $(C_3H_4O_2)_n$ is chemically synthesized from lactic acid (formula is $C_3H_6O_3$), which is produced using microbial fermentation of the biodegradable part of food-processing waste.

Question 27.25

The composition of wastewater from a fish-canning factory is as follows (in mg/L):

BOD (in mg O_2 L^{-1})	1400
Proteins (empirical formula is C_2H_2ON)	900
Lipids (empirical formula is $C_{51}H_{92}O_6$)	300
Sulfate (SO_4^{2-})	600

(a) Propose at least three types of microbial technologies to treat this wastewater and describe the microbial groups that can be used in these technologies.

(b) Calculate the following parameters:

 (i) The empirical formula of organic matter of the fish-canning factory wastewater. Consider that organic matter of wastewater consists of proteins and lipids.

 (ii) Theoretical COD in g O_2 g^{-1} of organic matter. Oxidation of nitrogen is not accounted for in the calculation.

(c) Calculate the percentage of BOD that can be removed from this wastewater using only anaerobic bacterial sulfate reduction. Use the balanced equation of sulfate bioreduction in this calculation.

(d) Explain the reasons for competition between sulfate-reducing and methanogenic prokaryotes during anaerobic digestion of the organic wastes. Describe how to diminish the activity of sulfate-reducing bacteria during anaerobic digestion of the organic wastes.

Question 27.26

(a) Characterize every microbial group from the list below. Select and justify at least one microbial group from this list that is suitable for the biodegradation of cellulose-containing waste:

- Methylotrophs
- Denitrifying bacteria
- Cyanobacteria
- Clostridia
- Nitrifiers

(b) Calculate the yield of biomass (empirical formula of microbial biomass is $CH_2O_{0.5}$) during aerobic growth in a medium with methanol (CH_3OH) as the source of carbon and energy. Production of CO_2, measured during growth, was 0.6 g CO_2 g^{-1} of consumed methanol.

(c) Characterize every microbial group from the list below. Select and justify at least one microbial group from this list that is suitable for the production of biofuel:
  (i) Anammox bacteria
  (ii) Fermenting bacteria
  (iii) Actinomycetes
  (iv) Anoxygenic phototrophs
  (v) Sulfate-reducing bacteria

Question 27.27

An effluent after anaerobic treatment of an industrial wastewater contains 1 mg/L of trichloroethylene, 20 mg/L of ammonia, and 690 mg/L of ethanol.

(a) Define "co-metabolism" and explain how co-metabolism of trichloroethylene is possible for the treatment of this industrial wastewater effluent.

(b) Write down the balanced equation describing the oxidation of ammonia to nitrate. Explain why nitrification is used in environmental engineering.

(c) Calculate the required intensity of aeration (kg O_2 per m^3 of working volume of aeration tank h^{-1}) for oxidation of ethanol (C_2H_6O) to CO_2 under the following conditions:
  - Hydraulic retention time (HRT) of wastewater in aeration tank = working volume of aeration tank/flow of wastewater = 2.5 h.
  - Efficiency of aeration = (consumed oxygen/supplied oxygen) × 100% = 12%.

(d) What could be the negative environmental consequences if an effluent containing ammonia is discharged to the aquatic system?

Question 27.28

Activated sludge produced in the aerobic treatment of a wastewater is concentrated and treated anaerobically to reduce the quantity of disposed waste sludge. The products of anaerobic digestion of the activated sludge (empirical formula $CH_{1.8}O_{0.5}N_{0.12}$) in the municipal wastewater treatment plant are as follows:

- Dewatered biomass of anaerobic microorganisms (empirical formula of dry matter is $CH_{1.8}O_{0.55}N_{0.07}$). Concentration of biomass of anaerobic microorganisms before dewatering is 20 g (dry weight)/L.
- Biogas containing 60% (v/v) of methane (CH_4) and 40% (v/v) of carbon dioxide (CO_2).
- Reject water, which is the liquid after dewatering of biomass of anaerobic microorganisms. This wastewater contains organic acids with empirical formula $CH_2O_{0.4}$ and ammonia (NH_3). Concentration of organic acids in the reject water is 1 g/L.

(a) Write down the balanced equation describing the anaerobic digestion of the activated sludge in the municipal wastewater treatment plant. Note that water is used for hydrolysis of biopolymers of the activated sludge and the growth yield of biomass of anaerobic microorganisms during anaerobic digestion of the activated sludge is 0.12 g of biomass of anaerobic microorganisms per 1 g of activated sludge.

(b) If there is production of H_2S in the anaerobic digester, propose at least three technologies to reduce the content of H_2S in biogas.

(c) List at least three microbiological methods for sewage-sludge management.

Question 27.29

Bioremediation of a site, which was polluted with polycyclic aromatic hydrocarbons (PAHs), could be an effective engineering solution. Laboratory tests demonstrated that aeration alone would not lead to either a significant reduction in the content of total petroleum hydrocarbons or stimulation of microbial activity in the polluted soil.

(a) Describe the advantages of bioremediation over other soil remediation techniques.

(b) Introduction of the biomass of the strain *Pseudomonas putida* 1QA7 into a soil polluted with 2400 mg of PAHs kg^{-1} of soil increased the average rate of PAHs biodegradation. The experimental data are shown in the table below.

Dosage of introduced biomass, g/kg of polluted soil	0	0.1	0.5	1.0	5.0	10.0
Average rate of PAHs degradation, mg/kg of soil day^{-1}	1.2	19.2	91.2	181.2	251.2	301.2

(i) What is the optimal dosage of the introduced biomass? Assume that the optimal dosage ensures both the highest specific biodegradation activity of introduced biomass in polluted soil (g of PAHs g^{-1} of biomass day^{-1}) and the highest average rate of PAHs degradation (mg of PAHs kg^{-1} of soil day^{-1}).

(ii) Calculate the minimum dosage of biomass of *Pseudomonas putida* 1QA7 to be introduced into a polluted soil with 2850 mg of PAHs kg^{-1} of soil to reach the targeted end point of bioremediation of 10 mg of PAHs kg^{-1} of soil for the period not longer than 30 days.

(c) Describe the major reactions as well as the biotic and abiotic conditions for the biodegradation of chlorinated aromatic hydrocarbons.

(d) Calculate the amount of NH_4NO_3 needed to support the bioremediation of a polluted soil using the following conditions:

- The initial level of pollution is 2850 mg of PAHs kg^{-1} of soil.
- The targeted end point of bioremediation is 10 mg of PAHs kg^{-1} of soil.
- The growth yield of *Pseudomonas putida* 1QA7 is 0.4 g of dry biomass g^{-1} of degraded PAHs.
- The content of N in microbial biomass is 6% (w/w).

Question 27.30

Calculate generation time (T), specific growth rate (μ), and hydraulic retention time (HRT) under steady-state conditions of chemostat cultivation of bacterial strain *Pseudomonas aeruginosa* FTTC2839 using the following conditions:

- The complete genome of *Pseudomonas aeruginosa* FTSC2839 contains 3749 genes.
- The average length of one gene sequence = 1050 bp.
- Time between cell division and start up of DNA replication = 17 min.
- Time between finishing of DNA replication and cell division = 20 min.
- DNA replication rate = 2400 bp/s.

Note: $\mu = \ln 2/T$; HRT = $1/\mu$.

Question 29.31

The storage of diesel fuel at an industrial site has led to the contamination of the area due to spillage. Laboratory tests demonstrated that aeration alone would not lead to significant reductions in levels of total petroleum hydrocarbons (TPH) or stimulation of microbial activity.

a. Bioremediation of the polluted site using different biotechnological processes could be an effective engineering solution for this polluted site:
   (i) Describe the advantages of bioremediation over other soil remediation techniques.
   (ii) Propose a bioremediation program for this polluted site.
b. Evaluate the applicability of in situ or ex situ bioremediation from the following conditions:
   (i) The expected duration of bioremediation must not be longer than 3 months.
   (ii) The cost of bioremediation must not be more than \$0.04 kg^{-1} of soil.
   (iii) The initial level of TPH (S_0) is 12,000 mg of TPH kg^{-1} of polluted soil.
   (iv) The targeted end point of bioremediation is 100 mg TPH kg^{-1} of soil.
   (v) The bioremediation can be described by the first order equation $S = S_0 e^{-kt}$, where S is the current content of TPH in soil, t is the time, and k is a constant.
   (vi) Values of k are 0.02 and 0.06 for in situ or ex situ bioremediations, respectively.

(vii) The costs of the biotreatment for in situ or ex situ bioremediations are \$0.02 and 0.03 kg^{-1} of soil, respectively.
(viii) The cost of soil excavation, transportation, piling for ex situ treatment, and return to site is \$0.006 kg^{-1} of soil.

c. Calculate the amount of $(NH_4)_3PO_4$ needed to support the bioremediation of this site using the following conditions:
   (i) The initial level of TPH is 12,000 mg of TPH kg^{-1} of polluted soil.
   (ii) The targeted end point of bioremediation is 100 mg TPH kg^{-1} of soil.
   (iii) The growth yield of microbial biomass is 0.6 g dry biomass g^{-1} of degraded TPH.
   (iv) The contents of N and P in microbial biomass are 6% (w/w) and 1% (w/w), respectively.

   Is $(NH_4)_3PO_4$, suitable for the supplementation of the bioremediation process? If this salt is not suitable, give the examples of at least three other supplements that can be used to enhance the bioremediation process.
d. Describe at least six physiological groups of microorganisms that could participate in the bioremediation of this polluted site.

Gram-positive fermenting bacteria	Gram-positive anoxic bacteria	Gram-positive microaerophilic and facultative anaerobic bacteria	Gram-positive aerobic bacteria
Fermenting archaea	Anoxic archaea	Microaerophilic and facultative anaerobic archaea	Aerobic archaea

Question 27.32

Biotreatment of municipal wastewater includes both aerobic treatment with production of activated sludge and its anaerobic digestion.

(a) The equation for the aerobic treatment of the organic-matter-containing wastewater is as follows:

$$CH_2O + X_1O_2 \rightarrow X_2CH_2O_{0.5} + X_3CO_2 + X_4H_2O$$

where CH_2O and $CH_2O_{0.5}$ represent chemical formulas of biodegradable matter of wastewater and biomass of activated sludge, respectively. Consider that yield of biomass of activated sludge produced from biodegradable matter of wastewater is 0.13 g of dry biomass/g of dry biodegradable matter. Write down the balanced equation for aerobic treatment of the organic-matter-containing wastewater.

(b) Activated sludge, which is produced by aerobic treatment of wastewater, is digested anaerobically to diminish the quantity of disposed sludge. Consider that the substrates of anaerobic digestion are biomass of activated sludge and water, and the products of anaerobic digestion are methane (CH_4) and carbon dioxide (CO_2). Write down the balanced chemical equation for the anaerobic digestion of activated sludge.

Question 27.33

A wastewater stream has a flow rate of 4000 m^3/day and an average ammonium concentration of 200 mg NH_4^+-N L^{-1}. The ammonium can be removed from wastewater using sequential nitrification and denitrification.

(a) Write down the balanced equation for nitrification, that is, the oxidation of ammonium (NH_4^+) to nitrate (NH_3^-), and calculate the daily consumption of oxygen for complete oxidation of ammonium in the wastewater stream.

(b) Write down the balanced equation for denitrification with methanol (CH_3OH) as electron donor and calculate the daily consumption of methanol for complete removal of nitrate from the wastewater stream.

(c) Describe nitrifying microorganisms. Explain how they can be used for the co-metabolism.

(d) Describe denitrifying microorganisms and how they can be used for soil bioremediation.

Question 27.34

Steady state of continuous cultivation in a chemostat with suspended biomass can be described by the following equation:

$$\mu - D = 0$$

where

μ is the specific growth rate

D is the dilution rate

The microbial strains A and B have different maximum specific growth rates (μ_{max}) and half-saturation constants (K_s).

$$\text{For strain A:} \mu_A = 0.16 \text{ h}^{-1}\left[\frac{S}{(S + 10 \text{ mg/L})}\right] \quad \text{and}$$

$$\text{For strain B:} \mu_B = 0.20 \text{ h}^{-1}\left[\frac{S}{(S + 25 \text{ mg/L})}\right],$$

where S is concentration of growth-rate-limiting substrate in the chemostat.

(a) Which strain will dominate at steady state during simultaneous cultivation of both strains in chemostat at $D = 0.15\,h^{-1}$ and $S = 75\,mg/L$?
(b) What are the dilution rates and concentrations of growth-rate-limiting substrate at steady state when both strains have the same specific growth rate?
(c) Describe the types of reactors for continuous cultivation of microorganisms.
(d) How would ecological interactions between two groups of microorganisms affect their simultaneous cultivation?

Index

B

C

R

S

T

U

V

W

X